Managing Innovation
Internationalization of Innovation

Series on Technology Management

Series Editor: J. Tidd (*University of Sussex, UK*)   ISSN 0219-9823

The Technology Management Series is dedicated to the advancement of academic research and management practice in the field of technology and innovation management. The series features titles which adopt an interdisciplinary, multifunctional approach to the management of technology and innovation, and includes work which seeks to integrate the management of technological, market and organisational innovation. All titles are based on original empirical research, and includes research monographs and multiauthor edited works. The focus throughout is on the management of technology and innovation at the level of the organisation or firm, rather than on the analysis of sectoral trends or national policy.

Published

Vol. 34 *Managing Innovation: Internationalization of Innovation*
edited by Alexander Brem (Friedrich-Alexander-Universität Erlangen-Nürnberg, Germany & University of Southern Denmark, Denmark), Joe Tidd (University of Sussex, UK) & Tugrul Daim (Portland State University, USA)

Vol. 33 *Managing Innovation: What Do We Know About Innovation Success Factors?*
edited by Alexander Brem (Friedrich-Alexander-Universität Erlangen-Nürnberg, Germany & University of Southern Denmark, Denmark), Joe Tidd (University of Sussex, UK) & Tugrul Daim (Portland State University, USA)

Vol. 32 *Managing Innovation: Understanding and Motivating Crowds*
edited by Alexander Brem (Friedrich-Alexander-Universität Erlangen-Nürnberg, Germany & University of Southern Denmark, Denmark), Joe Tidd (University of Sussex, UK) & Tugrul Daim (Portland State University, USA)

Vol. 31 *Innovation Heroes: Understanding Customers as a Valuable Innovation Resource*
by Fiona Schweitzer (Grenoble École de Management, France & University of Applied Sciences Upper Austria, Austria) & Joe Tidd (University of Sussex, UK)

Vol. 30 *Innovation Discovery: Network Analysis of Research and Invention Activity for Technology Management*
edited by Tugrul Daim (Portland State University, USA) & Alan Pilkington (University of Westminster, UK)

Vol. 29 *Exploiting Intellectual Property to Promote Innovation and Create Value*
edited by Joseph Tidd (University of Sussex, UK)

Vol. 28 *Promoting Innovation in New Ventures and Small- and Medium-Sized Enterprises*
edited by Joseph Tidd (University of Sussex, UK)

Vol. 27 *The Role of Creativity in the Management of Innovation: State of the Art and Future Research Outlook*
edited by Alexander Brem (University of Southern Denmark, Denmark), Rogelio Puente-Diaz (University Anáhuac, Mexico) & Marine Agogué (HEC Montréal, Canada)

More information on this series can also be found at http://www.worldscientific.com/series/stm

(Continued at the end of the book)

SERIES ON TECHNOLOGY MANAGEMENT – VOL. 34

Managing Innovation
Internationalization of Innovation

Editors

Alexander Brem
Friedrich-Alexander-Universität Erlangen-Nürnberg, Germany
University of Southern Denmark, Denmark

Joe Tidd
Science Policy Research Unit (SPRU), University of Sussex, UK

Tugrul Daim
Portland State University, USA

World Scientific

NEW JERSEY · LONDON · SINGAPORE · BEIJING · SHANGHAI · HONG KONG · TAIPEI · CHENNAI · TOKYO

Published by

World Scientific Publishing Europe Ltd.
57 Shelton Street, Covent Garden, London WC2H 9HE
Head office: 5 Toh Tuck Link, Singapore 596224
USA office: 27 Warren Street, Suite 401-402, Hackensack, NJ 07601

Library of Congress Cataloging-in-Publication Data
Names: Brem, Alexander, editor. | Tidd, Joseph, 1960– editor. | Daim, Tugrul Unsal, 1967– editor.
Title: Managing innovation : internationalization of innovation / edited by
　Alexander Brem (Friedrich-Alexander-Universität Erlangen-Nürnberg, Germany &
　University of Southern Denmark, Denmark), Joe Tidd (University of Sussex, UK) and
　Tugrul Daim (Portland State University, USA).
Description: New Jersey : World Scientific, [2019] | Series: Series on technology management ; Volume 34
Identifiers: LCCN 2018058273 | ISBN 9781786346544 (hc : alk. paper)
Subjects: LCSH: Technological innovations--Management. | Technological innovations--
　Management--Case studies.
Classification: LCC HD45 .M32612 2019 | DDC 658.4/063--dc23
LC record available at https://lccn.loc.gov/2018058273

British Library Cataloguing-in-Publication Data
A catalogue record for this book is available from the British Library.

Copyright © 2019 by World Scientific Publishing Europe Ltd.

All rights reserved. This book, or parts thereof, may not be reproduced in any form or by any means, electronic or mechanical, including photocopying, recording or any information storage and retrieval system now known or to be invented, without written permission from the Publisher.

For photocopying of material in this volume, please pay a copying fee through the Copyright Clearance Center, Inc., 222 Rosewood Drive, Danvers, MA 01923, USA. In this case permission to photocopy is not required from the publisher.

For any available supplementary material, please visit
https://www.worldscientific.com/worldscibooks/10.1142/Q0195#t=suppl

Desk Editors: Anthony Alexander/Jennifer Brough/Shi Ying Koe

Typeset by Stallion Press
Email: enquiries@stallionpress.com

Printed in Singapore

To My Family, Yonca, Tolga and Eda

— Tugrul Daim

About the Editors

Alexander Brem holds the Chair of Technology Management at the Friedrich-Alexander-Universität Erlangen-Nürnberg (FAU), Germany, which is located at the Nuremberg Campus of Technology (NCT). He is a CCSR International Research Associate at DeMontfort University (UK) and a Visiting Professor at the EADA Business School in Barcelona (Spain) and HHL Leipzig Graduate School of Management (Germany). Moreover, he serves as an Academic Committee Member of the Center of Technological Innovation, Tsinghua University, Beijing (China). In addition, he was appointed as Honorary Professor at the University of Southern Denmark (SDU) in May 2017. Further information can be found here: https://www.tm.rw.fau.eu.

Joe Tidd is a Physicist with subsequent degrees in Technology Policy and Business Administration. He is a Professor of Technology and Innovation Management at the Science Policy Research Unit (SPRU), and is a Visiting Professor at University College London, and was previously teaching at Cass Business School, Copenhagen Business School, and Rotterdam School of Management. Dr. Tidd was previously the Deputy Director of SPRU, and Head of the Innovation Group and Director of the Executive MBA Programme at Imperial College. He has worked as a Policy Adviser to the CBI (Confederation of British Industry), and is a Founding Partner of Management Masters LLP. He was a Researcher for the five-year International Motor Vehicle Program of the Massachusetts Institute of Technology (MIT), and has worked on Technology and Innovation Management Projects for Consultants Arthur D. Little,

CAP Gemini and McKinsey, and numerous technology-based firms, as well as UNESCO in Africa. He has written nine books and more than 60 papers on the Management of Technology and Innovation, including *Managing Innovation* (2018, 6th edition), has more than 19,000 research citations, and is the Managing Editor of the *International Journal of Innovation Management*.

Tugrul Daim is a Professor and the Director of the Technology Management Doctoral Program in the Maseeh College of Engineering and Computer Science at Portland State University (PSU). He is also the Director of the Research Group on Infrastructure and Technology Management. He is a Faculty Fellow at the Institute for Sustainable Solutions.

The US Department of Energy, National Science Foundation, National Cooperative Highway Research Program, and many other regional, national and international organizations have funded his research. He has published over 200 refereed papers in journals and conference proceedings. He edited more than 20 special issues in journals. He edited more than 20 books and conference proceedings. He was the adviser for 11 PhD graduates who are now in leading positions in government, industry and academia.

He is the Editor-in-Chief of *IEEE Transactions on Engineering Management*. He has been at various editorial roles in journals including *International Journal of Innovation and Technology Management, Technological Forecasting and Social Change, Technology in Society, Foresight, Journal of Knowledge Economy* and *International Journal of Innovation and Entrepreneurship*.

Prior to joining PSU, he had worked at Intel Corporation for over a decade in varying management roles. At Intel, he managed product and technology development. During his tenure at PSU, he has acted as a Strategic Consultant to the Chief Technology Innovation Officer of Bonneville Power Administration, a part of the US Department of Energy. He helped develop regional and national technology roadmaps in the energy sector. He is a member of the R&D Advisory Board for TUPRAS, the largest industrial firm in Turkey. He was also the consultant for many other international, national and regional organizations including Elsevier, Biotronik, NEEA, Energy Trust of Oregon, EPRI, ETRI, Koc Holding, Arcelik, Tofas, Ford Otosan, Kirlangic, Siemens, Mark and Spencer, and Castrol.

He is also a Visiting Professor with the Northern Institute of Technology at the Technical University of Hamburg, Harburg. He has given keynotes and distinguished speaker lectures at conferences, companies, universities and research centers around the world including Iamot, Euromot, Samsung, Helmut Schmidt University, Kuhne Logistics University, Seoul National University, Bogazici University, Koc University, University of Gaziantep, Izmir Institute of Technology, University of Pretoria, Tampere University of Technology, STEPI, EPIC at UNCC, etc.

He was given the Research Publication Award by the International Association of Management of Technology (IAMOT) and Fellow Award by the Portland International Center for Management of Engineering and Technology (PICMET) both in 2014.

Dr. Daim was the President of Omega Rho, International Honor Society in Operations Research and Management Science between the years of 2014 and 2016.

He received his BS in Mechanical Engineering from Bogazici University in Turkey, MS in Mechanical Engineering from Lehigh University in Pennsylvania, MS in Engineering Management from Portland State University, and PhD in Systems Science: Engineering Management from Portland State University in Portland, Oregon.

List of Contributors

Buse, S	Hamburg University of Technology
Dauth, T	HHL Leipzig Graduate School of Management
Fischer, S	Deutsche Telekom/Berlin Institute of Technology
Freitag, F	Friedrich-Alexander-Universität Erlangen-Nürnberg (FAU)
Herstatt, C	Hamburg University of Technology
Kraus, S	University of Liechtenstein
Lemminger, R	University of Southern Denmark (SDU)
Liu, Y	Queen's University Belfast
Mahdjour, S	Telekom Innovation Laboratories
Niemand, T	University of Liechtenstein
Nils, JF	Westfälische Wilhelms-Universität Münster
Piening, EP	Johannes Gutenberg-University Mainz
Podmetina, D	Lappeenranta University of Technology
Rasmussen, ES	University of Southern Denmark (SDU)
Salge, T-O	TIME Research Area, RWTH Aachen University
Sato, CEY	University of Sussex, Science Policy Research Unit (SPRU)
Schuessler, F	University of Liechtenstein
Schuessler, M	University of Liechtenstein
Shi, Y	Cambridge University
Smirnova, M	St. Petersburg State University
Svendsen, LL	University of Southern Denmark (SDU)
Tanev, S	Carleton University
Tiwari, R	Hamburg University of Technology
Torkkeli, M	Lappeenranta University of Technology
Väätänen, J	Lappeenranta University of Technology
Velamuri, VK	HHL Leipzig Graduate School of Management
Zhou, W	HHL Leipzig Graduate School of Management
Zijdemans, E	University of Southern Denmark (SDU)

Contents

About the Editors	vii
List of Contributors	xi
Introduction — Managing Innovation: Understanding International Innovation Alexander Brem, Joe Tidd and Tugrul Daim	xv

Part 1 Understanding International Innovation — 1

Chapter 1 Understanding International Product Strategy in Multinational Corporations Through New Product Development Approaches and Evolution
Yang Liu and Yongjiang Shi — 3

Chapter 2 Internationalisation of New Product Development and Research & Development: Results from a Multiple Case Study on Companies with Innovation Processes in Germany and India
Alexander Brem and Florian Freitag — 27

Chapter 3 Platform Leadership of Incumbent Telecommunications Operators: The Case of BT 21st Century Network (BT21CN)
Carlos Eduardo Yamasaki Sato — 59

Chapter 4 Changing Innovation Roles of Foreign Subsidiaries from the Manufacturing Industry in China
Wenqian Zhou, Vivek K. Velamuri and Tobias Dauth — 97

Chapter 5	Innovativeness and International Operations: Case of Russian R&D Companies *Daria Podmetina, Maria Smirnova, Juha Väätänen and Marko Torkkeli*	129
Chapter 6	Don't Get Caught on the Wrong Foot: A Resource-Based Perspective on Imitation Threats in Innovation Partnerships *Foege J. Nils, Erk P. Piening and Torsten-Oliver Salge*	153

Part 2 Lean and Global Innovation — **197**

Chapter 7	Lean and Global Technology Start-ups: Linking the Two Research Streams *Stoyan Tanev, Erik Stavnsager Rasmussen, Erik Zijdemans, Roy Lemminger and Lars Limkilde Svendsen*	199
Chapter 8	Global Innovation: An Answer to Mitigate Barriers to Innovation in Small and Medium-Sized Enterprises? *Stephan Buse, Rajnish Tiwari and Cornelius Herstatt*	241
Chapter 9	International Corporate Entrepreneurship with Born Global Spin-Along Ventures — A Cross-Case Analysis of Telekom Innovation Laboratories' Venture Portfolio *Sarah Mahdjour and Sebastian Fischer*	257
Chapter 10	Innovative Born Globals: Investigating the Influence of Their Business Models on International Performance *Sascha Kraus, Alexander Brem, Miriam Schuessler, Felix Schuessler and Thomas Niemand*	275

Index — 329

Introduction — Managing Innovation: Understanding International Innovation

Alexander Brem, Joe Tidd and Tugrul Daim

Phenomenon[1]

Globalization is the main economic phenomenon of the last centuries. Since the late 1980s, multinational companies (MNCs) from Europe and North America are shifting their production to Asian countries, especially to China and other Asian countries. Due to the growing markets in Asia, R&D was shifted there as well to appropriately adapt new products to local needs, but at a very low level (Edler *et al.*, 2003). However, this is changing in recent years through frugal and reverse innovation (Agarwal *et al.*, 2017). At the same time, these countries earned a huge amount of know-how, based on these technology transfers, and unbelievable amounts of foreign currency reserves (Klossek *et al.*, 2012).

In recent years, before the background of internationalization of these companies, the situation turned upside down (Schueler-Zhou and Schueller, 2009). Global companies invest in Asian R&D capabilities not only to adopt products on local needs but rather to use foreign knowledge sources and scientific capabilities (Veugelers, 2005; Brem and Freitag, 2015). Because on a macroeconomic level, from 2009 to 2010, Chinese expenses for R&D, which are already at a comparatively high level, increased by 38.5%. The average of this study was about 9.3% (Jaruzelski *et al.*, 2011). Another study by the United Nations revealed that 62% of all interviewed persons named China as their favorite new R&D site; 41% stated the USA, 29% India (UNCTAD, 2005).

On the microeconomic level, there is another trend of growing Research and Development (R&D) activities in Asian countries, to develop their own innovation capabilities, which are not dependent on foreign companies anymore (Jiatao and Rajiv, 2009; Gerybadze and Reger, 1999; Kumar, 1997; Brem and Moitra, 2012). Consequently, the number of innovations originated in China is growing, especially

---

[1]Based on Brem and Moitra, 2012.

by China-based MNCs (Zeschky *et al.*, 2011), and there is a high level of support by the Chinese government (Di Minin and Zhang, 2010). However, these companies must further develop their innovation capabilities to enhance technical innovation. For this, the focus must change from process to product innovation and from imitation to indigenous innovation (Li and Kozhikode, 2009).

Particularly, emerging Asian companies invest their money into strategic know-how all over the world, e.g., not only in technical rights like patents or licenses but in joint ventures and acquisitions as well (Schueler-Zhou and Schueller, 2009). In 2006, 11 billion US dollars were invested from European and American companies in Indian companies. On the contrary, Indian companies spent about 23 billion US dollars for 168 acquisitions in other countries (Rybak, 2007). In Germany, about 1,300 active Chinese companies spent about 1.7 trillion US dollars in German companies; in 2011 alone, these firms spent about 230 billion US dollars (Brüggmann, 2011). Until 2020, more than 2 trillion euros are expected to be invested in Germany.

These companies are using a variety of establishment modes to gain access to new markets.

Some prominent examples from such acquisitions by Chinese companies in 2011 are

- Lenovo bought Medion (Germany)
- CITIC bought KSM Castings (Germany)
- Wolong bought ATB Antriebstechnik (Austria)
- Youngman und Pang Da bought Saab (Sweden)

These examples show how far the internationalization of innovation already went within a comparably short timeframe. Hence, it is interesting to analyze in more detail how we can foster a common understanding of these developments. For this, current phenomena like lean and global innovation are also discussed.

Understanding International Innovation

International innovation mainly happens within the context of MNCs. These firms usually have enough resources to foster innovation at different locations all over the world, with distributed competences and organizational units. As Liu and Shi (2017) outline, factors like the development of Information and Communication Technologies (ICT), competitive pressure, brand awareness and technological capabilities determine the choice of how the company develops new products and allocate these resources. This is usually achieved through the application of New Product Development (NPD) approaches, which can be multi-local as well as adaptation- or platform-based. To achieve such a competitiveness at an international scale, certain innovation capabilities are needed. Based on further strategic, organizational and operational aspects, Brem and Freitag (2016) argue that some of the general management approaches can be applied universally, while others differ a lot in cultural setups. In their case of India, different expectations must be considered in order to be successful, e.g., reward systems are quite different, but also the average age of NPD teams. Only an appropriate awareness of these factors will lead to the expected success.

Sato (2014) uses the example of telecommunication operators to explain when platform-based innovation approaches work best. Zhou *et al.* (2016) argue in this context that innovation capabilities, organization structures and the interaction between the subsidiary and the headquarter determine the overall innovation success. For this, he found that the reusability of components and its sub-systems as well as the openness of platforms to externals drive innovation success in this industry in an international context. This phenomenon has become even more prominent in times of the sharing economy and their related platform strategies (Richter *et al.*, 2017). However, Sato (2014) also highlights that the low level of internationalization (in the case of telecom operators) might also harm the dynamics of such platform-based strategies. Such internationalization efforts can also be measured with export intensity. This path was chosen by Podmetina *et al.* (2009) who found that there is a significant impact of innovation activities, level of competition and NPD activities on this intensity, based on their Russian sample. However, there is also always an inherent risk of being imitated, as the success is then also visible on a global scale. Based on the results of Foege *et al.* (2017), there is an indication that the probability of being imitated increases with the partnership variety. In such a situation, companies need to carefully select whom to work with and how the relationship can be secured.

Lean and Global Innovation

Following research avenues on the internationalization of innovation in recent years inevitably ended up with reading about lean and global innovation setups. One of these approaches ended in the definition of Lean Global Start-Ups (LGS) by Tanev *et al.* (2015), which faces the specific combination of business development, new product development and early internationalization. They further distinguish between lean-to-global and lean-and-global firms, whereby both have high levels of complexity, uncertainty and risk on a global scale. Hence, knowledge sharing and Intellectual Property (IP) protection have an even higher importance, but also an even higher complexity, as they also happen on a global level. Buse *et al.* (2010) already identified such barriers to innovation in their study on SMEs in Western countries.

As a result, they recommend that firms should decide on a specific form of internationalization: captive offshoring, offshore cooperation or offshore outsourcing. As a consequence of such internationalization efforts, spin-offs may occur. Early internationalization of such spin-offs using the spin-along idea (Rohrbeck *et al.*, 2007) can avoid or even reduce the main problems they are facing (Mahdjour and Fischer, 2014). A main dimension for such "born global" firms is the ideal business model design, which they have in the best case from the start of their firm. If this setup is strategically chosen, such a business model design can also have a positive impact on the international firm performance (Kraus *et al.*, 2016).

These results show the high dynamics of research on the internationalization of firms, which continues till today. With this overview, we would like to stimulate further reading in this edited book which contains main research milestones from the *International Journal of Innovation Management* as well as from the *International Journal of Technology and Innovation Management* on the internationalization of innovation. Finally, we also hope to encourage further research on this interesting phenomenon.

Book Structure

Table 1 below summarizes what to expect in each part and chapter. The expectations are separated for a practitioner versus a researcher or a student. The readers can review this table to identify the chapters they should focus on.

Table 1. Overview of results for practitioners and researchers/students

Parts	Chapters	Practitioner focus	Researcher focus/Student focus
1: Understanding International Innovation	1	A review of four cases from multinational corporations analyzing new product development.	Case analysis identified several topics that can be developed into propositions to test with a field study.
	2	A review of four international cases from the R&D world.	Case analysis identified several topics that can be developed into propositions to test with a field study.
	3	Review of the impact of platform-based strategy in the telecommunication industry.	A comprehensive case study analysis including a series of interviews. The results can be used to develop propositions to test with a larger sample.
	4	A review of the innovation activities of foreign subsidiaries in China.	Data from a set of interviews provide a platform for a future research involving a larger sample and a survey.

(*Continued*)

Table 1. (*Continued*)

Parts	Chapters	Practitioner focus	Researcher focus/Student focus
	5	A review of the innovation activities in Russia.	A set of validated and tested hypotheses that can be tested in other settings.
	6	Review of imitation threats in collaborative innovation.	A set of validated and tested hypotheses that can be tested in other settings as well as a list of additional aspects that can be studied within the same setting.
2: Lean and Global Innovation	7	A review of strategies for technology start-ups.	Case study results provide a future research opportunity to test propositions out of this study.
	8	A review of innovation barriers in Germany.	A set of preliminary areas for future research on innovation success factors.
	9	A case of corporate entrepreneurship.	A case study which can be used to develop propositions for a future expanded study.
	10	Review of business models and their impact on international success.	A set of validated and tested hypotheses that can be tested in other settings as well as a list of additional aspects that can be studied within the same setting.

References

Agarwal, N, M Grottke, S Mishra and A Brem (2017). A systematic literature review of constraint-based innovations: State of the art and future perspectives. *IEEE Transactions on Engineering Management*, 64(1), 3–15.

Ale Ebrahim, N, S Ahmed, Z Taha, M Personal, R Archive, A Ebrahim and K Lumpur (2010). SMEs; Virtual research and development (R&D) teams and new product development: A literature review. *International Journal of the Physical Sciences*, 5(7), 916–930.

Brem, A and F Freitag (2015). Internationalisation of new product development and research & development: Results from a multiple case study on companies with innovation processes in Germany and India. *International Journal of Innovation Management*, 19(1). doi: 10.1142/S1363919615500103.

Brem, A and D Moitra (2012). Learning from failure: Case insights into a UK–India technology transfer project. In: *Technology Transfer in a Global Economy*, pp. 253–275. Springer, US.

Brüggmann, M (2011, Oktober). China und Katar auf Einkaufstour. In: *Handelsblatt*, 25, Nr. 206, S. 15.

Buse, S, R Tiwari and C Herstatt (2010). Global innovation: An answer to mitigate barriers to innovation in small and medium-sized enterprises? *International Journal of Innovation and Technology Management*, 7(3), 215–227. doi: 10.1142/S0219877010001970.

Di Minin, A and J Zhang (2010). An exploratory study on international R&D strategies of Chinese companies in Europe. *Review of Policy Research*, 27(4), 433–455.

Economist (2010). *First Break all the Rules*. 395(8678), pp. 6–8.

Edler, J, R Döhrn and M Rothgang (2003). Internationalisierung industrieller Forschung und grenzüberschreitendes Wissensmanagement. Eine empirische Analyse aus der Perspektive des Standortes Deutschland, Essen — Karlsruhe.

Foege, JN, EP Piening and T Salge (2016). Don't get caught on the wrong foot: A resource-based perspective on imitation threats in innovation partnerships. *International Journal of Innovation Management*. doi: 10.1142/s1363919617500232.

Gerybadze, A and G Reger (1999). Globalization of R&D: Recent changes in the management of innovation in transnational corporations. *Research Policy*, 28(2–3), 251–274.

Jaruzelski, B, J Loehr and R Holman (2011, Winter). The global innovation 1000 — Why culture is key. *Strategy and Business*, Issue 65.

Jiatao, L and KK Rajiv (2009). Developing new innovation models: Shifts in the innovation landscapes in emerging economies and implications for global R&D management. *Journal of International Management*, 15, 328–339.

Klossek, A, BM Linke and M Nippa (2012). Chinese enterprises in Germany: Establishment modes and strategies to mitigate the liability of foreignness. *Journal of World Business*, 47(1), 35–44.

Kraus, S, A Brem, M Schuessler, F Schuessler and T Niemand (2016). Innovative born globals: Investigating the influence of their business models on international performance. *International Journal of Innovation Management*. doi: 10.1142/S1363919617500050.

Kumar, N (1998). Technology generation and technology transfers in the world economy: Recent trends and implications for developing countries. *Science, Technology and Society*, 3(2), 265–306.

Li, J and RK Kozhikode (2009). Developing new innovation models: Shifts in the innovation landscapes in emerging economies and implications for global R&D management. *Journal of International Management*, 15(3), 328–339.

Liu, Y and Y Shi (2017). Understanding international product strategy in multinational corporations through new product development approaches and evolution. *International Journal of Innovation Management*, 1750057.

Mahdjour, S and S Fischer (2014). International corporate entrepreneurship with born global spin-along ventures — A cross-case analysis of telekom innovation laboratories' venture portfolio. *International Journal of Innovation Management*, 18(3). doi: 10.1142/S1363919614400076.

Podmetina, D, M Smirnova, J Väätänen and M Torkkeli (2009). Innovativeness and international operations: Case of Russian R&D companies. *International Journal of Innovation Management*, 13(2), 295–317. doi: 10.1142/S1363919609002303.

Rohrbeck, R, M Dohler and HM Arnold (2007). Combining spin-out and spin-in activities — The spin-along approach. Paper presented at ISPIM 2007 Conference: *Innovation for Growth: The Challenges for East & West*, pp. 1–12, June 17. Warsaw, Poland.

Richter, C, S Kraus, A Brem, S Durst and C Giselbrecht (2017). Digital entrepreneurship: Innovative business models for the sharing economy. *Creativity and Innovation Management*, 26(3), 300–310.

Rybak, A (2007). Available at http://www.ftd.de/unternehmen/industrie/:agenda-indische-welteroberer/182300.html#z.

Sato, CEY (2014). Platform leadership of incumbent telecommunications operators: The case of BT 21ST century network (BT21CN). *International Journal of Innovation Management*, 18(2), doi: 10.1142/S1363919614500157.

Schueler-Zhou, Y and M Schueller (2009). The internationalization of Chinese companies — What do official statistics tell us about Chinese outward foreign direct investment?*Chinese Management Studies*, 3(1), 25–42.

Tanev, S, ES Rasmussen, E Zijdemans, R Lemminger and LL Svendsen (2015). Lean and global technology start-UPS: Linking the two research streams. *International Journal of Innovation Management*, 19(3). doi: 10.1142/S1363919615400083.

UNCTAD (2005). WORLD INVESTMENT REPORT 2005 WIR 2005 UNCTAD/WIR/2005. Available at: http://unctad.org/en/Docs/wir2005_en.pdf [Accessed May 29, 2017].

Veugelers, R (2005). *Internationalisation of R&D in the UK — A Review of the Evidence*. Leuven — Cambridge.

Zeschky, M, B Widenmayer and O Gassmann (2011). Frugal Innovation in emerging markets: The case of mettle. *Research-Technology Management*, 54(4), 38–45.

Zhou, W, VK Velamuri and T Dauth (2016). Changing innovation roles of foreign subsidiaries from the manufacturing industry in China. *International Journal of Innovation Management*. doi: 10.1142/S1363919617500086.

Part 1
Understanding International Innovation

Part I

Understanding
International Innovation

Chapter 1

Understanding International Product Strategy in Multinational Corporations Through New Product Development Approaches and Evolution*

Yang Liu and Yongjiang Shi

International product strategy regarding global standardisation and local adaptation is one of the challenges faced by multinational corporations (MNCs). Studies in this area have tested the antecedents and consequences of standardisation/adaptation, but lack a new product development (NPD) perspective. In this study, we explore how product standardisation/adaptation is determined in the NPD context. Through a qualitative case study of four MNCs, we found three NPD approaches: multi-local, adaptation-based and platform-based. We analysed the advantages and challenges of each approach. In addition, we reveal how the factors (development of information and communication technology, competition pressure, brand awareness and technical capability) could influence the choice of a certain NPD approach. We draw implications on the paths to ensuring full leveraging of the benefits of a platform-based approach.

Keywords: Multinational corporations; global standardisation; local adaptation; new product development; platforms.

Introduction

Multinational corporations (MNCs) are believed to be at the forefront of organisational and managerial innovations (Bélanger *et al.*, 1999), as they face the challenges of global competition and the management of worldwide activities resulting in greater complexity than that faced by domestic firms (Bartlett and Ghoshal, 2000; Yip, 2003). One of the challenges in MNCs is the development of international product strategy regarding global standardisation and local adaptation (Katsikeas *et al.*, 2006; Kotabe, 1990). In many industries, there are still different market requirements across countries in terms of customer tastes, local conditions

---

*Originally published in *International Journal of Innovation Management*, 21(7), pp. 1–23.

and regulations (Gooderham, 2012; Rugman and Hodgetts, 2001). MNCs need to identify commonalities and differences in requirements and offer products accordingly (Kotler, 1986; Levitt, 1983).

The strategic importance of new product development (NPD) lies in the cross-functional nature of this task (Wheelwright and Clark, 1992). Studies have shown interactions between product strategy and NPD activities. Many product-related decisions are actually made in the NPD process (Cooper, 1994; Schmidt and Calantone, 2002; Ulrich and Eppinger, 2012). Therefore, it is essential to understand product strategy in the context of NPD.

Whereas previous studies have tested the antecedents and consequences of international product strategy (Calantone *et al.*, 2006; Cavusgil and Zou, 1994; Katsikeas *et al.*, 2006), very few studies have explored international product strategy from an NPD perspective. However, NPD activities could significantly influence the product form. For example, when some requirements are not considered early on in an NPD project, significant redesign is needed afterwards (Gunzenhauser and Bongulielmi, 2008). Such NPD activities are likely to affect the competitiveness of MNCs and therefore it is important to understand how MNCs choose an NPD approach under certain conditions. For NPD approaches in MNCs, we focus on the way of organising NPD activities that affects global standardisation and local adaptation of products.

This study aims to contribute to the understanding of international product strategy and NPD in several ways. To be specific, we identify several NPD approaches in MNCs and explain their advantages and challenges. We show how product standardisation/adaptation is determined in these NPD approaches. In addition, we reveal how certain factors could influence the adoption of a certain NPD approach.

This paper is structured as follows. In the next section, Theoretical Background, relevant studies related to international product strategy and NPD are analysed and research gaps are identified. In the section of Research Methods, the research design is presented including the choice of case companies, data collection and data analysis. In the Case Studies section, we present a detailed description of cases. In the Findings section, we show key findings of three NPD approaches and relevant influencing factors across four MNCs. In the Discussion section, we highlight the theoretical contributions and practical implications of this study, and explore the research limitations and future research avenues.

Theoretical Background
International product strategy in MNCs

Global standardisation and local adaptation have been the subject of discussion for a long time. With standardisation, firms can achieve economies of scale and therefore offer high quality products at a low price (Levitt, 1983). With adaptation, products may be more appealing to customers in terms of the desired functions or aesthetics (Kotler, 1986). Cooper and Kleinschmidt (1985) suggested that international product strategy can be viewed on a continuum and firms position themselves somewhere between standardisation and adaptation. It is measured as the level of product or component sharing across countries (Calantone *et al.*, 2004; Zou and Cavusgil, 2002).

Empirical studies have examined the antecedents and consequences of global standardisation and local adaptation. Such studies have explored product strategy either directly or as an element of marketing strategy. Regarding antecedents, Samiee and Roth (1992) argue that the rate of technological change and the frequency with which competitors change products will influence the emphasis on global standardisation in a firm. Katsikeas *et al.* (2006) found that the degree of standardisation is related to the similarity between markets in six respects: regulatory environment, technological intensity and velocity, customs and traditions, customer characteristics, the stage of the product in its life cycle and competitive intensity. Zou and Cavusgil (2002) found international experience, global orientation and external globalising conditions to be antecedents of global marketing standardisation. These studies have been undertaken for MNCs.

In terms of studies on export firms, Cavusgil *et al.* (1993) tested the influence of three factors on the degree of product adaptation in export ventures: company characteristics (a firm's international experience, export sales goal and entry scope), product/industry characteristics (technological orientation of the industry, product uniqueness, cultural specificity of the product and type of product) and export market characteristics (similarity of legal regulations, competitiveness of the export market and product familiarity of export customers). Cavusgil and Zou (1994) identified six significant antecedents in a study of export marketing strategy: international competence, product uniqueness, the cultural specificity of the product, export market competitiveness, a firm's experience with the product and the technological orientation of the industry. Calantone *et al.* (2004) conducted research on the product adaptation of US and South Korean export firms. They identified three antecedents of product adaptation: similarity in the legal environment of the home and export markets, relevant experience of the business unit in international marketing, responsive marketing organisation and customer-orientated practices. Calantone *et al.* (2006) identified another three antecedents: export dependence, industry adaptation and market similarity.

Studies have tested the consequences of standardisation/adaptation, including strategic and financial performance. Zou and Cavusgil (2002) showed that global standardisation has a positive relationship with the strategic and financial performance of MNCs. In contrast, Samiee and Roth (1992) study revealed that the emphasis on global marketing standardisation in MNCs is not significantly related to financial performance. Some studies draw the conclusion that the degree of product adaptation is positively related to export performance for export firms (Calantone et al., 2004; Cavusgil and Zou, 1994; Leonidou et al., 2002).

Overall, prior studies suggest that a standardisation/adaptation strategy is contingent on many factors such as industry and company characteristics, and there is no single optimal strategic position for all firms (Schmid and Kotulla, 2011). In addition, choosing the right strategy is essential for the superior performance of the firm (Katsikeas et al., 2006).

Global product development

Previous studies on global product development have predominantly examined cross-border collaboration and there are several streams of research. One stream concerns the behavioural environment, defined as the firm's organisational culture and management commitment (de Brentani and Kleinschmidt, 2004). Studies have tested the direct (de Brentani and Kleinschmidt, 2004; Salomo et al., 2010) and indirect (de Brentani et al., 2010; Kleinschmidt et al., 2007) relationships between the behavioural environment (innovation/globalisation culture, resource commitment and top management involvement) and global NPD performance, emphasising the importance of the behavioural environment in facilitating cross-border collaboration.

Also, there are studies exploring the challenges of global NPD teams. By integrating globally dispersed members into a global NPD team, MNCs could leverage talents worldwide and increase cultural sensitivity (Eppinger and Chitkara, 2006; Graber, 1996; Salomo et al., 2010). However, as global NPD team members are culturally diverse, they are likely to lack shared beliefs, experiences and expectations, which diminishes trust (Barczak and McDonough, 2003; Bierly III et al., 2009; McDonough III et al., 2001). Team members may even encounter conflicts as a result (Tavcar et al., 2005). Bierly III et al. (2009) argue that increasing the frequency of face-to-face communication is one approach to enhancing trust, but the team members also need to overcome the communication barrier caused by cultural differences (Hansen and Ahmed-Kristensen, 2011; Jarvenpaa and Leidner, 1999).

Other studies have examined knowledge management in global NPD. MNCs have the advantage of acquiring local knowledge. However, to reap the benefits, dispersed knowledge needs to be integrated, and how to integrate knowledge in NPD is a challenge for MNCs (Söderquist, 2006). Subramaniam (2006) found that the cross-national collaboration climate is the key to integrating knowledge globally. Subramaniam and Venkatraman (2001) argued that increased frequency of communication in project teams is effective for processing tacit overseas information. Tavcar et al. (2005) found that there is an optimum level of communication which fosters creativity, while too much or too little communication reduces creativity.

The development of information and communication technology (ICT) has changed the NPD approach in many aspects. One important aspect is promoting cross-border collaborations (Chang, 2006; Howells, 1995). ICT tools are useful for global knowledge integration, especially when knowledge is not highly tacit. For virtual global teams, videoconferencing makes intense communication possible and when videoconferencing is combined with face-to-face meetings, communication in NPD can be highly effective while maintaining low travel costs (Tavcar et al., 2005). Kleinschmidt et al. (2010) argued that ICT infrastructure can increase the firm's ability to access, integrate and transform widely dispersed information and skills. They empirically tested the positive relationship between ICT infrastructure and NPD performance. In addition, Nambisan (2003) finds that ICT could influence process management (making the process more comprehensive or flexible) and project management (better resource monitoring and control) of NPD. Ozer (2000) founds that ICT can influence the speed of NPD.

NPD perspective of product strategy

NPD is an important source of a firm's competitiveness (Bessant and Francis, 1997; Eslami and Lakemond, 2016; Millson and Kim, 2015). It is a complex task calling for collaboration across the functions of R&D, marketing, manufacturing, finance, etc. (Griffin and Hauser, 1993; Wheelwright and Clark, 1992). In practice, product strategy is formulated and adjusted in the NPD context, with many product-related decisions being made throughout the NPD process (Luchs and Swan, 2011; Muffatto, 1999; Shibata and Kodama, 2015). For example, in the idea generation phase, firms need to identify business opportunities and conduct business case analysis, through which the firms will determine what products to develop (Kim and Wilemon, 2002; Ulrich and Eppinger, 2012; Verworn, 2006). Another example is the Stage-Gate® process, through which firms review an NPD project at each gate meeting (Cooper, 1994; Hart et al., 2003; Schmidt and Calantone, 2002; Tzokas et al., 2004). Based on the available information at that time point, the project can pass the gate or be killed off. Therefore, discussions concerning product strategy should not be separated from the NPD context.

More studies have confirmed this NPD perspective on product strategy. Bloch (1995) indicated that in the product design process, the requirements of consumers and distributors, regulations, production equipment, the marketing programme and designers jointly influence the form of products. Hauser *et al.* (2006) suggested that firms need to choose the right technology to develop in the early phase of NPD. Seidel (2007) showed how initial product concepts are changed in the NPD process when new technical or market information becomes available. Ulrich (1995) showed how product modularity is realised in NPD. Moorman and Miner (1998) defined the term 'improvisation' and suggest that formulating strategies and performing activities could happen simultaneously in NPD. Trappey *et al.* (2009) suggested firms strive to optimise product portfolios with limited NPD resources.

Studies on international product strategy have predominantly explored the antecedents and consequences of standardisation/adaptation. However, very few studies have addressed the issue of international product strategy from an NPD perspective, that is, how standardisation/adaptation decisions are made in the NPD context. An NPD perspective is needed because various issues of NPD (e.g., how NPD activities are organised) are likely to interact with standardisation/adaptation. We also know less about how different NPD approaches in MNCs can influence the standardisation/adaptation of product form, and why an approach is chosen over others. Studies on global NPD have tended to focus on cross-border collaboration and knowledge management, falling short of discussions on standardisation/adaptation issues. This study aims to contribute to the understanding of international product strategy and NPD in this regard.

Research Methods

In this research, we chose the qualitative case study method to explore product standardisation/adaptation in the NPD context. A case study is suitable when the boundaries between concepts and the context are not clear (Yin, 2009). In this study, we argue that the NPD context is especially important for understanding international product strategy. Case study research is also appropriate for explorative study (Yin, 2009), as in this case we had limited knowledge regarding the NPD perspective in relation to international product strategy. A multiple-case design allows us to make comparisons across cases to broaden our insights (Eisenhardt, 1989; Eisenhardt and Graebner, 2007).

When choosing case companies, we adopted a diverse sampling approach (Eisenhardt, 1989). This approach is widely used, because firms with different settings could broaden the insights generated from the case study (Andriopoulos and Lewis, 2009; Bohnsack *et al.*, 2014; Lawrence and Dover, 2015). Also, replication can be realised through studying firms with different backgrounds (Yin, 2009). In this research, we chose four automotive MNCs headquartered in

Table 1. Summary of Cases.

Case	Headquarters	Sales volume*	Market
Case 1	USA	155 billion	Global
Case 2	Japan	101 billion	Global
Case 3	China	3 billion	Mainly developing countries
Case 4	China	91 billion	Global except the USA

Note: *Sales data in 2013, converted to US dollars according to the average exchange rate in 2013.

different countries as case companies (see Table 1). We focused on their NPD approaches in the global context, how the NPD approaches evolved over time and their influence on product forms (standardisation/adaptation). We intentionally chose MNCs which adopted different NPD approaches for standardisation/adaptation at the time of the study or in history. To collect data, we interviewed R&D managers and engineers in the case companies. In total, we conducted 12 semi-structured interviews for all case firms. Each interview lasted for 1–1.5 h. The interview protocol is shown in appendix. Also, we collected secondary data (mainly press releases and news articles) from the internet to complement our understanding of the case companies.

This study focuses on the category of passenger cars (i.e., not including commercial vehicles). Based on our study, we found that while the same sizes (such as C-segment) are needed globally (except A-segment), market requirements differ evidently across countries and regions. For example, American people need greater power whereas European people require higher efficiency of engines. Another example is the hardness of the suspension due to different road conditions. Also, the crash test standards are different region by region.

We conducted data analysis in the way suggested by Yin (2009). First, we compiled information and wrote case reports for each company including: the background of the company, the companies' historical NPD approach, evolution in the NPD approach and influencing factors. At this stage, we conducted within-case analysis. We then compared the findings across cases to explore common patterns and variations. Through this comparison, we aimed to explain the findings in the four cases and deepen our understanding. For example, the Chinese MNCs (Cases 3 and 4) adopted different NPD approaches from developed country MNCs (Cases 1 and 2). We tried to find explanations for such variation from different aspects. In addition, we found that even for Chinese MNCs (Cases 3 and 4), the NPD approaches could be somewhat different and certain factors could influence that. The findings in the four cases allowed us to build a theory explaining the choice of NPD approaches in MNCs.

Case Studies

Case 1

The company in Case 1 is headquartered in the US. It has R&D centres in North America (the US), Europe (Germany) and Asia Pacific (Australia). Historically, each R&D centre developed cars independently for the regional market. In NPD, the cars were tailored to regional requirements (e.g., styling) and regulations (e.g., crash tests). This was actually a natural outcome as this company entered Europe and the Asia Pacific region through acquisitions. The acquired companies had NPD capability and strong brands. Case 1 largely maintained the existing operations at that time. In an era in which ICT was less developed and used, this global NPD approach was a reasonable choice as R&D centres were closer to the market so they understood what the customers really wanted. However, under this approach it was very difficult to share components globally as the R&D centres were too independent in terms of power, processes and design habits. The firm suffered from duplication in product design, which made it less and less competitive over time due to high costs.

Starting from the late 1990s, the company changed its NPD approach. Global platforms (referred to as architectures in the company) were created based on car categories: A, B, C, CD and D. With the platform approach, the international product strategy focused on "simplifying the platforms and diversifying the products". When developing the global platform, the requirements of different regions and feedback from those regions were taken into consideration from the beginning. This became easier with the use of ICT tools. To facilitate the change, the firm shifted the organisational structure of R&D. For example, the German centre was nominated as the lead centre for the B-car development to facilitate collaboration across regions. Other R&D centres were controlled by the German centre for NPD. Under this approach, products were still differentiated to accommodate differences in local requirements, but significant parts (defined as the platforms) were shared globally. The level of sharing could vary case by case. In most cases only the chassis was shared; in other cases, the car body was also shared, but with modified bonnets, bumpers and doors.

With the new NPD approach, for each car the cost was lower due to shared parts, which gave the company a cost advantage. Also, the company was able to develop more car models to attract different customer groups.

Case 2

Case 2 is a Japanese automotive company. Historically, its NPD capability was concentrated in Japan. The traditional approach was that the cars were developed based on Japanese requirements only, as communication across borders was not very convenient at that time. Then, after their launch on the Japanese market, some of the cars were introduced in other countries. These cars were then adapted in local subsidiaries, mainly to take account of local regulations, production facilities and suppliers. While this strategy was intended to minimise costs, economies of scale were never fully leveraged as such local adaptation needed considerable redesign work. The costs were still relatively low as the basic design was the same globally. However, the car, although cheaper, did not fully meet local customer requirements in terms of dimensions and aesthetics.

In the late 2000s, as the company found it was less and less competitive in the global market, it moved away from the traditional NPD approach. Different local requirements were taken into consideration early on in NPD projects, so that the developed cars could appeal to more customers in different countries. The NPD team separated the global and local parts during the project and made different versions of cars accordingly for different markets. To achieve this, the firm created a culture of collaboration. All the local subsidiaries were highly involved in the NPD process through providing feedback. ICT tools made this task easier. The firm also worked to abandon the old mind set of central authority and be more accommodating to local subsidiaries' views on car design.

With the new global NPD approach, the company attained better economies of scale. Significant numbers of parts were shared globally, not needing redesign. Also, the cars were more suited to customers' tastes in different markets.

Case 3

Case 3 is a Chinese automotive company. Under the government's "going-out" strategy, it sought to internationalise operations and sell its cars aboard in the early 2000s. In terms of an NPD approach, it followed the same approach as Case 2, developing cars for Chinese customers only and then introducing some cars in other countries with adaptations.

In the mid-2000s, this company attempted to develop a global car for different markets, but the project ultimately failed. In the NPD process, the engineering team found it very difficult to pass the regulations in developed countries. It finally managed to comply with the regulations with many revisions, but the cost was very high. What was worse, the product was not at all attractive in developed countries, partly because few people recognised the brand. It was not even attractive in China as it deviated from the low-cost position of the firm.

After the failure of this project, the company stayed focused on developing countries which did not have very high standards. The customers in developing countries were more price sensitive, making it easier for the company to sell its cars. Also, the company reverted to its original NPD approach to focus on Chinese requirements in projects as the NPD team could not determine the volume of potential car sales in a certain country. This was also because the company was more established in China and the volume was more likely to be sufficient to justify such an investment. Then, after a product was launched in China, the local marketing team promoted it in the local market and then evaluated the local market volume. If there was sufficient volume, the company would introduce that car with adaptations.

Case 4

Case 4 is a Chinese automotive company, too. Unlike Case 3, in the mid-2000s it obtained technologies, brands and distribution channels through the acquisition of a Western firm. This provided an opportunity for the firm to expand its overseas market. Although it was able to meet the standards in developed countries with the acquired technologies, the quality of the car and brand awareness were still much lower than the industry leaders. Under such conditions, the company largely followed the same approach as Case 2, focusing on Chinese requirements in NPD projects and then introducing the cars in sufficient volume to other countries with adaptations.

The company was seeking the opportunity for a global car and it developed and launched one for the global market (except in the USA) in the late 2000s. For this car, the requirements of all markets were taken into consideration early on in the NPD project. Compared to Case 3, this car was not a total failure. However, the overseas sales were still below the target. The company chose to continue the global car approach for the next generation of this model and included a diesel engine option to stimulate sales. It remained to be seen if this model could reach the target sales volume.

For other cars, the company still defined them as Chinese cars in NPD projects although the engineering team started to consider a few requirements (such as left-wheel and right-wheel drive) in other countries for the convenience of later modification. As the overseas market volume was not certain in the early stage of NPD, the firm could hardly benefit from developing global cars. In comparison, the market volume was much higher and stable in the Chinese market.

Findings

NPD approaches and evolution

In the four cases, three types of NPD approaches in MNCs were identified: multi-local, adaptation-based and platform-based. In the multi-local approach, a company

Table 2. NPD approaches in MNCs.

NPD approach	Definition	Advantages	Challenges
Multi-local	R&D units are located in different countries developing products for local markets.	Products are fully tailored to local customers. Developing local products is less complex.	There is duplication in product design. R&D costs are higher at the global level.
Adaptation-based	Products are developed for the domestic market initially and are then modified for sale in other markets.	There is less duplication. NPD is less complex as only local requirements are considered.	Significant redesign work is needed as some requirements are not considered initially.
Platform-based	Different requirements are considered in the global NPD project. Global and local parts are determined in projects.	Economies of scale are fully leveraged. The different requirements across countries are fulfilled. More variety can be generated based on a common platform.	Global projects are more complex than local projects to manage. Specific target markets must be determined from the beginning of platform development.

has R&D centres in different areas of the world and each of them develops products for the local market. In the adaptation-based approach, the products are developed mainly in one location with only domestic market requirements considered in NPD projects, the products then being introduced later in other countries incorporating local adaptations. The platform-based approach is different from the above two in that the NPD projects are defined as global from the beginning, with requirements from all the countries taken into consideration. In NPD, the global (common) parts and local (special) parts are determined as early as possible. Table 2 shows a summary of three NPD approaches in MNCs.

The four cases show different patterns of evolution in NPD approaches (see Fig. 1). Case 1 evolved from a multi-local approach to a platform-based approach. Case 2 changed from an adaptation-based approach to a platform-based approach. The two cases show a converging trend in NPD approaches. In contrast, Case 3 attempted to change from an adaptation-based towards a platform-based approach, but failed and then returned to the initial approach. Case 4 launched a global car as a sign of moving towards a platform-based approach.

The multi-local approach and the adaptation-based approach used to be common in the past in developed countries (Bartlett and Ghoshal, 1998). Each has

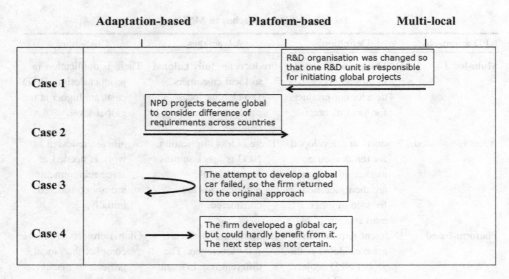

Fig. 1. NPD approach: Evolution of cases.

certain advantages and disadvantages. In the multi-local approach, R&D centres are located close to the markets, and products are tailored to local customers in terms of aesthetic and functional requirements. NPD capability in multiple locations could be the result of a firm's growth by acquisition. However, with this approach, it is very difficult to share components as coordination is difficult across countries. As a result, duplication raises the R&D costs. For example, the car chassis may be developed in totally different ways in different places, which can actually be avoided. When competition becomes more intense, the profitability of a company employing such a strategy will decline.

In the adaptation-based approach, NPD capability is usually centralised in one location. Different versions of products in different countries are derived from one basic design and therefore, duplication is significantly lower. The adaptation of an existing product is much cheaper than developing a new product. However, with this approach, economies of scale are not fully realised. Although adaptation is usually intended to be minimal, mainly arising from differences in regulations and manufacturing equipment, the modifications often require the redesign of many interrelated components. Such modification is easier if requirements are considered early on in the NPD project (Gunzenhauser and Bongulielmi, 2008; Halman et al., 2003). Also, due to specifications that are not easily modified (such as the size and appearance of the product), the products may not suit local customers' preferences.

The platform-based approach combines the advantages of the above two approaches and therefore denotes the trend in the NPD approach for MNCs.

As parts are shared globally at significant levels, duplication is eliminated and economies of scale are fully realised. Also, as requirements from all over the world are considered early on in NPD projects, the products will be suitable for different markets. The products still have different versions in different countries, but by adopting a platform-based approach, it is clear which parts are global and which are local, so that no significant redesign is needed after product launch. Firms with a platform-based NPD approach can be more competitive in the global market by offering higher quality products at a lower price and still fulfilling local requirements. However, this approach is more complex to implement in practice. Firms have to determine the target markets (countries) at the beginning and handle a larger amount of information in product design, which can drive up the cost of a single project. Considering the challenges, such an approach may not be successfully implemented in every firm. We will elaborate this point next.

The motivation for change towards a platform-based approach

Change of NPD approaches takes significant effort and cost. For example, firms need to change organisational structures and NPD processes. Despite that, development of ICT tools and competition pressure lead to the motivation for change. For the platform-based NPD approach, a much larger amount of information is needed from local subsidiaries, including local requirements and feedback at different stages of projects. Such operations would not be desirable without the development and use of ICT tools. Two decades ago, when ICT tools were less developed and not as widely used as today, engineering and marketing personnel had to travel frequently to facilitate the flow of information needed for the platform-based approach, adding to the R&D costs. Under such conditions, multi-local and adaptation-based approaches would be reasonable choices, as they do not need an intensive flow of information from all over the world.

Modern ICT tools, such as NPD process management software and teleconferencing systems, make cross-border communication much easier. For example, in the NPD process, engineers need to receive feedback from marketing personnel regarding whether products can meet local requirements. With teleconferencing systems, virtual meetings can be held globally in an efficient way. In addition, the documents of product design and market information can be transmitted easily through NPD process management software.

With more MNCs moving to a platform-based approach using ICT tools, firms that do not change (as in Case 2) will become less competitive over time. Many firms, by employing a platform-based approach, offer products of higher quality that meet local requirements at lower costs. As a result, unchanged firms will lose either market share or profit. Feeling the pressure of competition and survival, firms will seek to change towards a platform-based approach, even though firms will incur significant costs of organisational change.

The benefit of change towards a platform-based approach

However, benefiting from the platform-based approach has certain prerequisites. Therefore this approach may not be desirable for every firm. Cases 3 and 4 offer good examples for this. Unlike Cases 1 and 2, they lack strong brand awareness overseas and technical capability, hindering them from achieving sufficient market volume to yield the benefits of the more complex platform-based approach. In comparison, Cases 1 and 2 can benefit from platform-based approach due to their strong brand awareness and technical capability.

Brand awareness plays a vital role in car purchase as the brand is usually linked to safety for customers. Without a strong brand, the expected (average) sales volumes of cars are low. Though some cars can be popular in certain countries, as we found in the case studies, it is difficult to figure out at the beginning of NPD whether the market volume overseas worth the investment. There is a risk that after a global car is developed within a platform-based approach at a greater cost, it will not sell in overseas markets. Cases 3 and 4 illustrate this risk very well. Under such conditions, the benefits of the adaptation-based approach are very clear: after domestic product launch and overseas promotions, the company can choose the cars which have a higher success rate based on the market feedback. Any such prediction is more precise after domestic product launch and overseas promotions.

Technical capability is another factor contributing to the expected sales volumes. With low technical capability, the quality, functionality and aesthetics of a product will probably fall behind what competitors offer. Like in Cases 3 and 4, their cars fell behind competitor offerings in terms of fuel efficiency and the power of the engine, for instance, which made their cars unattractive even to customers who knew the brands or would like to consider cars with these brands.

The reinforcing cycles

Based on brand awareness and technical capability, MNCs are likely to be in a reinforcing cycle (virtuous or vicious) influencing the choice of an NPD approach (see Fig. 2). Firms such as Cases 1 and 2 have a high level of brand awareness and technical capability. When they start to develop a product, they expect the sales volume of the product to be high in their global markets. Therefore, with the use of ICT tools and competition pressure, they will adopt the platform-based approach

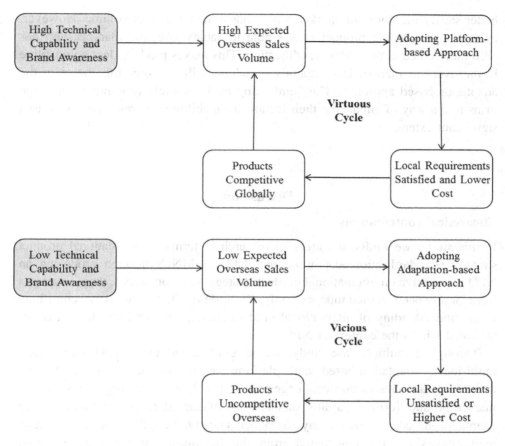

Fig. 2. Global NPD approach and competitiveness.

to define a global car at the beginning of a project. Although developing a global car is more expensive, they can achieve economies of scale through high volume globally, so firms are willing to make such an investment. Knowing the requirements of all countries, the product can readily satisfy local customers at a low cost, which contributes to product competitiveness in the global market. Product competitiveness further contributes to the expected sales volume, therefore reinforcing the choice of the platform-based approach.

Firms with low brand awareness and technical capability, such as Cases 3 and 4, are in a vicious cycle. When they develop a product, they expect the overseas volume to be low. In fact, there are variations across products, but the average volume is low. Under such conditions, firms adopt an adaptation-based approach due to uncertain benefit from a platform-based approach which costs more. Another reason is that through an adaptation-based approach, they can focus on and better satisfy the domestic market, which has a higher market volume. However, in this approach, the product is less able to satisfy overseas markets, and has high incurred costs of product modification. This makes products less competitive in the overseas market. Uncompetitive products will reinforce the choice of the adaptation-based approach. This reinforcing cycle is likely to continue until the firms find a way of enhancing their technical capability and brand awareness to a significant extent.

Discussion

Theoretical contributions

In this study, we bridge the areas of research in terms of international product strategy (standardisation/adaptation) and NPD in MNCs through proposing an NPD perspective on international product strategy. In prior studies, these two areas have largely been studied independently. By bridging the two areas, we contribute to the understanding of international product strategy, in particular how it is determined within the context of NPD.

Through a multiple-case study, we have identified three NPD approaches: multi-local, adaptation-based and platform-based. We have discussed their definitions, advantages and challenges. We show how NPD approaches can influence product forms regarding standardisation/adaptation. For example, for the adaptation-based approach, many components need to be redesigned as numerous requirements are not considered from the beginning. In the platform-based approach, much redesign work can be avoided as the requirements are considered from the beginning. Prior studies mainly addressed whether products need to be standardised/adapted (Calantone *et al.*, 2006; Katsikeas *et al.*, 2006; Samiee and Roth, 1992; Zou and Cavusgil, 2002). However, our findings show how the products should be standardised/adapted in NPD for the global market, which has been discussed to a lesser extent in prior studies.

More importantly, we argue that certain factors could influence the choice of an NPD approach in MNCs. Prior studies identified the advantages of the platform-based approach (Chai *et al.*, 2012; Gunzenhauser and Bongulielmi, 2008; Robertson and Ulrich, 1998), which are confirmed in this study. However, prior studies have not addressed under what circumstances MNCs will choose this approach over others. In this study, we find the influencing factors including ICT development, competition pressure, brand awareness and technical capability. ICT development and competition pressure lead to the motivation for change towards a platform-based approach, which denotes the trend. Prior studies show that ICT tools make cross-border collaboration easier (Chang, 2006; Howells, 1995; Kleinschmidt et al., 2010). We show that with the convenience of cross-border communication, manufacturers tend to change their NPD approaches under the competition pressure. Therefore, the competition pressure promotes the use of ICT tools. MNCs by adopting the platform-based NPD approach can be more competitive than those with the adaptation-based or multi-local approach.

Brand awareness and technical capability determine the benefit of the platform-based approach. Firms that have low technical capability and brand awareness are not likely to benefit from a platform-based approach due to their low expected market volume. As the literature shows that there is no single optimal position between standardisation and adaptation (Delene et al., 1997; Katsikeas et al., 2006; Schmid and Kotulla, 2011), this study suggests that the same applies to the choice of an NPD approach in MNCs. In this regard, we contribute to research on contingencies affecting the choice of NPD approaches (Pasche et al., 2001). In addition, we reveal the reinforcing cycles (virtuous or vicious) based on brand awareness and technical capability. This offers an additional explanation of case firms and highlights the role of the two factors.

Limitations and future research

As with all studies, this research has some limitations. In this study, only four case companies are examined, therefore, its generalisability needs further confirmation. In the future, more case companies could be studied to see if our findings could be replicated. Future studies could also test our findings through quantitative methods with a large sample. In addition, in this study we only explored the automotive industry. Future research could examine other manufacturing industries to see what findings can be generalised to other industries and what cannot. For example, brand awareness may be less important in some industries than in the automotive industry, which may influence the choice of the NPD approach. It will be important to identify key industrial factors that could influence the results.

Also, this study explores NPD as a whole without observing the role of relevant functions (e.g., R&D, sourcing, production and marketing) involved in NPD. Future research could explore how these functions interact with each other in each NPD approach we have found. These functions may exert some influences on the evolution and the choice of the NPD approach. For example, the production function can be centralised globally or independent locally, which will influence NPD because manufacturing issues need to be taken into account during NPD. There is likely the co-evolution between each function and the NPD approach, which needs to be studied.

Managerial implications

This study can draw implications for practices not only in the automotive industry, as focused on in this study, but also other manufacturing industries in which different requirements across countries remain. The platform-based NPD approach is superior in terms of global competition and therefore, firms should aim to move towards this approach in the long run. However, in the short term, such a path may not be suitable for every company depending on the current condition. Based on our case studies, there are certain paths that companies can follow to break the vicious cycle in relation to technical capability and brand awareness and be more competitive. The first path is acquisitions. Through acquisitions, the company can obtain stronger brands and advanced technologies, thus attaining a higher expected sales volume overseas. The NPD approach may be multi-local immediately after acquisition. The challenge is how to integrate R&D centres to facilitate the platform-based approach. Change of power and processes of NPD is needed for integration. ICT tools should be used to support the platform-based approach which needs a large amount of information flows across countries. The second path is to gradually accumulate technical knowledge and brand awareness overseas. One tactic could be forming strategic alliances with leading manufacturers, which may be helpful for learning technologies and enhancing brand awareness overseas (Doz, 1996). For learning technologies, the firm could collaborate with the partner in NPD of cars. For enhancing brand awareness, the firm could advertise this alliance relationship whenever possible in overseas markets. In practice, a company can also mix the two paths in a flexible way in the growing process, for example accumulating technical knowledge through strategic alliances and the acquisition of brands.

Appendix Interview Protocol

Company background and market requirement

- What are the product lines of the company?
- How are market requirements similar or different across countries?
- What is the market share?
- What are the target markets?

NPD approaches related to product standardisation/adaptation

- How are products developed for different countries?
- How are products standardised/adapted for different requirements across countries?

- How are NPD activities organised across R&D centres?
- Why are NPD activities organised this way?

Change of NPD approaches and influencing factors
- How were the NPD approach changed over the years?
- What is the story behind the change?
- Was the change successful?
- If the old approach could not work well, why?

References

Andriopoulos, C and MW Lewis (2009). Exploitation-exploration tensions and organizational ambidexterity: Managing paradoxes of innovation. *Organization Science*, 20(4), 696–717.

Bélanger, J, C Berggren, T Björkman and C Köhler (1999). *Being Local Worldwide: Abb and The Challenge of Global Management*. Ithaca, NY: Cornell University Press.

Barczak, G and EF McDonough (2003). Leading global product development teams. *Research-Technology Management*, 46(6), 14–18.

Bartlett, CA and S Ghoshal (1998). *Managing Across Borders: The Transnational Solution*. 2nd edn. Boston, Massachusetts: Harvard Business School Press.

Bartlett, CA and S Ghoshal (2000). *Transnational Management: Text, Cases, and Readings in Cross-border Management*, 3rd edn. Boston: Irwin McGraw-Hill.

Bessant, J and D Francis (1997). Implementing the new product development process. *Technovation*, 17(4), 189–197.

Bierly III, PE, EM Stark and EH Kessler (2009). The moderating effects of virtuality on the antecedents and outcome of NPD team trust. *Journal of Product Innovation Management*, 26(5), 551–565.

Bloch, PH (1995). Seeking the ideal form: Product design and consumer response. *Journal of Marketing*, 59(3), 16–29.

Bohnsack, R, J Pinkse and A Kolk (2014). Business models for sustainable technologies: Exploring business model evolution in the case of electric vehicles. *Research Policy*, 43(2), 284–300.

Calantone, RJ, ST Cavusgil, JB Schmidt and G-C Shin (2004). Internationalization and the dynamics of product adaptation — An empirical investigation. *Journal of Product Innovation Management*, 21(3), 185–198.

Calantone, RJ, D Kim, JB Schmidt and ST Cavusgil (2006). The influence of internal and external firm factors on international product adaptation strategy and export performance: A three-country comparison. *Journal of Business Research*, 59(2), 176–185.

Cavusgil, ST and S Zou (1994). Marketing strategy-performance relationship: An investigation of the empirical link in export market ventures. *Journal of Marketing*, 58(1), 1–21.

Cavusgil, ST, S Zou and GM Naidu (1993). Product and promotion adaptation in export ventures: An empirical investigation. *Journal of International Business Studies*, 24(3), 479–506.

Chai, KH, Q Wang, M Song, JIM Halman and AC Brombacher (2012). Understanding competencies in platform-based product development: Antecedents and outcomes. *Journal of Product Innovation Management*, 29(3), 452–472.

Chang, CM (2006). Web-based tools for product development. *International Journal of Product Development*, 3(2), 167–180.

Cooper, RG (1994). Third-generation new product processes. *Journal of Product Innovation Management*, 11(1), 3–14.

Cooper, RG and EJ Kleinschmidt (1985). The impact of export strategy on export sales performance. *Journal of International Business Studies*, 16(1), 37–55.

de Brentani, U and EJ Kleinschmidt (2004). Corporate culture and commitment: Impact on performance of international new product development programs. *Journal of Product Innovation Management*, 21(5), 309–333.

de Brentani, U, EJ Kleinschmidt and S Salomo (2010). Success in global new product development: Impact of strategy and the behavioral environment of the firm. *Journal of Product Innovation Management*, 27(2), 143–160.

Delene, LM, MS Meloche and JS Hodskins (1997). International product strategy: Building the standardisation-modification decision. *Irish Marketing Review*, 10(1), 47–54.

Doz, YL (1996). The evolution of cooperation in strategic alliances: Initial conditions or learning processes? *Strategic Management Journal*, 17(S1), 55–83.

Eisenhardt, KM (1989). Building theories from case study research. *Academy of Management Review*, 14(4), 532–550.

Eisenhardt, KM and ME Graebner (2007). Theory building from cases: Opportunities and challenges. *Academy of Management Journal*, 50(1), 25–32.

Eppinger, SD and AR Chitkara (2006). The new practice of global product development. *MIT Sloan Management Review*, 47(4), 22–30.

Eslami, MH and N Lakemond (2016). Internal integration in complex collaborative product development projects. *International Journal of Innovation Management*, 20(1).

Gooderham, P (2012). The transition from a multi-domestic to globally integrated multinational enterprise–in an industry where local taste matters. *European Journal of International Management*, 6(2), 175–198.

Graber, DR (1996). How to manage a global product development process. *Industrial Marketing Management*, 25(6), 483–489.

Griffin, A and JR Hauser (1993). The voice of the customer. *Marketing Science*, 12(1), 1–27.

Gunzenhauser, M and L Bongulielmi (2008). A value chain oriented approach for the development of global platforms in the systems business. *Journal of Engineering Design*, 19(6), 465–487.

Halman, JI, Hofer, AP and W van Vuuren (2003). Platform-driven development of product families: Linking theory with practice. *Journal of Product Innovation Management*, 20(2), 149–162.

Hansen, ZNL and S Ahmed-Kristensen (2011). Global product development: The impact on the product development process and how companies deal with it. *International Journal of Product Development*, 15(4), 205–226.

Hart, S, E Jan Hultink, N Tzokas and HR Commandeur (2003). Industrial companies' evaluation criteria in new product development gates. *Journal of Product Innovation Management*, 20(1), 22–36.

Hauser, J, GJ Tellis and A Griffin (2006). Research on innovation: A review and agenda for marketing science. *Marketing Science*, 25(6), 687–717.

Howells, JR (1995). Going global: The use of ICT networks in research and development. *Research Policy*, 24(2), 169–184.

Jarvenpaa, SL and DE Leidner (1999). Communication and trust in global virtual teams. *Organization Science*, 10(6), 791–815.

Katsikeas, CS, S Samiee and M Theodosiou (2006). Strategy fit and performance consequences of international marketing standardization. *Strategic Management Journal*, 27(9), 867–890.

Kim, J and D Wilemon (2002). Focusing the fuzzy front-end in new product development. *R&D Management*, 32(4), 269–279.

Kleinschmidt, EJ, U de Brentani and S Salomo (2007). Performance of global new product development programs: A resource-based view. *Journal of Product Innovation Management*, 24(5), 419–441.

Kleinschmidt, E, U de Brentani and S Salomo (2010). Information processing and firm-internal environment contingencies: Performance impact on global new product development. *Creativity & Innovation Management*, 19(3), 200–218.

Kotabe, M (1990). Corporate product policy and innovative behavior of European and Japanese multinationals: An empirical investigation. *Journal of Marketing*, 54(2), 19–33.

Kotler, P (1986). Global standardization — courting danger. *Journal of Consumer Marketing*, 3(2), 13–15.

Lawrence, TB and G Dover (2015). Place and institutional work: Creating housing for the hard-to-house. *Administrative Science Quarterly*, 60(3), 371–410.

Leonidou, LC, CS Katsikeas and S Samiee (2002). Marketing strategy determinants of export performance: A meta-analysis. *Journal of Business Research*, 55(1), 51–67.

Levitt, T (1983). The globalization of markets. *Harvard Business Review*, 61(3), 92–102.

Luchs, M and KS Swan (2011). The emergence of product design as a field of marketing inquiry. *Journal of Product Innovation Management*, 28(3), 327–345.

McDonough III, EF, KB Kahn and G Barczak (2001). An investigation of the use of global, virtual, and colocated new product development teams. *Journal of Product Innovation Management*, 18(2), 110–120.

Millson, MR and J Kim (2015). A moderation study of organisational integration and NPD process proficiency in the us and Korean heavy construction equipment industries. *International Journal of Innovation Management*, 19(5).

Moorman, C and AS Miner (1998). The convergence of planning and execution: Improvisation in new product development. *Journal of Marketing*, 62(3), 1–20.

Muffatto, M (1999). Platform strategies in international new product development. *International Journal of Operations & Production Management*, 19(5/6), 449–459.

Nambisan, S (2003). Information systems as a reference discipline for new product development. *MIS Quarterly*, 27(1), 1–18.

Ozer, M (2000). Information technology and new product development: Opportunities and pitfalls. *Industrial Marketing Management*, 29(5), 387–396.

Pasche, M, M Persson and H Lofsten (2011). Effects of platforms on new product development projects. *International Journal of Operations & Production Management*, 31(11), 1144–1163.

Robertson, D and K Ulrich (1998). Planning for product platforms. *Sloan Management Review*, 39(4), 19–31.

Rugman, A and R Hodgetts (2001). The end of global strategy. *European Management Journal*, 19(4), 333–343.

Söderquist, KE (2006). Organising knowledge management and dissemination in new product development: Lessons from 12 global corporations. *Long Range Planning*, 39(5), 497–523.

Salomo, S, EJ Kleinschmidt and U de Brentani (2010). Managing new product development teams in a globally dispersed NPD program. *Journal of Product Innovation Management*, 27(7), 955–971.

Samiee, S and K Roth (1992). The influence of global marketing standardization on performance. *Journal of Marketing*, 56(2), 1–17.

Schmid, S and T Kotulla (2011). 50 years of research on international standardization and adaptation - from a systematic literature analysis to a theoretical framework. *International Business Review*, 20(5), 491–507.

Schmidt, JB and RJ Calantone (2002). Escalation of commitment during new product development. *Journal of the Academy of Marketing Science*, 30(2), 103–118.

Seidel, VP (2007). Concept shifting and the radical product development process. *Journal of Product Innovation Management*, 24(6), 522–533.

Shibata, T and M Kodama (2015). Managing the change of strategy from customisation to product platform: Case of Mabuchi motors, a leading DC motor manufacturer. *International Journal of Technology Management*, 67(2–4), 289–305.

Subramaniam, M (2006). Integrating cross-border knowledge for transnational new product development. *Journal of Product Innovation Management*, 23(6), 541–555.

Subramaniam, M and N Venkatraman (2001). Determinants of transnational new product development capability: Testing the influence of transferring and deploying tacit overseas knowledge. *Strategic Management Journal*, 22(4), 359–378.

Tavcar, J, R Zavbi, J Verlinden and J Duhovnik (2005). Skills for effective communication and work in global product development teams. *Journal of Engineering Design*, 16(6), 557–576.

Trappey, CV, AJC Trappey, C Tzu-An and K Jen-Yau (2009). A strategic product portfolio management methodology considering R&D resource constraints for engineering-to-order industries. *International Journal of Technology Management*, 48(2), 258–276.

Tzokas, N, EJ Hultink and S Hart (2004). Navigating the new product development process. *Industrial Marketing Management*, 33(7), 619–626.

Ulrich, K (1995). The role of product architecture in the manufacturing firm. *Research Policy*, 24(3), 419–440.

Ulrich, KT and SD Eppinger (2012). *Product Design and Development*, 5th edn. New York: McGraw-Hill.

Verworn, B (2006). How german measurement and control firms integrate market and technological knowledge into the front end of new product development. *International Journal of Technology Management*, 34(3/4), 379–389.

Wheelwright, SC and KB Clark (1992). *Revolutionizing Product Development: Quantum Leaps in Speed, Efficiency, and Quality*. New York: The Free Press.

Yin, RK (2009). *Case Study Research: Design and Methods*. 4th edn. Los Angeles, CA: Sage.

Yip, GS (2003). *Total Global Strategy II: Updated for the Internet and Service Era*. Upper Saddle River, N.J.: Pearson Education International.

Zou, S and ST Cavusgil (2002). The GMS: A broad conceptualization of global marketing strategy and its effect on firm performance. *Journal of Marketing*, 66(4), 40–56.

Chapter 2

Internationalisation of New Product Development and Research & Development: Results from a Multiple Case Study on Companies with Innovation Processes in Germany and India*

Alexander Brem and Florian Freitag

A rich body of literature has emerged from research on Western new product development (NPD). However, the impact of country- and culture-specific influences on these processes has not been examined in detail yet. Hence, this study identifies the differences in NPD practices between the Indian and German research and development (R&D) subsidiaries of multinational companies (MNCs). Data have been generated by interviews with R&D executives in both countries across multiple cases. The study samples strategic, organisational, and operational aspects and indicates differences in process coordination, reward systems, NPD creativity techniques, market orientation, and the average age of NPD teams. Other aspects, such as top management support, the use of structured NPD processes, and the use of heterogeneous NPD teams, show no substantial differences between the countries. Our findings suggest that, while some aspects are universally applicable across cultural frontiers, Western companies must understand Indias different expectations regarding NPD and adjust their practices accordingly.

Keywords: Innovation process; R&D Management; New Product Development; Internationalization; India; Germany; Emerging markets.

Introduction

In the last 10 years, the face of innovation and its management has changed. The main driver for this development is the changing role of Asian countries like India and China (e.g., Agarwal and Brem, 2012). Multinational companies (MNCs) used to rank emerging market economies primarily as low-cost locations for routine operations, while most of their research and new product and service development was carried out in the home country, as the creation of new technology was

---

*Originally published in *International Journal of Innovation Management*, 19(1), pp. 1–32.

geographically sticky to a company's headquarters (Patel and Pavitt, 1991). However, research and development (R&D) activities began to shift to developed countries outside of the home countries, producing a few preeminent centres of innovation within the triad of North America, Japan, and Western Europe (Ernst, 2005; Karlsson, 2006; Bruche, 2009). Since around the turn of the millennium, foreign direct investment (FDI) from Western multinational firms into R&D in developing countries has increased progressively at the expense of R&D investments in developed countries (UN, 2005).

Now, MNCs see enormous business and market opportunities in those markets, and have located operations of higher value — such as sales, marketing and, more recently, R&D — in these economies (Deloitte, 2007). As a result, today's R&D map is far more geographically dispersed across the globe, with more centres for innovation than ever (e.g., Cantwell, 1995).

This paper examines the implications of this worldwide spreading of R&D processes on the innovation processes of MNCs, focusing on the example of India. First, we analyse the literature on R&D and its management in India and introduce the three main dimensions of analysis, namely strategy, organisation, and operations. Our empirical section presents the results of the analysis of four companies pursuing R&D activities in Germany and India. Finally, we discuss our findings as well as the study's limitations; we then provide an outlook on future research.

India attracts foreign R&D operations

The number of innovations developed by MNCs in emerging economies have significantly increased, and India and China have emerged as prominent countries in MNCs' global R&D map (Bruche, 2009; Zeschky et al., 2011). A survey conducted in the manufacturing industry by Deloitte in 2007 presented the projected directions for internationalising R&D. As shown in Fig. 1, China and India come first and second.

The most frequently cited motivations for relocating R&D to emerging economies are the wish to obtain a better understanding of the market, a faster time to market, lower costs, government incentives, and new ideas (Deloitte, 2006; see also Fig. 2). The wish to access the most talented R&D personnel and exploit pools of skilled labour emerged as further important drivers (Manning et al., 2008; Economist Intelligence Unit, 2004). However, labour cost advantages begin to lessen as competition for skilled employees rises (Deloitte, 2006).

A survey on the changing importance of the strategic drivers of offshoring decisions, as shown in Fig. 3, describes a notable shift over the years 2004, 2005, and 2006. The lower costs of labour still ranked in first place; however, the access

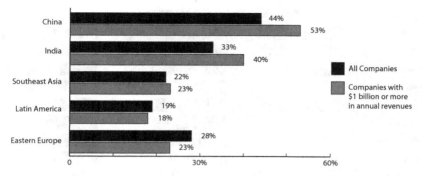

Fig. 1. Expected types of future investments in R&D operations.

Note: Percent extremely or very likely to establish or significantly expand operations within the next five years. Base: Companies, not headquartered in market, which are at least somewhat likely to invest.

Source: Deloitte (2007: 5).

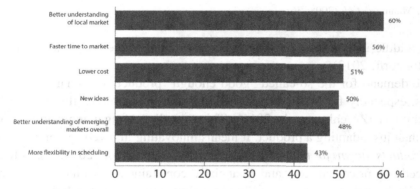

Fig. 2. Key benefits from local R&D.

Note: Percentage of executives rating the benefits of locating R&D in emerging markets as extremely or very important.

Source: Deloitte (2006: 13).

to qualified personnel gained importance. In the short span of just two years, this driver's importance increased by 26%.

A total of 60% of consulted executives rated a better understanding of the local market as the most decisive factor for relocating R&D to emerging countries (Deloitte, 2007). India's market incorporates features that vary greatly from Western economies. Its nominal gross domestic product is the ninth largest in the world and it is ranked third in the world by purchasing power parity. Despite being one of the fastest growing economies in the world, India is only rated a

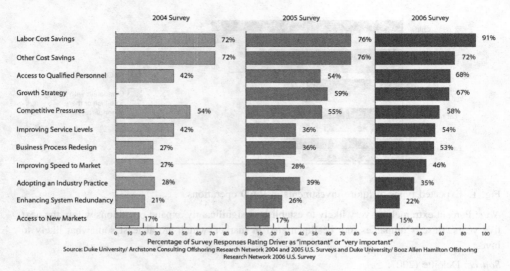

Fig. 3. Changing importance of strategic drivers of offshoring decisions.
Source: Manning *et al.* (2008: 36).

lower-middle income economy by the International Monetary Fund (Sithemsetti and Borstorff, 2012).

The demand for the so-called "good enough" products is a challenge for producers, especially for Western multinationals and their differently oriented business models (Zeschky *et al.*, 2011). This so-called "frugal innovation" means more than just adapting a product. It means innovating in reverse in order to "*strip the products down to their bare essentials*" (Economist, 2010: 7), reflecting the needs of price-sensitive and financially constrained consumers, who would otherwise be non-consumers. At the same time, the face of the Indian market is changing with the rise of a new middle class that has become an interesting part of a market that offers great business potential (Agarwal and Brem, 2012; Zeschky *et al.*, 2011).

This potential, manifested in the emergence of new clusters in India, is providing talent and up-stream services (e.g., software development, product design, engineering) to MNCs. Bangalore, a metropolis in the State of Karnataka in South India, has become a cluster for science and engineering (S&E), which attracts a large number of MNCs searching for specialised skills and S&E talent (Manning, 2008). The city of Bangalore has become a centre of excellence for information and communication technology and is one of India's software hubs (Caniels and Romijn, 2003). By sharing in those clusters, companies can benefit from the dynamics of those S&E regions and harness technology developed by other companies (Karlsson, 2006).

National culture affects organisational processes

> "You cannot simply take a North American version of a business practice, move it to China or India, and just flip the switch. It won't work."

Director of Global Process Optimization at a major U.S.-based manufacturer (Deloitte, 2007: 4).

Many scholars have contributed to the literature on innovation practices and success factors in Western and developed economies (Barczak et al., 2009; Grinstein, 2007; Cooper et al., 2004a, 2004b, 2004c; Griffin, 1997). However, literature on innovation practices in India published in English is scarce. Studies in this discipline have produced contradictory findings on aspects of innovation as support for organisational structures within firms. Some organisational shapes condemned as repressive towards innovation in the Western literature have been found to have positive effects on innovation in the Indian business environment (Prakash and Gupta, 2007).

This finding is in line with research conducted by the Economist Intelligence Unit (2004) on the internationalisation of R&D activities. On the one hand, the standardisation of the R&D approach is mentioned as one of ten principles of international R&D success. However, another principle is described as *"Don't underestimate cultural differences"* (Economist Intelligence Unit, 2004: 16). Thus, cultural differences should be embraced and taken into consideration when determining an offshore R&D subsidiary's space. As different cultural backgrounds are in play, the challenge is to create the right balance between independence and similarity in the R&D activities of different countries. These principles are endorsed by other scholars, who state that the implementation of standardised processes is inefficient in culturally inconsistent markets. Griffith et al. (2000) found that standardisation is applicable across nations featuring similar cultural characteristics but is inappropriate across nations featuring different cultural types. In another context, Lindholm (2000) researched the possibility of standardising human resource management practices across nations. Again, he found that practices must be modified before they are applied to other cultural contexts. Newman and Nollen (1996) stated that business performance improves if management practices are congruent with the values of the external national culture. This is in line with research in the context of internationalisation of R&D (Brem and Ivens, 2013).

A major work on national cultures by Hofstede (2001) claimed that the cultures of Germany and India are significantly distinct. The cultural dimensions Power Distance, Uncertainty Avoidance, Individualism versus Collectivism, Masculinity versus Femininity, and Long-term versus Short-term Orientation were evaluated across more than 50 countries. In all categories, India and Germany scored substantially differently; thus, one can legitimately claim that Germany and India are characterised by big cultural differences.

A more recent publication by Hofstede *et al.* (2010) shows that cultural differences on the national level affect firms' organisational processes. The affiliation of an individual to a company and therefore to an organisational culture influences daily practices rather than underlying cultural values. Hence, beyond the superficial organisational culture, which manifests in shared practices, lies the deeper level of culturally conditioned values. Interviews with project managers have shown that Hofstede's cultural dimensions have implications for running scientific projects within different cultures in terms of management style and decision-making (Shore and Cross, 2005).

Research on the constructs culture and new product development (NPD) has suggested interference. The findings have underlined the importance of national culture to MNCs in a globalized world and to NPD in particular (Nakata and Sivakumar, 1996). The simultaneously articulated need for further empirical study is taken as an occasion to examine the processes implemented in NPD at German and Indian sites of three MNCs and analyse cases of practices that trace back to different national cultures.

Aspects of Inquiry

Following Ozer (2011), the aspects under study are grouped into three dimensions for NPD: strategy, organisation, and operations. Relevant subordinate aspects of these dimensions have been identified by a literature review focusing on Western NPD practices. Relevance was assigned according to whether the studies indicated the existence of cultural differences or expressed an expectation that culture has significant effects. As mentioned, there are few English studies on Indian NPD. Hence, literature on Chinese and Hong Kong NPD was sometimes consulted as a proxy[1] for Indian NPD.

Figure 4 illustrates how the strategic dimension covers the framework in which NPD is embedded. The second dimension is the organisational span within which NPD is integrated into a company. Depth development processes are covered by the operational dimension.

---

[1]Geographic and relative cultural proximity, as defined by Hofstede (2001), encouraged the decision to use China and Hong Kong as proxy countries for India. The purpose of using a proxy here is to outline the already detected cultural differences in NPD between Western and Asian countries.

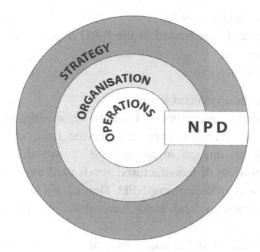

Fig. 4. Dimensions of corporate NPD.

The first dimension, strategy, comprises the strategic dimensions of NPD and the involvement of top management in the NPD process.

Technology and innovation strategy should play a vital role in a company's efficient integration of market-pull and technology-push (Brem and Voigt, 2009). Leiponen and Helfat (2010) state that the choice of innovation objectives also plays a significant role. Their empirical study found that in Western companies a greater breadth of innovation objectives correlates with greater new product success in terms of sales revenues. Another study found that Western firms tend to follow a balanced choice across several innovation objectives (Cooper *et al.*, 2004a). Chinese firms use a less balanced set of innovation objectives; cultural differences concerning risk-adversity are mentioned as having an effect here (Ozer, 2011).

Cooper *et al.* (2004a) and Elenkov (2005) found that the degree of senior management support has noticeable effects on a firm's NPD performance; senior management must be actively involved in designing and focusing NPD projects, show high commitment to new products, and provide support and empowerment to NPD project teams while avoiding micro-management. However, as suggested in Elenkov (2005), effective management for innovation is dependent on the ambient national culture. Ozer (2011) supports this thesis by pointing out that the perceived top management support in Western firms (in which 79% of managers are "best performers" (Cooper *et al.*, 2004a) far exceeds the perception of top management support in Chinese firms (in which 23% of managers are in that category).

The organisation dimension refers to the

- functional organisation of NPD,
- integration of marketing and sales,

- team structures in R&D and NPD,
- hierarchical structures implemented in the R&D department,
- forms of rewarding, and
- degree of formalisation.

These criteria will be explained below.

The literature describes many possibilities for organising NPD responsibilities in a firm. Research on manufacturing and service firms conducted by Griffin (1997) found that firms engage in an average of two different structures for organising NPD; producers of manufactured goods used even more. Accordingly, there is no one best way of organising NPD. Hence, this study examines whether NPD processes in Germany and India have different functional organisation preferences.

Ernst et al. (2010) elucidate the importance of a cross-functional integration of sales and marketing into the NPD process. The marketing inflow of customer and market information into the NPD process and participation in decisions about product positioning and features increase the success of new products (Wren et al., 2000; Griffin and Hauser, 1996). Involving sales in the conceptual and product development phases is a driver for NPD success (Ernst et al., 2010). Therefore, this study examines the integration of both functions into the NPD process in both countries to determine how connected these processes are.

The use of multifunctional teams is considered a best practice factor in Western NPD and is thus widely practiced in Western firms (Cooper et al., 2004a) and is positively related to NPD success (Griffin, 1997). This has been confirmed by studies on Chinese, Japanese, and Korean firms (Song and Thieme, 2006; Song and Noh, 2006). Hence, this study analyses the diversity and average age of NPD teams in both countries.

The Western literature states that minimising vertical complexity increases a firm's innovative capabilities. Hierarchy and thus inequality amongst members of an organisation hinder change (Burns and Stalker, 1994). By contrast, Prakash and Gupta (2007) found that, in Indian manufacturing firms, hierarchy is seen to support innovation and has a significantly positive relationship with the number of innovations due to the variety of national cultures and their respective *"preference for hierarchy"* (Shane, 1992).

Western best practice firms use only non-financial rewards for successful NPD projects. Financial rewards are seldom used and are therefore not popular in Western firms (Barczak et al., 2009; Griffin, 1997). Another study, conducted against a different cultural background (Ozer and Chen, 2006), found that both financial and non-financial rewards are used in Hong Kong firms to motivate employees in NPD teams.

Formalisation refers to *"the degree of work standardisation and the amount of deviation that is allowed from standards"* (Aiken and Hage, 1966: 499). Formalisation is used to exert control over what is to be done and what is to be refrained from (Bodewes, 2002). From a Western perspective, an extensive and concrete set of rules in the workplace restricts an employee's autonomy to define much of his or her own work, blocking employee initiative concerning how tasks could be executed differently or even better. Hence, a work environment of freedom gives employees the space to make decisions and share information in order to conceive new methods based on their own perspective (Ekvall, 1996; Grønhaug and Fredriksen, 1981). The Indian perspective again differs: formalisation is an important aspect of innovation, as Indian employees greatly value discipline and coordination through rules (Prakash and Gupta, 2007).

The last dimension, operations, comprises the NPD process, the idea sources, and creativity techniques as well as the market orientation of NPD processes. A formal, structured NPD process helps to move through the development of innovative products and is associated with best-practice firms (Barczak *et al.*, 2009). A study found that, in 2004, about 70% of all participating firms had implemented a formal NPD process (Cooper *et al.*, 2004c). Cooper *et al.* (2004c) stress that, though most businesses have such a process, it is the quality of the execution that drives NPD performance. While firms in both the United States and in Hong Kong use formal NPD processes, Asian companies use them less than their Western counterparts (Ozer and Chen, 2006).

Even in Western best-practice firms, almost 50% of ideas for new products come from random, informal sources. Such ideas tend to lack strategic fit and hence lack potential for realisation. Formally generated ideas arising from a strategic need are more likely to be successful in the marketplace (Barczak *et al.*, 2009). As idea sources account for a great deal of NPD success, this study examines which and how many sources are used for NPD in Germany and India.

A study by Cooper and Edgett (2009) looked at 160 firms in both the business-to-business (B2B) and business-to-consumer (B2C) markets, and examined how extensively the methods of NPD idea generation are applied. Among the sample firms, a multitude of techniques had been used extensively. The use of formal techniques has positive impacts on NPD processes (Barczak *et al.*, 2009). Chinese firms, in particular, seem to focus on a smaller set of NPD techniques (Ozer, 2011). Grinstein (2008) found that the use of customer and competitor orientation has a positive effect on innovative capability in Western firms. Culture has impacts here as well; in countries with a high power distance or a high degree of individualism, the effect of market orientation on new product performance was found to be even stronger.

Methodology

This study employed a multiple case study based on nine face-to-face interviews across different multinational corporations who carry out research activities in Germany and India.[2] The case study follows an inductive and primarily interpretive logic, as it aims at generating a descriptive and explanatory framework of how national culture influences the practice of NPD in a globally dispersed R&D environment.

Study design

Considering social phenomena, case studies are regularly applied as a research method (Yin, 2003a, b). This method provides rich data and is particularly suited to research questions requiring a grounded understanding of social or organisational processes (Hartley, 2004). This study satisfies all the criteria listed by Yin (2003a): the researcher requires no control over the events, the research focuses on contemporary events, and it poses "how" and "why" questions. Thus, a case study strategy is particularly appropriate for this context. Further, a multiple case study was applied that treats the cases as independent units and thus forms a set of multiple holistic case studies (Yin, 2003a). The multiple case study method is less vulnerable to uniqueness than is a single case study (Yin, 2003a); it produces *"more robust, generalizable, and testable theory than single-case research"* (Eisenhardt and Graebner, 2007: 27) and thus provides stronger evidence for theory building (Leonard-Barton, 1990). The replication of results across cases, either direct or literal, produces robust data and strong support for the theory derived (Yin, 2003a). A mixed-methods approach is used in order to utilise both qualitative and quantitative research and thus gain most complete understanding of the researched phenomenon (Johnson *et al.*, 2007; Creswell, 2002). As the study is aimed primarily at gaining inductive insights, it uses qualitative research to discover and understand the procedural contrasts between German and Indian NPD. Quantitative research elements are utilised to investigate and test certain factors considered important in the extant Western literature in the context of a different cultural background (Johnson and Onwuegbuzie, 2004). Eisenhardt (1989) explains that case studies can provide both, quantitative data to build a theorys foundation or to test theory, as well as qualitatively acquired data to build theory or to explain phenomena.

---

[2]These data were used for a comparative study of Chinese and Indian innovation processes as well (published in the proceedings of the XXIV ISPIM Conference held in Helsinki, Finland, 2013).

As *"interviews are one of the most commonly recognized forms of qualitative research"* (Mason, 2002: 63), this study conducted problem-centred interviews (PCIs). This method combines listening with interposing questions (Witzel, 2000). The PCI consists of a short questionnaire that collects the social characteristics of the interviewee; it breaks the ice between interviewer and interviewee, and provides guidelines that build an orientation framework and assure comparability among the interviews. A framework of pre-formulated lead questions forms a guideline for the interview (Witzel, 2000). On the quantitative side, standardised questionnaires with seven-item Likert scaling were incorporated into the dimensions at issue (Witzel, 2000). The questionnaire consists of open questions, and questions that can be analysed with descriptive statistics. For each case company, at least two interviews took place to ensure validity of the given information. All questions in the PCI were derived from literature. The full interview guideline can be found in the appendix of this paper.

The interviews were recorded and transcribed. Following Mayring (2000), the data were then analysed step-by-step. The applied categories are partially inductive and deductive. For the purpose of orientation, categories are grouped into sections following the aspects discussed in the interviews. This coding agenda was put together on a spreadsheet, and passages of the transcribed interviews were assigned to corresponding categories in order to make the interviews formally comparable. Finally, a cross-case synthesis was performed to determine similarities and patterns in the interviews. Case studies were grouped by country and examined for intra-group similarities and inter-group differences (Eisenhardt, 1989).

Units of analysis

This study's aim is to gain insights into how NPD processes at different R&D sites of MNCs in Germany and Bangalore (India)[3] differ and to what extent differences in national cultural backgrounds account for these differences. Hence, the case study units were selected according to the following criteria:

- The MNCs must have a Western origin (i.e., Europe or USA).
- There must be fully qualified R&D as well as NPD activities at both locations in Germany and India from the same business unit.

---

[3]As shown in Chan *et al.* (2010), subnational regions have greater influence on management and performance in emerging countries than they do in developed countries. Hence, we decided to ensure comparability across cases by looking at only one region in India. Further, Manning (2008) indicates that S&E clusters like Bangalore feature a unique environment.

Based on these criteria, we identified 17 companies fulfilling these conditions. Two German technology-oriented MNCs and one US-based technology-oriented MNC, a direct competitor to one of the German companies, took part in our study.

Interviews were conducted personally on the correspondent companies' premises in Germany and India. Interviewees were chosen according to the following criteria:

- Interviews can be conducted with at least one leading person in R&D.
- Participants must be full time employees of the company (no external staff).

Case A

This case involves a German tech company with several business segments and more than 100,000 employees worldwide. Its R&D is conducted at many locations worldwide, with operative R&D in Germany and India.

Case B

Case B examines the same MNC as Case A, but was accomplished in a completely different business segment.

Case C

This case involves a US tech company with subsidiaries in several countries worldwide. It also employs more than 100,000 employees and conducts R&D in centres around the world, with Germany and India hosting such centres. The same business segment used in Case A is the background of this analysis.

Case D

This case involves a German tech company with over 10,000 employees that primarily operates in the automotive industry. Operations and R&D are carried out globally as well, with subsidiaries in Germany and India.

Nine out of 10 interviews took place in person; one interview was conducted via telephone, as the personal appointment had to be postponed. Each interview lasted 90 min to 120 min. Table 1 shows an overview of the times and locations of interviews in Germany and India.

Results and Discussion

Strategy

NPD in both German and Indian R&D locations follow a broad and balanced set of strategic aspects. Figure 5 depicts the arithmetic means per aspect and country and the rating of the strategic dimension of the NPD process in both countries.

Table 1. Timing and location of interviews.

Case	Position	Location	Date of interview	Mode of interview
A	Global Technology Leader	India	July 6, 2012	face-to-face
B	Research Group Leader	India	July 6, 2012	face-to-face
B	R&D Group Leader	Germany	June 24, 2012	face-to-face
B	R&D Group Leader	Germany	June 24, 2012	face-to-face
A	Research Group Head	Germany	July 27, 2012	face-to-face
C	R&D Program Leader	India	July 9, 2012	face-to-face
C	Head of R&D	Germany	July 19, 2012	face-to-face
C	R&D Lab Manager	Germany	July 19, 2012	face-to-face
D	Head of R&D	India	August 1, 2012	telephone
D	Director Product Development	Germany	July 18, 2012	face-to-face

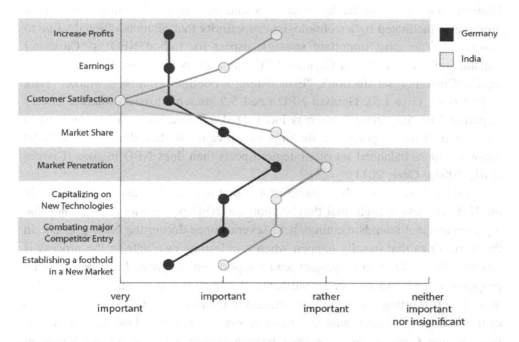

Fig. 5. Rating of strategic aspects of NPD in both countries.

Note: Median Scores per Country.

In the arithmetic averages, significant deviations between both countries can be detected in "Increase Profits" (deviation: 0.83), "Customer Satisfaction" (deviation: 0.75), "Market Penetration" (deviation: 0.58), and "Capitalising on New Technologies" (deviation: 0.42).

Hence, Indian NPD is more focused on delivering to the buyer's satisfaction than on increasing profitability, while the opposite is the case for Germany. Increasing market penetration is more important to NPD in India, consistent with India's Long-Term Orientation and the corresponding objective of strengthening their own market position (Hofstede, 2001). As India's markets are developing and growing at a much faster pace than Germany's, there is greater possibility that participating companies take advantage within as well as with those markets (CIA, 2012a, 2012b). The ability to capitalise on new technologies shows a reversed image. In India, the still growing segment of resource-constrained consumers with little excess income to spend calls for no-frills products. Products are only considered as the required *"value for money"* (Zeschky et al., 2011: 39); product development takes this demand into consideration and attempts to deliver the same benefit with cheaper material or technologies or by excluding additional features. However, Germany can be described as a saturated market in which selling new products is facilitated by a technological superiority that offers additional value to customers. The most important strategic aspect for Indian NPD is "Customer Satisfaction" ($\hat{x} = 1.25$). In German NPD, the aspects "Increase Profits", "Earnings", "Customer Satisfaction", "Establishing a Foothold in a New Market" rank in first place ($\hat{x} = 1.5$). German NPD rated 5.5 strategic aspects (median) "very important" or "important," whereas India rated 4.5 strategic aspects (median) in this range. This supports the findings of previous studies that Western NPD follows a more balanced set of strategic aspects than does NPD in Asia (Cooper et al., 2004a; Ozer, 2011).

All interviewees in both countries stated that top management involvement in the NPD process is high, that they support the NPD process in general, and that top management attends meetings held several times during the NPD process. In those meetings that usually happen when a milestone or a gate in the process is reached, the NPD team or project leader reports on progress. Management then proposes a "go," "no-go," or amendments. While all Indian contacts declared that those report meetings happen in determined intervals of $\bar{x} = 6.2$ times a year, two of the four German cases said that those reports are scheduled on-demand. In the two residual German cases, reviews happen in frequent and comparably short intervals of $\bar{x} = 8.6$ or 12 reviews per year. However, no investigated case in the present study indicates micro-management in the NPD processes by the top management.

Indian NPD tends to have determined intervals in its top management reviews. This is consistent with the observations made by Prakash and Gupta (2007), who found that employees in the Indian manufacturing sector prefer a formalised work environment, as it fosters discipline and coordination. However, this result diverges from India's low preference for Uncertainty Avoidance (Hofstede, 2001). The finding that 50% of the German cases have no determined NPD review schedules also partly confirms the results in Ekvall (1996) and Grønhaug and Fredriksen (1981), who found a formulated work environment to be harmful to innovation processes in Western firms.

Organisation

Both Indian and German R&D subsidiaries mainly use project-based teams in their functional organisation of firm NPD; four of the five German and two of the four Indian contacts commented accordingly. Figure 6 shows the responses given in each country.

In two German cases and one Indian case, the NPD was organised in more than one way, congruent with the finding in Griffin (1997) that there is not one but several ways to organise NPD within a firm. Still, building project-based teams seems to be most common methodology in both countries.

Given the data generated, it is not possible to determine a country-specific practice for integrating the marketing and sales functions into the NPD process. In three of four cases, both countries' contacts said that marketing and/or sales are integrated. Interestingly, the extent to which those departments are integrated into NPD is congruent across countries and cases, implying that if marketing is integrated throughout the NPD process in Germany, it would also be integrated in India and if marketing or sales is only integrated at one specific stage (e.g., identifying market needs) in India, it would be the same in Germany. In only one of the four cases were mismatching answers given by the interviewees, suggesting that this facet of NPD is determined by corporate culture rather than by national culture.

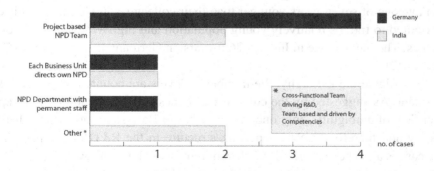

Fig. 6. Functional organisation of NPD.

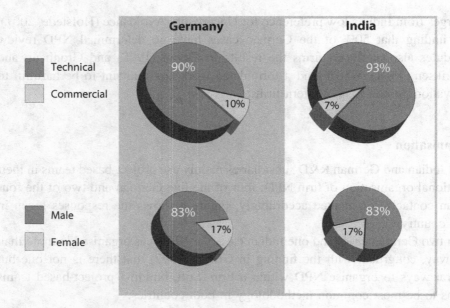

Fig. 7. Make-up of NPD teams in Germany and India.

Note: Percentage of technical versus commercial background of team members/female and male team members.

With the exception of one R&D subsidiary per country, a heterogeneous workforce is employed in the R&D departments. As a result, NPD team structures in both countries and across all cases are functionally heterogeneous. As NPD is mostly a technical discipline, engineers account for the largest part of NPD team members. Moreover, technical disciplines are typically dominated by men, and female NPD team members are rare. Figure 7 illustrates that, in both aspects, there is little difference between German and Indian NPD teams.

However, we recognized that the average age of NPD team members is significantly different when comparing India and Germany. Team members in India are on average of seven years younger than their colleagues in Germany. This can be attributed to India's relatively young population and high availability of young graduates. The median age in India is 26.6 years, in Germany it is 44.9 years (CIA, 2012a, 2012b).

To investigate the hierarchy, the number of levels are counted for every R&D department. As suggested by the cultural willingness of Indian culture to accept a greater deal of ambiguity, only one of four cases in Bangalore indicated a distinct hierarchy and therefore a clear reporting structure in the R&D department. This single case is also the youngest R&D location in Bangalore, led by a German Head of R&D, which may account for the difference. This result contradicts the findings

Prakash and Gupta (2007) and India's high index value on Power Distance (Hofstede, 2001), suggesting that Indian NPD favours high vertical complexity.

In Germany and in the respective R&D subsidiaries, two of five respondents stated having a distinct hierarchy, while three interviewees said that they had a flat hierarchy, illustrating Germany's position in the middle field (ranked 35 out of 53 countries) in Power Distance (Hofstede, 2001).

All polled R&D subsidiaries in both countries use a system of both monetary and non-monetary rewards for successful NPD projects (Fig. 8), yet a significant difference exists concerning the non-monetary rewards given to the whole team. All four Indian cases use team dinners to motivate or reward the whole team, whereas only one respondent in Germany used such practices. This result is in line with India's collectivistic culture and the emanating "we" consciousness (Hofstede, 2001). Only one case in India grants special monetary rewards to project leaders; in all other cases, project leaders and team members are rewarded the same way. Examples of non-monetary rewards include team lunches or dinners and more responsibility for successful project leaders. Monetary rewards are given either directly after contributing a significant part to a project or are included in annual performance rewards.

The extensive use of individual monetary rewards in Germany is inconsistent with the findings of Barczak *et al.* (2009) and Griffin (1997), who conclude that monetary rewards are not popular in Western firms. Ozer and Chen (2006) state that Hong Kong firms likewise use financial and non-financial rewards for NPD projects.

As R&D employees did not participate in this survey section, the "degree of formalisation" aspect could not be examined in this context.

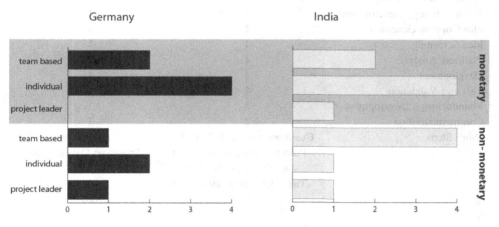

Fig. 8. Rewards given for successful NPD projects.

Operations

A structured and determined NPD process is used by all the R&D subsidiaries examined. However, there is a structural difference in how rigidly it is adhered to. All German interviewees remarked that there is flexibility for process adaptation, whereas their Indian colleagues stated that the NPD process in use is to be followed strictly — or even *religiously*, as one interviewee put it. Adaptations to the process could be made or a less extensive process used because of time pressure, low (financial) importance of the project, and special process technology requirements. While companies in both countries use formal NPD processes, Indian NPD seems to follow them more rigidly than German NPD.

The NPD processes in both countries include almost the same number of steps. With a median of nine, German NPD processes include one step more than India's. Which steps are included in the NPD processes is shown in Table 2. The process steps in this table are based on Griffin (1997).

These findings do not replicate the results of Ozer and Chen (2006). Whereas Hong Kong firms use their NPD process less than Western firms do, Indian NPD seems to use the NPD process more rigidly than their Western equivalents do. These findings for Germany and India diverge widely from what is suggested by

Table 2. Steps of NPD processes in Germany and India.

Process step	Number of NPD processes per country including this step (out of the four cases)	
	Germany	India
Product Line Planning	3	2
Project Strategy Development	3	2
Idea/Concept Generation	4	4
Idea Screening	4	4
Business Analysis	4	4
Development	4	4
Test and Validation	4	4
Manufacturing Development	3	2
Commercialisation	2	2
Other Steps	Customer Survey, Filing of Patents, Maturation, Virtual Development, Supplier Talks, Quality Management	Engineering Reviews, Prototype Development
$\bar{X}$	9	8.67
$\hat{X}$	9	8

their cultural Uncertainty Avoidance index scores (Hofstede, 2001), which point to contrary results on the rigidity of the processes used in these countries.

Interesting results are found regarding idea sources for NPD. On average, German R&D executives named 3.5 idea sources and Indian R&D executives 3.25 for NPD. In all cases, the importance of the creative potential of the employees was stressed. Three of the four cases in both Germany and India use their customers for idea input. Two of the four German cases also analyze competitors for ideas, whereas only one Indian R&D subsidiary utilises this source. Suppliers are widely consulted by all four cases in Germany, yet no Indian R&D department in the sample involves their suppliers in the process of generating or sourcing ideas. Figure 9 gives an overview of the distribution of idea sources per country.

In two of the four cases, German R&D executives named the company's employees as the most important idea source. The remaining two cases consider all idea sources, internal and external, to be equally important. Three of the four Indian cases place equal weight on all the idea sources they use. One case, the R&D subsidiary with the German head, places emphasis on their own employees as the most important idea source.

Across the cases, the median number of creativity techniques used for NPD in Germany is 6 and 4.5 in India. In two cases, Germany scored higher than its Indian equivalent (Case A: +3, Case B: +6); in one case, the number of used creativity techniques was equal in both countries, and one Indian case scored higher by one technique. Figure 10 depicts the distribution across cases.

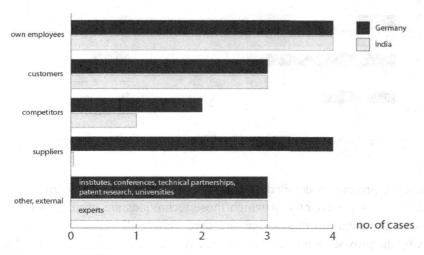

Fig. 9. Idea sources of NPD.

Note: Use of different idea sources for NPD by the examined four cases.

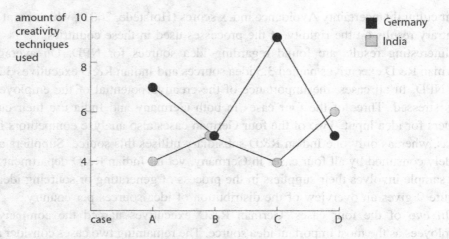

Fig. 10. Creativity techniques used per case.

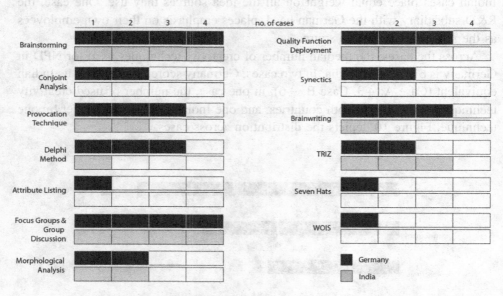

Fig. 11. Creativity techniques in NPD across cases and per country.

Figure 11 presents a detailed array of the used creativity techniques and the number of cases per country in which those techniques are used. Brainstorming and focus groups/group discussions are used across all cases, followed in frequency by the provocation technique, quality function deployment, and the Theory of Inventive Problem Solving (TIPS resp. TRIZ). NPD conducted in Germany uses a wider set of creativity tools to generate ideas for new products. Again, this

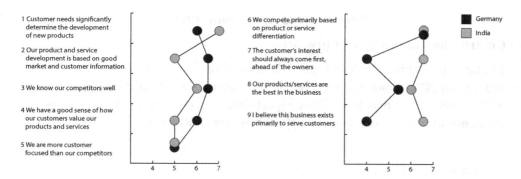

Fig. 12. Market orientation of NPD in Germany and India.

Note: R&D executives were asked to rate how far the following statements apply; median scores (1 = "strongly agree," 2 = "disagree," 3 = "rather disagree," 4 = "neither agree nor disagree," 5 = "rather agree," 6 = "agree," and 7 = "strongly agree"); as all ratings were between 4 to 7, only that rating scale is shown in the figure.

matches the findings in Ozer (2011) for China and indicates that Indian NPD also focuses on a smaller set of NPD techniques.

Two cases in Germany said that they use creativity techniques at several points during the NPD process, whereas this is true for only one case in India. The remaining cases apply those techniques solely in the front end of NPD.

In three cases and in total, Indian R&D executives are more market-oriented than their German equivalents. Figure 12 presents the scores per country and aspect. German NPD seems to be based on a better understanding of the market (including customers and competitors), as German R&D executives scored higher in Categories 2 and 3. A substantial country difference in this regard can be found in the degree of customer valuation, with Indian R&D executives scoring 2.5 points higher in Categories 7 and 9.

The Indian contacts constantly stressed the high demands and special requirements of the Indian customer, whose satisfaction requires extra effort. This may account for the high scores in 1, 7, and 9. Another driving factor is the relatively high combined score of Power Distance and Individualism in India.[4] Those two cultural dimensions add weight to the importance of market orientation (Grinstein, 2007). India's Long-Term Orientation quality explains why the customers' interests outweigh the focus on the business' bottom line. Long-term orientated cultures prefer to forge longer lasting relationships with business partners (Hofstede, 2001).

---

[4]The combined index scores on Power Distance and Individualism are 102 for Germany and 125 for India.

Implications and Limitations

Contribution to theory and practice

The findings of this study indicate that there are several differences between German and Indian NPD practices, the underlying reasons for which could be cultural, market-specific, or manifold. Other aspects and success factors of Western NPD seem, however, to be universally applicable to Indian R&D subsidiaries.

Contribution to theory

This study performs the first holistic comparison of German and Indian NPD practices and thus contributes to the body of literature that compares Western and Asian NPD practices. The findings highlight the aspects of Indian NPD that differ from Western NPD practices and present cultural and market-related interpretations. Differences between the countries have been detected, consistent with Griffith *et al.* (2000), who stated that standardisation is not appropriate for countries with different cultural types. This study also confirms the interference of national culture and NPD in MNCs (Nakata and Sivakumar, 1996), as divergences are ascribed to discordant national cultures.

According to our results, Hofstede's dimensions cannot be applied by default to detect divergence in NPD practice. As seen above, our findings do not reflect Hofstede's scores of both countries in several cases.

Implications for practice

Although the R&D department in India is *just another* corporate subsidiary, differences in processes and practices can be grave. The desire for a rigid process to regulate NPD and schedule top management confirmation or feedback poses an especially sharp contrast to German NPD practices. Contrary to that degree of process coordination, Indian R&D subsidiaries have a flat hierarchy. The younger NPD workforce (by almost seven years) and a balanced use of financial and non-financial rewards demand a different kind of personnel management. German executives coming to India to set up or lead NPD will have to rethink the role of the customer, as this factor is more highly valued there than in German NPD.

However, some aspects were shown to be applicable to both Western and Indian NPD. Both countries display support for NPD by top management across all cases. All cases across both countries show a structured process for NPD, with a similar number of steps taken by a heterogeneous NPD team. Both countries also display an equal degree of integration of their marketing and sales departments into NPD, though the shape of this aspect is related to the company's preference.

Evidently, India offers great potential and talent for carrying out R&D and NPD, but Western firms locating development facilities in India's industrial clusters must acknowledge the differences in order to understand and be able to deal with the behaviour of partners and employees. Whereas some practices may be transferred in a standardised matter, others will need adjustment if the Indian potential is to be used most efficiently. This knowledge will allow Western MNCs to amend their expectations and perceptions about NPD organisation in India.

Limitations and further research

The primary aim of this study is to examine the differences in NPD practices between German and Indian R&D subsidiaries and provide a starting point for an explanation of these. However, due to the study's small sample, more extensive research will be needed to extend and refine the foundation laid by this paper. A potential bias might be the fact that not all MNCs were from the same country. Through the fact that all interviews were conducted in person and only one via telephone, we ensured a high openness regarding the innovation processes, and how they are really lived in each country. As pointed out by Nakata and Sivakumar (1996), Hofstede predicts aggregate behaviour but does not take into account individual behaviour. Thus, a larger case study sample may be better able to replicate the cultural predictions made by Hofstede (2001). We suggest examining country- and culture-specific NPD practices with regard to their NPD efficiency in order to provide better guidance as to whether the preferable option is to adjust or standardise NPD practices. Moreover, future research could also include the problem of coordinating between the companies' headquarters and its subsidiaries. That kind of research will be in line with the seven major fields of future research in innovation management theory and practice (Horn and Brem, 2013). In particular, the relationship among frugal and reverse innovation created in India, corporations' sustainability management, and their performance constructs should be further investigated to determine their influence on corporate success (Brem and Ivens, 2013). The authors gratefully acknowledge the research support by Hans-Frisch-Stiftung (Nürnberg, Germany).

Appendix

PCI Interview Guideline

I Which job position do you have within the R&D department?

II What are your main tasks?

III How many years have you been in this company?

IV What do you think about the current market situation?

- Which strategy is applied to meet customers and competitors?
- Would you consider the strategy to be a pioneer, follower or late follower strategy?
- What is more prevalent in NPD processes: Market pull/Technology push?

STRATEGY

1. How important are the following aspects for the strategic direction of NPD?

	Very important	Important	Rather important	Neither important nor insignificant	Rather insignificant	Insignificant	Very insignificant
Increase profits	1	2	3	4	5	6	7
Earnings	1	2	3	4	5	6	7
Customer satisfaction	1	2	3	4	5	6	7
Market share	1	2	3	4	5	6	7
Market penetration	1	2	3	4	5	6	7
Capitalising on new technologies	1	2	3	4	5	6	7
Combating major competitive entry	1	2	3	4	5	6	7
Establishing a foothold in a new market	1	2	3	4	5	6	7

2. The value assigned to NPD internally is high enough.

Strongly disagree	Disagree	Rather disagree	Neither agree nor disagree	Rather agree	Agree	Strongly agree
1	2	3	4	5	6	7

3. Investments made in NPD are high enough.

Strongly disagree	Disagree	Rather disagree	Neither agree nor disagree	Rather agree	Agree	Strongly agree
1	2	3	4	5	6	7

- How did the investments made in R&D develop over the last years?

4. Newly developed products significantly contribute to the company's turnover.

Strongly disagree	Disagree	Rather disagree	Neither agree nor disagree	Rather agree	Agree	Strongly agree
1	2	3	4	5	6	7

5. How high is the involvement of the Top Management in NPD?

- Does Top Management provide strong support for, empowerment to and authority over NPD team members?
- To what extent is the Top Management involved in day-to-day business and decisions of NPD?
- Does the Top Management make Go, No-Go decisions?
- How often do you have to report to the Top Management?

ORGANISATION

6. Is NPD rather steered centrally or decentrally?

Absolutely central	Quite central	Rather central	Nor central neither decentral	Rather decentral	Quite decentral	Absolutely decentral
1	2	3	4	5	6	7

7. How is NPD set up in the firm?

- NPD with permanent staff members
- Project based NPD teams
- Each business unit's general manager directs their own NPD efforts
- Other, please specify

8. Is the organisational structure in the R&D department rather functionally heterogeneous or homogeneous? (e.g., are colleagues from other departments present?)

- And what structure exists within the NPD project team?

9. Are Marketing and Sales integrated into NPD processes?

- If not, why?
- If yes, how?

10. Team structure

- How many people are working in R&D? And how many in the company? How many people work in NPD teams?
- What is the average age of people working in NPD teams?
  20–25 25–30 30–35 35–40 40–45 45–50
- What is the gender allocation in NPD teams?
- What is the allocation of employees with a technical and employees with a business background?
- According to whom is the team leader appointed and what aspects are desired for such an individual?

11. How is the hierarchal structure within the R&D department?

12. Are rewards given for successful NPD projects?

- If yes, which?
- If not, why not?
- Are there distinct monetary/non-monetary rewards?
- Are rewards given to the project only — or also to the members of the NPD team?
- Are rewards given individually or to the NPD team as a whole?

13. Is your organisation structured by strict rules or loose guidelines?

- Formalisation Inventory (with R&D employees, questionnaire, 7-item Likertscale)

OPERATION

14. Does your organisation follow a well-defined, structured process for the development of innovative new products?

- If yes, please make a quick draft.
- In the case of something missing or no draft possible:
  Which steps are included?
    - Product line planning
    - Project strategy development
    - Idea/Concept generation
    - Idea screening
    - Business analysis
    - Development

- Test and validation
- Manufacturing development
- Commercialisation
- Other activities, please specify

- Is this process to be followed strictly or is it understood as a guideline?
- Which steps are the most important?
- Where do the ideas for NPD come from? Are they from internal and/or external sources?
- Which are the most important sources?

15. Which creativity techniques are applied?

- Which techniques are mainly used?
- Are you familiar with the following techniques/which do you use?
    - brainstorming
    - conjoint analysis
    - provocation technique
    - Delphi methods
    - attribute listing
    - focus groups and group discussions
    - morphological analysis
    - quality function deployments
    - synectics
    - brain writing
    - others, please specify

16. In which phase of NPD are those techniques applied?

17. How market-oriented are processes that are carried out in NPD?

	Strongly disagree	Disagree	Rather disagree	Neither agree nor disagree	Rather agree	Agree	Strongly agree
Customer needs significantly determine the development of new products	1	2	3	4	5	6	7
Our product and service development is based on good market and customer information	1	2	3	4	5	6	7
We know our competitors well	1	2	3	4	5	6	7
We have a good sense of how our customers value our products and services	1	2	3	4	5	6	7

(Continued)

(Continued)

	Strongly disagree	Disagree	Rather disagree	Neither agree nor disagree	Rather agree	Agree	Strongly agree
We are more customer focused than our competitors	1	2	3	4	5	6	7
We compete primarily based on product or service differentiation	1	2	3	4	5	6	7
The customer's interest should always come first, ahead of the owners	1	2	3	4	5	6	7
Our products/services are the best in the business	1	2	3	4	5	6	7
I believe this business exists primarily to serve customers	1	2	3	4	5	6	7

References

Agarwal, N and A Brem (2012). Frugal and reverse innovation — Literature overview and case study insights from a German MNC in India and China. In *IEEE Xplore Proc. the 2012 18th Int. Conf. Engineering, Technology and Innovation*, B Katzy, T Holzmann, K Sailer and KD Thoben (Eds.), pp. 1–11.

Aiken, M and J Hage (1966). Organizational alienation: A comparative analysis. *American Sociological Review*, 31(4), 497–507.

Barczak, G, A Griffin and KB Kahn (2009). PERSPECTIVE: Trends and drivers of success in NPD practices: Results of the 2003 PDMA best practices study. *Journal of Product Innovation Management*, 26(1), 3–23.

Bodewes, WEJ (2002). Formalization and innovation revisited. *European Journal of Innovation Management*, 5(4), 214–223.

Brem, A and K-I Voigt (2009). Integration of market pull and technology push in the corporate front end and innovation management — Insights from the German software industry. *Technovation*, 29(5), 351–367.

Brem, A and B Ivens (2013). Do frugal and reverse innovation foster sustainability? Introduction of a conceptual framework. *Journal of Technology Management for Growing Economies*, 4(2), 31–50.

Bruche, G (2009). The emergence of China and India as new competitors in MNCs' innovation networks. *Competition & Change*, 13(3), 267–288.

Burns, T and GM Stalker (1994). *The Management of Innovation*, 3rd Edition. New York, NY: Oxford University Press.

Caniels, MCJ and HA Romijn (2003). Dynamic clusters in developing countries: Collective efficiency and beyond. *Oxford Development Studies*, 31(3), 275–292.

Cantwell, J (1995). The globalization of technology: What remains of the product life cycle model? *Cambridge Journal of Economics*, 19(1), 155–174.

Chan, CM, S Makino and T Isobe (2010). Does subnational region matter? Foreign affiliate performance in the United States and China. *Strategic Management Journal*, 31(11), 1226–1243.

CIA (2012a). The World Factbook, India. Available at https://www.cia.gov/library/publications/the-world-factbook/geos/in.html.

CIA (2012b). The World Factbook, Germany. Available at https://www.cia.gov/library/publications/the-world-factbook/geos/gm.html.

Cooper, BR and S Edgett (2009). Ideation for product innovation: What are the best methods? *Development*, 1(March 2008), 12–17.

Cooper, RG, SJ Edgett and EJ Kleinschmidt (2004a). Benchmarking best NPD practices — II. *Technology*, 47(3), 50–59.

Cooper, RG, SJ Edgett and EJ Kleinschmidt (2004b). Benchmarking best NPD practices — I. *Research-Technology Management*, 47(1), 31–43.

Cooper, RG, SJ Edgett and EJ Kleinschmidt (2004c). Benchmarking best NPD practices — III. *Research-Technology Management*, 47(6), 43–55.

Creswell, JW (2002). *Research Design: Qualitative, Quantitative, and Mixed Methods Approaches*, 2nd Edition, *Organizational Research Methods*, Vol. 6, p. 246. Thousand Oaks, CA: Sage Publications, Inc.

Deloitte (2006). Innovation in emerging markets: Strategies for achieving commercial success.

Deloitte (2007). Innovation in emerging markets: Annual report. Annual Study.

Economist Intelligence Unit (2004). Scattering the seeds of invention: The globalisation of research and development.

Economist (2010). First break all the rules: The charms of frugal innovation. *Economist 2010*, 395, 6–8.

Eisenhardt, KM (1989). Building theories from case study research. *Academy of Management Review*, 14(4), 532–550.

Eisenhardt, KM and ME Graebner (2007). Theory building from cases: Opportunities and challenges. *Academy of Management Journal*, 50(1), 25–32.

Ekvall, G (1996). Organizational climate for creativity and innovation. *European Journal of Work and Organizational Psychology*, 5(1), 105–123.

Elenkov, DS (2005). Top management leadership and influence on innovation: The role of sociocultural context. *Journal of Management*, 31(3), 381–402.

Ernst, D (2005). The complexity and internationalization of innovation: The root causes. In *Globalization of R&D and Developing Countries: Proceedings of an Expert Meeting*, K Kalotay (Ed.). Switzerland: United Nations Publications.

Ernst, H, WD Hoyer and C Rübsaamen (2010). Sales, marketing and R&D cooperation across new product development stages: Implications for success. *Journal of Marketing*, 74(5), 1–43.

Griffin, A (1997). PDMA research on new product development practices: Updating trends and benchmarking best practices. *Journal of Product Innovation Management*, 14(6), 429–458.

Griffin, A and JR Hauser (1996). Integrating R&D and marketing: A review and analysis of the literature. *Journal of Product Innovation Management*, 13(3), 191–215.

Griffith, DA, MH Hu and JK Ryans Jr. (2000). Process standardization across intra- and intercultural relationships. *Journal of International Business Studies*, 31(2), 303–324.

Grinstein, A (2007). The effect of market orientation and its components on innovation consequences: A meta-analysis. *Journal of the Academy of Marketing Science*, 36(2), 166–173.

Grønhaug, K and T Fredriksen (1981). Resources, environmental contact and organizational innovation. *Omega International Journal of Management Science*, 9(2), 155–162.

Hartley, J (2004). Case study research. In *Essential Guide to Qualitative Methods in Organizational Research*, C Cassel and G Symon (Eds.). Thousand Oaks, CA: Sage Publications.

Hofstede, G (2001). *Culture's Consequences: Comparing Values, Behaviors, Institutions, and Organizations Across Nations*, 2nd Edition. Thousand Oaks, CA: Sage Publications.

Hofstede, G, GJ Hofstede and M Minkov (2010). *Cultures and Organizations: Software of the Mind*, 3rd Edition. New York, NY: McGraw-Hill.

Horn, C and A Brem (2013). Strategic directions on innovation management — a conceptual framework. *Management Research Review*, 36(10), 930–954.

Johnson, RB and AJ Onwuegbuzie (2004). Mixed methods research: A research paradigm whose time has come. *Educational Researcher*, 33(7), 14–26.

Johnson, RB, AJ Onwuegbuzie and LA Turner (2007). Toward a definition of mixed methods research. *Journal of Mixed Methods Research*, 1(2), 112–133.

Karlsson, M (Ed.) (2006). *The Internationalization of Corporate R&D: Leveraging the Changing Geography of Innovation*. Stockholm, Sweden: Elanders.

Leiponen, A and CE Helfat (2010). Innovation objectives, knowledge sources, and the benefits of breadth. *Strategic Management Journal*, 31(2), 224–236.

Leonard-Barton, D (1990). A dual methodology for case studies: Synergistic use of a longitudinal single site with replicated multiple sites. *Organization Science*, 1(3), 248–266.

Lindholm, N (2000). National culture and performance management in MNC subsidiaries. *International Studies*, 29(4), 45–66.

Manning, S (2008). Customizing clusters: On the role of western multinational corporations in the formation of science and engineering clusters in emerging economies. *Economic Development Quarterly*, 22(4), 316–323.

Manning, S, S Massini and A Lewin (2008). A dynamic perspective on next-generation offshoring. *Academy of Management Perspectives*, 22(3), 35–55.

Mason, J (2002). *Qualitative Researching*. Thousand Oaks, CA: Sage Publications.

Mayring, P (2000). Qualitative Inhaltsanalyse. *Forum: Qualitative Social Research*, 1(2).

Mir, R and A Watson (2000). Strategic management and the philosophy of science: The case for a constructivist methodology. *Strategic Management Journal*, 21(9), 941–953.

Nakata, C and K Sivakumar (1996). National culture and new product development: An integrative review. *Journal of Marketing*, 60(1), 61–72.

Newman, KL and SD Nollen (1996). Culture and congruence: The fit between management practices and national culture. *Journal of International Business Studies*, 27(4), 753–779.

Ozer, M (2011). Strategic, organizational, and operational challenges of product innovation in China. *Research Technology Management*, 54(4), 46–52.

Ozer, M and Z Chen (2006). Do the best new product development practices of US companies matter in Hong Kong? *Industrial Marketing Management*, 35(3), 279–292.

Patel, P and K Pavitt (1991). Large firms in the production of the worlds technology: An important case of "non-globalisation". *Journal of International Business Studies*, 22(1), 1–21.

Prakash, Y and M Gupta (2007). Relationship between organisation structure and firm-level innovation in the manufacturing sector of India. *The Indian Journal of Industrial Relations*, 43(2), 191–216.

Shane, SA (1992). Why do some societies invent more than others? *Journal of Business Venturing*, 7(1), 29–46.

Shore, B and B Cross (2005). Exploring the role of national culture in the management of large-scale international science projects. *International Journal of Project Management*, 23(1), 55–64.

Sithemsetti, NR and PC Borstorff (2012). Is right now the right time to choose India as a business location? *Proceedings of the International Academy for Case Studies*, 19(1), 57–62.

Song, M and J Noh (2006). Best new product development and management practices in the Korean high-tech industry. *Industrial Marketing Management*, 35, 262–278.

Song, M and RJ Thieme (2006). A cross-national investigation of the R&D — marketing interface in the product innovation process. *Industrial Marketing Management*, 35(3), 308–322.

The PDMA Glossary for New Product Development (2012). *Product Development & Management Association.* Available at http://www.pdma.org/npd_glossary.cfm.

UN (2005). *World Investment Report 2005 "Transnational Corporations and the Internationalization of R&D."* Switzerland: United Nations Publications.

Witzel, A (2000). The problem-centered interview. *Forum: Qualitative Social Research*, 1(1).

Wren, BM, W Souder and D Berkowitz (2000). Market orientation and new product development in global industrial firms. *Industrial Marketing Management*, 611(6), 601–611.

Yin, RK (2003a). *Applications of Case Study Research*. Thousand Oaks, CA: Sage Publications.

Yin, RK (2003b). *Case Study Research: Design and Methods*, 3rd Edition. Thousand Oaks, CA: Sage Publications.

Zeschky, M, B Widenmayer and O Gassmann (2011). Frugal innovation in emerging markets: The case of Mettler Toledo. *Research Technology Management*, 54(4), 38–45.

Chapter 3

Platform Leadership of Incumbent Telecommunications Operators: The Case of BT 21st Century Network (BT21CN)*

Carlos Eduardo Yamasaki Sato

This paper addresses the problem of survival and growth of incumbent telecommunication operators. In particular, this paper investigates the extent to which the platform-based approach is appropriate for the internationalisation strategy (platform leadership) of incumbent telecommunications operators in the context of the transition to the Next Generation Network (NGN). It examines the case of BT in the UK, as a large-scale first mover in this transition, exploring the major transformation project BT 21st Century Network (BT21CN). The case demonstrates how the platform-based approach can be used to promote business transformation of an incumbent telecom operator in a turbulent environment. The platform architecture integrates two main features that are usually treated in different contexts and time frames in the literature: (i) the reusability of components and sub-systems (typically in the automotive industry); and (ii) the openness of the platform to external actors in order to drive innovation in the industry (typically in the ICT industry). BT21CN emerged as an industry platform, made possible by the maturation of the Internet Protocol (IP) as a "common technology" able to transport not only data, but also real time voice and video, with an acceptable quality of service (QoS). Reusability can help reducing costs and decreasing time-to-market for new products and services. However, openness of the platform to external actors has still limited impact due to the limited success of BT (and incumbent telecom operators in general) in their process of internationalisation. Thus, in the context of BT (and of potentially other incumbent telecom operators), the limitations in their process of internationalisation have a negative impact on the evolutionary dynamics of the platform-based approach.

Keywords: Platform; platform innovation; platform leadership; ICT services; next generation network; BT 21st Century Network.

---

*Originally published in *International Journal of Innovation Management*, 18(2), pp. 1–37.

Introduction

The concept of platforms has been increasingly used in the Information and Communication Technology (ICT) sector. The success of Facebook, Apple, and Amazon popularised the idea of platform as an underlying common entity that facilitates the interaction with customers, developers and business partners in general. Interest in platform research has also been growing (Muffato, 1999; Muffato and Roveda, 2000; Olleros, 2008; Ottoson, 2003; Standing and Kiniti, 2011) and platforms are being increasingly associated with industry innovation (Gawer, 2009a; Gawer and Cusumano, 2002). Baldwin and Woodard (2009) have suggested three waves of platform development focusing on products, technical systems, and transactions. In practice, the concept of platform is being applied to various contexts and situations. Platforms can be embedded in products, projects, firms, and markets. When a concept becomes so widespread, it can become confusing, as a wide range of phenomena can be seen through the lenses of platforms. And when a concept is used in a wide range of contexts, its explanatory power may be lost. Thus, trying to minimise this confusion, Gawer (2009a) proposed a typology and an evolutionary process to provide some guidance on the use of platforms and their development. These were based on well known "products as platforms" such as PCs, microprocessors, automobiles, and smart phones. Less research has been done on "firms as platforms," i.e., on designing or redesigning firms and their underlying infrastructure following principles of the platform-based approach.

This paper explores the extent to which the platform-based approach is adequate for the internationalisation strategy of incumbent telecom operators. With the rise of the Internet and within a fierce competitive market, incumbent telecom operators needed to consider strategies and approaches to improve their performance in order to survive and grow. It has been become increasingly clear that, in order to survive and grow, incumbent telecom operators need to internationalise. In this context, BT was chosen for being, at the time of the research, an incumbent or traditional firm in the ICT sector in urgent need of innovation. BT aimed at transforming their business due to declining performance in their traditional lines of business, and adopted a platform-based approach to reorganise their operations. As pointed out by Gawer (2009a) and Suarez and Cusumano (2009), research on platforms in the service sector is still underdeveloped. Currently the benchmark for platforms at the firm level seems to be Amazon, opening its "platform" to third parties which can sell the same type of products as Amazon is selling.

This paper integrates two strands of literature which are usually considered separately: (i) the internal approach of finding commonalities within the products and processes, largely based on the literature on product platform (Meyer, 2008;

Meyer and Dalal, 2002; Meyer and DeTore, 2001; Meyer and Mugge, 2001; Muffato and Roveda, 2000; Tatikonda, 1999; Wheelwright, 1992); and (ii) the external approach of opening up interfaces to interact with external stakeholders (customers, developers, business partners) in order to create variety and drive innovation (Gawer, 2009b; Gawer and Cusumano, 2002).

Platforms at the firm level are used as a strategic approach to simultaneously reduce costs (identifying internal commonalities in processes) and increase revenues (creating variety by opening up interfaces to interact with the external ecosystem of developers, customers, and other stakeholders). It is a very attractive approach being used by many successful firms in the ICT industry such as Google, Facebook, and Microsoft. For incumbent and very traditional firms such as BT which are highly dependent on network services (based on "hardware"), the problem is whether the platform-based approach is adequate for an incumbent telecom operator for its internationalisation strategy as it is for those successful firms (Google, Facebook, and Microsoft), mostly based on software products and services. Thus, the underlying research question for this paper is:

- *To what extent is the platform-based approach suitable to incumbent telecom operators for implementing its internationalisation strategy?*

In a very competitive environment, BT took the bold decision to replace its traditional network (based on several different types of technology) into a unified network based on Internet Protocol (IP)/Multi-Protocol Label Switching (MPLS) technology. This was done through the launch of a megaproject,[1] BT 21st Century Network (BT21CN). This new unified network is usually referred to as the Next Generation Network (NGN)[2] in the telecommunications industry.

The conditions under which the platform-based approach emerged in BT highlight the importance of the emergence of the IP as a "common technology" that made possible to transport not only data, but also real-time voice and video under a common infrastructure. For the suitability of the platform-based approach to BT, it is argued that the reusability helps them to decrease operational costs and time-to-market of new products and services. However, the openness of the platform to external actors (hence generating more revenues) is hampered by the limited success of BT (and incumbent telecom operators in general) in their process of internationalisation. The infrastructure as a platform makes the internationalisation strategy more difficult compared to firms based on product

---

[1] Flyvbjerg *et al.* (2003) define megaprojects as multibillion dollar mega infrastructure projects.
[2] For this research, NGN is viewed as "a multi-service network based on IP technology" (OECD, 2005:p. 7). It is based on the premise that voice, video, and data services are digitalised and transported using packet-switching technology based on the IP.

platform. Thus, the impact of the platform-based approach on incumbent telecom operators such as BT is still debatable, but promising.

To reach the conclusions above, the paper is structured as follows. The following section positions the platform-based approach, pointing out two main features: The internal- and external-oriented views of platforms. It also relates these features to current platform research, pointing out ways in which research in this area can be developed, especially in service industries and in terms of the co-evolution of the platform architecture and the environmental dynamics. It builds a framework of management analysis based on platform emergence, platform architecture, platform evolution, and platform leadership. This framework is used to structure the presentation of the results of the analysis. The succeeding section explains in more detail the qualitative research methodology chosen for this research, addressing issues of reliability and validity of the data collected and used. The next section develops the analysis and results, according to the framework of analysis elaborated in the preceding section. It explains the evolution of the telecommunications industry towards the adoption of IP, the unprecedented agreement in the telecom industry as a common technology to transport voice, video, and data, and the realisation of the NGN. It investigates the emergence of the platform-based approach in BT through BT 21st Century Network (BT21CN). This megaproject illustrates the building of the network as a platform for innovation based on the predominance of IP/MPLS technology. It highlights BT's business transformation, i.e., the transformative effects on BT's operations, organisational structure, and ways of doing business. It also provides a discussion on the implications of the platform strategy of BT21CN for BT and the telecommunications industry, putting forward the argument that the platform-based approach may not provide its full benefits to BT due to internationalisation issues typical of incumbent telecom operators. More specifically the reusability of components and sub-systems may help with lowering operational costs and improving time-to-market for new products and services, but the openness for external actors to drive industry innovation may not render the expected revenues as BT (and incumbent telecom operators in general) are struggling to internationalise. The final section summarises the main findings for the research question proposed, and refers to some implications and limitations of this research.

The Platform-Based Approach

Firms are constantly striving to devise ways of developing new products and services while reducing operational costs. There is also the issue of speed: To create value (in the form of products, services, and solutions) faster while being cost effective. The notion of platform seems, to a great extent, to fulfil those needs of lowering costs and increasing revenues at the same time, and this will be briefly reviewed in this section as part of the overall service concept being adopted by leading incumbent telecommunications operators.

Platform is defined in the Oxford Dictionary as "level surface raised above the surrounding ground or floor, especially one from which public speakers, performers, etc. can be seen by their audience" (Oxford, 1989: 946). This definition highlights an important feature of platforms: visibility to the audience. The visibility corresponds to the level of exposure to the audience, who can be customers or users in the telecommunications industry context. Thus, the concept of platform is preferred to system, as the latter does not highlight the visibility or exposure of the system to customers and users, in such a way that these customers and users can interact and influence its design and the products and services derived from the platform. Interestingly, Gawer and Cusumano (2002: 2, 3) define high-tech platform as "an evolving system made of interdependent pieces that can each be innovated upon." This definition seems to be still highly dependent on system and does not emphasise the visibility or exposure of the system to the "audience," i.e., to what is supposed to be considered as external to the system under consideration. It does however emphasise the interdependency of the various systems' parts and the evolution through innovation of each part. These are characteristics already emphasised in systems.

The concept of platform may also suggest the notion of something in transit, moving or ultimately changing: The launching platform for spaceships, for example, or the boarding platforms in train stations and airports, or petroleum platforms. Both the infrastructure and the service level depend on some degree of openness for the different actors to interact and integrate their efforts into new products and services. In this sense, the platform facilitates the interaction with the "moving" element, be it a spaceship, a train or an aircraft. More recently, with the popularisation of the Public Internet, the concept of platform has been transported to the software domain, where, for example, Ottoson (2003) discusses the development of an Intranet platform, and Standing and Kiniti (2011) proposed a wiki innovation platform for organisations to use wikis more effectively when a clear purpose is defined.

Baldwin and Woodard (2009: 41) propose a unified view of the platform architecture, suggesting that it is "a modularisation" that partitions the system into (1) a set of components whose design is stable and (2) a complementary set of components which are allowed — indeed encouraged — to vary. This section elaborates on these two aspects and how they can be used by firms to overcome the constraints of cost, speed-to-market, and customer experience at the same time. Based on the two aspects of the unified view of the platform architecture, on the discussion on the internal interdependence of the elements of the system of the platform, and on the exposure to external elements, there are two major approaches to the platform-based approach, the internally and externally focused strategy approaches for innovation.

Internal platform-based approach for product/service innovation

The internal (to the firm) approach of platform recognises "a subsystem or interface that is used in more than one product, system, or service" (Meyer, 2007:149). This is the product platform, where the reusability of components to improve time-to-market and cost reduction in product and service development is emphasised (see, for example, Tatikonda (1999); Muffato and Roveda (2000); Meyer and Mugge (2001); Meyer and DeTore (2001); Meyer and Dalal (2002); Meyer (2008)). This stream of literature is inspired by the automotive industry. For example, Wheelwright (1992) uses the concept of platform in the context of product development, and Meyer (2008) shows how Honda reuses its engines in different models of cars for different market segments. It is also applicable to IT (e.g., IBM) and services industries as shown in, for example, Meyer and Mugge (2001).

The concept of platform is a "common sense way for a firm to leverage technologies into new markets and, at the same time, reduce per-unit costs through more efficient production and procurement" (Meyer and Mugge, 2001:26). Here, the idea of platforms is applied to products from the supplier perspective (like IBM and SUN). The issue of product complexity in this instance is very generic and not well defined. Usually this literature of product platform is connected to manufacturing, and thus mass production. This is not the case for incumbent telecom operators that have outsourced their equipment development to specialised equipment providers. Also, the reduction in per-unit cost does not explore the potential of different forms of collaboration that the Internet culture is making possible and more popular. This leads to the more collaborative, external platform-based approach for innovation in the industry.

External platform-based approach for industry innovation

The notion of platforms emphasises the visibility or exposure of the internal system to the external environment. It also conveys the idea of flux or flow in the interfaces.

Gawer and Cusumano (2002) put forward the idea of platform leadership, using examples like Intel, Cisco Systems, and Microsoft. Their perspective, as well as of those from the product platform literature, is from the suppliers' perspective and usually the literature does not focus on how large users build their platforms in order to deliver new services. Telecom operators use Cisco Systems and Microsoft product and system platforms to build their network platform. The leadership (from the suppliers' perspective) consists of establishing market standards and architectures that are eventually adopted by large users and that are continuously advanced, providing the initiator with a sustained competitive advantage against rivals.

The discussions about platform in the literature usually concentrate on the product as the unit of analysis.[3] The notion of platform does not scale up to the large network platforms being implemented by incumbent network operators, like BT, France Telecom (FT), and Deutsche Telekom (DT). An exception is Gawer and Cusumano (2002), who used the example of NTT DoCoMo to illustrate how NTT is using different business models to create an environment where third parties are encouraged to develop applications for their mobile phones. This is part of the scope that this paper intends to cover. The platform being developed is for any device (mobile and fixed phone, PC, laptop, Blackberry, iPod, Palm, etc.).

It is important to take into account some characteristics of platforms in order to understand the platform-centric organisation and how platform innovation leads and facilitates service innovation in the telecom industry. Ciborra (1996:115) argued that

> The platform [-centric organisation] is far from being a specific organisational structure, where one can recognise a new configuration of authority and communication lines. Rather it is a virtual organising scheme, collectively shared and reproduced in action by a pool of human resources, where structure and potential for strategic action tend to coincide in highly circumstantial ways, depending upon the transitory contingencies of the market, the technology and the competitors' moves. Schematically, the platform can be regarded as a pool of schemes, arrangements, and human resources.

Firms organise differently in order to develop and implement capabilities to adopt a platform-centric organisation. The platform-based approach has significant implications for the way firms organise innovation in services (Meyer and DeTore, 2001). From the perspective of services, the platform-based approach

---

[3]See, for example, van de Paal and Steinmueller (1998) and Mansell and Steinmueller (2000) for a discussion on multimedia platforms, analysing DVD and CD-ROM; Gawer (2000) about Intel's microprocessor; and Gawer and Cusumano (2002) about Intel, Cisco, Microsoft, Palm, NTT DoCoMo, and Linux.

derives its underlying principles from the "high-variety strategy," where it is assumed that a higher variety of product or service line makes it more likely that customers will find what they are looking as they have more choices (Kahn, 1998). However, Schwartz (2004) argues that having more choice does not mean that customers will be necessarily more satisfied. Sometimes they will become more confused and/or unhappy. On the other hand, for firms to survive they must pursue the development of new products and services, and extend the high-variety strategy. Sawhney (1998:54) introduces the concept of "leveraged high-variety strategies — strategies that allow firms to achieve high variety and high growth, without a corresponding increase in costs or complexity." Sawhney (1998:54) argues that the success of the leveraged high-variety strategies is *"platform thinking* — the process of identifying and exploiting commonalities among firms' offerings, target markets, and the processes for creating and delivering offerings" (italics in the original). The platform thinking offers an interesting strategic approach for firms to survive and grow in a turbulent environment. It assumes the existence of a common core, which Olleros (2008) warns organisations to keep it lean in order not to compromise the platform's neutrality, scalability, and evolvability.[4] The platform strategy is supposed to include the totality of the firm which is pursuing the aims of reducing cost, decreasing the time-to-market of its products, services and solutions, and improving customer experience *at the same time*. Ultimately, the platform-based approach offers a way to balance fragmentation and wholeness (or totality), making sense of the interdependencies of the part and the whole within the organisation.

Evolving platform architecture to platform leadership

The concept of platform stems from the principle that stability and variability can go together (Baldwin and Woodard, 2009). And businesses yearn to have more stable aspects that may help predictability and control. Thus within the platform there is a "stable component" which can make sense of the variability around it. Business can eventually reduce costs by not having to "reinvent all the wheel," and by providing a more stable mental construct (e.g., an architecture) that can provide some rationale for actions and decision making.

As Baldwin and Woodard (2009) pointed out, the platform architecture is conceived in such a way that there is a core or common part which does not change or change very little over time, and a variable part which gives the flexibility of producing a new or different product or service without having to

---

[4]Olleros (2008) compared four digital platforms: Internet versus Minitel, Wikipedia versus Nupedia, Visa versus BankAmericard, and eBay versus OnSale.

"reinvent the wheel," keeping costs lower due to the existence and reuse of the common part.

The concept of platform can be analysed in various units of analysis. Platform-based approach can be embedded in products, projects, firms, networks, and markets. As much as the concept of fractal is concerned, the idea is that a closer look of certain complex objects can reveal the same type of structure, i.e., there are some elementary principles or rules upon which complexity is created. Platforms were first widely used in products, thus the so-called product platform. Another level of analysis is the network, as a set of interconnected nodes. The network itself is not sold to a customer, but it is the means to sell other products and services. These various units of analysis can be viewed in what Baldwin and Woodard (2009) identify as three waves of research on platforms. The first and second waves are primarily based on product platforms. The first one focusing on the reusability concept of platforms to create derivative products (e.g., Meyer and Lehnerd, 1997; Wheelwright and Clark, 1992). The second wave emphasised the power of the product platform to create an ecosystem and drive innovation in the industry, with the typical examples of Intel's microprocessors, Microsoft's operating system, and Cisco's systems (e.g., Gawer and Cusumano, 2002). The third wave emphasise the platform at the level of markets, networks, and governance structures (e.g., Rochet and Tirole, 2003). Intel, Microsoft, and Cisco are examples of "platform leaders" (Gawer and Cusumano, 2002) due to their extensive use of the platform-based approach and their success as global players in the ICT market.

In order to explain the evolution to platform leadership, Tiwana *et al.* (2010) suggests that the evolutionary dynamics of ecosystems and modules in platform settings is influenced by: (i) platform architecture, (ii) platform governance, and (iii) environmental dynamics. Platform architecture co-evolves with environmental dynamics. Taking this into account, it is possible to better evaluate to what extent the platform-based approach is suitable to firms engaging with this type of approach.

As noted before, so far most of the research on platforms has been focusing on tangible products (Intel, Microsoft, and Cisco are examples), and little research has been done on platforms in service industries. The existence of this deficiency is corroborated by Gawer (2009a) and Suarez and Cusumano (2009). Another deficiency is that the literature on platforms usually focuses on mass market products (in a business-to-consumer context) from the perspective of their suppliers (e.g., microprocessors, operating systems, and automobiles). Less attention is paid to the business-to-business context of platforms.

It is the aim of this paper to shed some light on these deficiencies of the literature on platforms, and to provide empirical evidence on the emergence and evolution of platforms towards platform leadership.[5] Another aspect that is explored in this paper is the use of the platform-based approach as a strategy in the context of business renewal (or transformation), with impact on the whole organisation, and not only focusing on the way particular products or services are designed. As suggested by Aerts *et al.* (2004), the ICT platform architecture is mutually influenced by and aligned with the business architecture and application architecture. And as highlighted by Tiwana *et al.* (2010) exogenous factors (i.e., the environmental dynamics) coupled with the endogenous factor of platform architecture might be important to explain the degree of success of the evolutionary dynamics to platform leadership. These interdependencies are important to consider when analysing the success of platform architectures.

Framework of management analysis

Based on the literature review above on platform-based approach, it is possible to draw a framework of management analysis based on the following elements: (i) platform emergence, (ii) platform architecture, (iii) platform evolution, and (iv) platform leadership. This is represented in Fig. 1.

Platform emergence and platform evolution are mostly discussed by Tiwana *et al.* (2010) and Gawer (2009a). Platform architecture is mostly discussed in Baldwin and Woodard (2009) with the support of other authors exploring the internal and external platform-based approaches as discussed earlier in this section. And platform leadership is mostly discussed by Gawer and Cusumano (2002) and Gawer (2000). The notion of platform leadership is much related to the internationalisation strategy which incumbent telecom operators are seeking to implement.

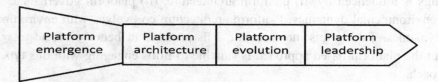

Fig. 1. Framework of management analysis based on platform literature.

Source: Author's own elaboration.

---

[5]Gawer (2009a) proposed an evolutionary typology of platforms, emerging from internal platforms, evolving to supply chain platforms, and eventually to industry platforms. She highlights the need to validate this hypothesis empirically. Although this paper provides empirical evidence of the impact of platforms on the service industry, it focuses on the emergence phase (initial stages) of the platform-based approach.

Research Methodology

This research was based upon a variant of participant observation in which the author's previous background as a telecommunication engineer and manager allowed him to be recognised by people in the industry as a fellow engineer rather than a social science researcher.[6] In seeking an understanding of telecommunication industry developments by attending trade conferences and interviewing specialists, it became apparent that the major issue for companies was defining the fundamental change needed within the industry and the organisations, namely the traditional telecommunication operators, in order to cope with the shifting competitive environment. More particularly, the fundamental change was concerned with the development of a more flexible infrastructure, and with the rethinking of the innovation processes to create and deliver new services. This change can be translated into a new dominant logic based on platform and solutions, where the customer and the service delivered to the customer are the centre of business practices. The question was not whether incumbent telecom operators needed to change their infrastructure and their innovation processes in services, but how to make these changes in an uncertain and competitive environment carrying a huge legacy system.[7]

The transition to NGN is a relatively recent phenomenon. NGN was legitimised and adopted by the main incumbent telecommunications operators like BT in the first half of the 2000s (OECD, 2005). At the time of this research, BT intended to complete the transition to NGN by 2011/12 while others, like DT and FT, would supposedly take longer (completion by 2015 or later).[8] The methodology is primarily qualitative, and the data collection involved conducting interviews and collecting documentation during the period between 2005 and 2008. An important element of the data collection was the attendance at trade conferences in order to

---

[6]The participant observation was variant in the sense that, although I was attending conferences as I normally did in my previous job, I was not employed by any of those firms, which helped me "to retain some critical subjectivity about the situation" (Maylor and Blackmon, 2005:236). Thus, the research objectives and the participants' objectives were not co-determined, and had a high level of independence. On the other hand, the participants may be less willing to cooperate or may give less information than expected. I address these issues and how I tried to avoid or overcome them in this section.

[7]Interview with DT Technical Manager, March 2005; interview with Lucent Technical Manager, March 2005; interview with Nortel Senior Technical Manager, March 2005.

[8]Interview with BT Senior General Manager, November 2005; interview with DT Project Manager, November 2005; and interview with FT Technical Manager, November 2005. These different approaches were also mentioned in the interview with KT (Korea Telecom) Business Development Manager, November 2005.

interview executives, attend their presentations and gain insights which would not have been possible (or would have taken much more time) by only analysing documents. The interaction between the information obtained through interviews (as primary sources) and through documentation and presentations (as secondary sources) helped to speed up the process and deepen the understanding of the phenomenon.

The research process (design and strategy)

Unit of analysis

BT 21st Century Network (BT21CN) was chosen as the most advanced case (at the time of the research in 2005) of deployment of the NGN technological change being undertaken in the telecommunications industry by the incumbent telecom operators.[9] This choice was made at the end of stage 1 of the research as shown in Table 1. BT21CN is a major project to change the infrastructure/network of BT, adopting IP on a massive scale.[10]

Observational strategy

The background of this research is the transition to the NGN in telecommunications, leading to business renewal at BT. At the time this research was starting (late 2004), the concept of NGN was still stabilising. As a contemporary phenomenon, it was difficult to find quantitative data about it. The research question turned out to be more related to the platform-based approach that was emerging as incumbent telecommunications operators revealed their plans to migrate to NGN. Thus a qualitative research approach was preferred over a quantitative approach for the data collection, and thematic analysis was preferred over statistical analysis. Ultimately, the transition to NGN is a contemporary phenomenon, unfolding as this research was undertaken, and this suggested an inductive rather than deductive approach.

Furthermore, when considering a quantitative approach, the number of incumbent telecommunications operators undertaking the transition to NGN was low for the purposes of conducting a quantitative analysis. At the time of starting the research (late 2004), agreement about implementing the principles of NGN

---

[9]Interview with DT Manager, March 2005; interview with Siemens Senior Manager, March 2005; interview with Heavy Reading Senior Analyst, June 2005; interview with Cisco Account Manager, June 2005.

[10]BT issued a press release on 9th June 2004 announcing its plan to build BT21CN.

Table 1. Overview of the research stages for the data collection and empirical sources being used.

	Stage 1: March 2005–July 2005 (Exploration)	Stage 2: August 2005–July 2006 (Exploitation)	Stage 3: August 2006–May 2008 (Exploitation and Confirmation)
Objectives	• Understanding industry structure, processes, and resources to deliver and build NGN; • Identifying main suppliers of NGN; • Identifying main fixed-line incumbent telecom operators building NGN; • Exploring the dynamics of capabilities development, disruption, and inter-firm collaboration.	• Exploring in detail the specifics of industry change in terms of innovation and capabilities development in order to deliver and build the NGN; • Exploring in detail the dynamics of innovation and capabilities development in the transition to NGN of BT21CN, and in BTGS.	• Finalising data collection about the innovation dynamics of the transition to NGN at industry level; • Finalising the data collection about the capabilities development in BT: BT21CN and BTGS; • Resolving remaining discrepancies.
Interviews	Interviews with suppliers, service providers, industry analysts, consultants, and regulators: • 7 interviews in CEBIT 2005; • 3 interviews in VON Europe 2005; • 3 interviews in Light Reading Carrier Class Ethernet; • 1 interview in IEE Course.	Interviews with suppliers, service providers, industry analysts, consultants, and regulators: • 2 interviews in Light Reading — The Future of Telecom; • 6 interviews in Carriers World 2005; • 8 interviews in Broadband World Forum Europe 2005; • 9 interviews in International Telecommunication Union-Telecommunication Standardisation Sector (ITU-T) NGN Focus Group and Industry Event; • 14 interviews in CEBIT 2006; • 16 interviews in 21st Century Communications World Forum 2006.	Interviews with suppliers, service providers, industry analysts, and consultants: • 3 interviews in The New Telco: Europe 2006; • 9 interviews in Broadband World Forum Europe 2006; • 5 interviews in IP Leaders 2007; • 14 interviews in C5 World Forum 2007; • 1 interview in Carrier Ethernet Expo 2007; • 3 interviews in ITU-T Kaleidoscope Academic Conference 2008.

(Continued)

Table 1. (*Continued*)

	Stage 1: March 2005–July 2005 (Exploration)	Stage 2: August 2005–July 2006 (Exploitation)	Stage 3: August 2006– May 2008 (Exploitation and Confirmation)
Secondary sources	• Annual reports; • Press releases; • Newspapers and magazine articles; • Official websites; • Trade Conference presentations.	• Annual reports; • Press releases; • Newspapers and magazine articles; • Official websites; • BT Technology Journal; • Trade Conference presentations.	• Annual reports; • Press releases; • Newspapers and magazine articles; • Official websites; • BT Technology Journal; • Trade Conference presentations.
Events involved in	• CEBIT 2005; • VON Europe 2005; • Light Reading — The Future of Carrier Class Ethernet 2005; • The IEE Annual Course on Telecoms NGN.	• Light Reading — The Future of Telecom — Europe 2005 (7–8 September 2005); • Carriers World 2005; • Broadband World Forum Europe 2005; • ITU-T Focus Group on NGN 2005; • CEBIT 2006; • 21st Century Communications World Forum 2006.	• The New Telco: Europe 2006; • Broadband World Forum Europe 2006; • IP Leaders 2007; • C5 World Forum 2007; • Carrier Ethernet Expo 2007; • ITU-T Kaleidoscope Academic Conference 2008.

Source: Author's elaboration.

was widespread in the industry, but the strategy of implementation (i.e., when and how) varied from one operator to another. In Europe, BT, DT and FT took the lead to deploy NGN with different strategies. Among these three, BT decided to deploy NGN most aggressively, committing a huge amount of money (about £10 billion) over roughly five years.

Initially, data were collected from BT, DT, and FT, as part of the observational strategy to look at large-scale changes in complex systems. However, along the way, it became evident that BT was committed to make the transition to NGN happen more quickly, while DT and FT would take longer. As BT is a large and complex company, it has made a major commitment to change. It is leading the other companies (hence the potential for learning and imitation from this experience), and it has moved to begin implementation in a more aggressive manner (providing an important observational opportunity to compare planning and implementation).

Selecting the case

The process described below indicates the effort to approach the problem of developing platform thinking from the viewpoint of applying theory in a context of technological change. This was accomplished by a careful examination of the central issue of NGN transition, leading to an empirical and inductive process that also led to the selection of BT as the best candidate for in-depth investigation.

After considering incumbent telecommunications operators in various continents (Europe, America, and Asia), the decision was taken to focus on Europe. Initial interviews and participation in trade conferences (during Stage 1 of Table 1, as mentioned before in this section) indicated that Europe was ahead of other continents (compared to North America/USA and Asia/Japan) in the deployment of NGNs.[11] The main three operators (in terms of size) in Europe are BT, DT, and FT. All three operators have the same background of having overinvested in 3G licences and having a huge amount of debt by the time of the downturn, and all three were suffering from declining revenues in their traditional fixed voice services. However, BT took the decision to sell its mobile business back in 2001, a decision not taken by DT and FT.[12]

In the following years, the mobile business grew, and DT and FT could offset to a certain extent their losses in revenues in the traditional fixed voice services. However, that was not the case with BT. In an audacious move, BT announced BT 21st Century Network (BT21CN) in 2004 as a £10 billion project to build its NGN in five years, switching off its PSTN network.[13] There was a general agreement in the telecommunications market that building the NGN was really needed, but there was no agreement on the deployment strategy. DT and FT (and others like NTT) had announced that they would be converting their network to NGN, but at a slower pace (not in the five years announced by BT).[14]

As mentioned before in this section, the research was initially conducted collecting data from all three incumbent operators: BT, DT, and FT. However, over time (and mostly based on the first stage of the data collection represented in Table 1), it became apparent that (i) by trying to lead the race BT was experiencing the challenges of the transition to NGN before the other incumbent operators, as DT and FT still had the mobile business to support their overall

---

[11]Interview with Siemens Senior Technical Director, March 2005; interview with DT Senior Manager, March 2005.
[12]mmO2, BT's mobile unit, was divested on 19th November 2001 (BT, 2002:9).
[13]BT issued a press release on 9th June 2004 announcing its plan to build BT21CN (BT, 2005).
[14]Interview with DT Project Manager, November 2005; and interview with FT Technical Manager, November 2005; interview with NTT Senior General Manager, October 2005.

business; (ii) BT was pointed out by many interviewees as the most advanced deployment of NGN in Europe; and (iii) collecting data from DT and FT was not proving advantageous, since they were moving at a slower pace. Thus, it was decided to focus on BT, as it was the most advanced case of the transition to NGN, and it could offer more relevant insights before events unfolded in DT and FT.

The public Internet is having a major impact on the incumbent telecom operators, regarding its network architecture and business models. The transition of BT was chosen because it seems to be, at the time of this research, the most influential and radical approach in the global telecommunications market when BT21CN was announced in 2004. The difference from other approaches, like DT and FT, is that BT is proposing a deadline to switch off the PSTN, a commitment that is not assumed by other operators.[15] BT was also chosen due to its innovation "leadership" in the transition. Innovation "leadership" occurs when "firms aim at being first to market based on technological leadership" (Tidd *et al.*, 2005:121), whereas innovation "followership" occurs when "firms aim at being late to market, based on imitating (learning) from the experience of technological leaders" (Tidd *et al.*, 2005:121). These definitions tend to be used in reference to technology suppliers, but in the context of this research the technological leadership occurs in the adoption of the technology on a large scale, where huge amount of resources are committed to its deployment.

Operationalising the research strategy

Being a recent phenomenon, an inductive approach was adopted in three stages. This is in line with what Eisenhardt (1989) calls grounded case study, where theory is built from case study research. Although the author identified some prospective literature in the beginning of the research, it was during and after the data collection that emerging literature could be identified to better explain the data and compare the findings. The research was conducted through interviews and analysis of documents such as reports, newspaper articles, and official Internet websites. The reports included annual reports of suppliers and incumbent service providers, and documents of regulators. The interviews were conducted with senior managers, managers and other practitioners of incumbent telecommunications service providers and suppliers, regulators, consultants, and market research analysts. An overview of the documentary and interview data used is shown in Table 1.

---

[15]Interview with BT Senior Manager, November 2005.

Stage 1 was the exploration phase where the context of the research problem and incumbent operators were investigated. One of the outcomes of this phase was to narrow the options down to BT as the main case study to be developed. Stage 2 was the phase of exploitation where more information about BT and the industry was gathered addressing the research question on three aspects: Platform, service innovation, and NGN. Stage 3 served to further exploit the insights and propositions reached in phase 2 and attempted to confirm (or not) those propositions.

The interviews were conducted during the trade conferences attended by the author. It was organised a questionnaire with several questions related to this research and during the trade conferences it was adopted the approach to make few questions very focused on the expertise of the interviewee, and wherever possible, pose the same question to many interviewees. All questions were supposed to be covered in one trade conference. Then, whenever possible, received answers were compared with documentary data, trying to confirm (or not) the information thus obtained in the following trade conference. Dubious or ambiguous information was either discarded or considered for a discussion topic. When necessary and possible, previous interviewees were contacted again (by telephone and/or e-mail) for clarification or to obtain more information.

The specific case of BT is to address the conditions under which the platform-based approach emerged in the transition to NGN of an incumbent telecommunications operator. The context of incumbent telecommunication operators provides an opportunity to examine traditional firms where business renewal and growth are harder to achieve compared to new entrant operators and emerging Internet-based communications firms such as Google, Yahoo, and Facebook. The incumbent telecom operators come from a traditional and secular voice-only service and they are competing with other incumbent operators as well as smaller operators, new entrants, cable TV operators, satellite and Internet firms, which are also able to provide communication services.

Analysis and Results

This section uses the framework of management analysis of Fig. 1 to structure the discussion and present the results. Each of the four elements of the framework (platform emergence, platform architecture, platform evolution, and platform leadership[16]) is elaborated in the following.

---

[16]Platform evolution and platform leadership are developed together in this section.

Platform emergence: IP and NGN

The incumbent telecommunications operators have been renewing themselves in order to remain competitive in the face of declining revenues of traditional services, fierce competition, major technological change, and the financial difficulties following the burst of the financial bubble in telecommunication- and Internet-related companies that occurred at the beginning of the 2000's. Incumbent telecommunication operators (such as BT, DT, and FT) are striving to create and deliver new services in the course of large-scale adoption of the IP as the core of their networks, migrating their infrastructure and transitioning to the IP NGN. In particular, BT in the UK is a significant representative of the transition to NGN. Within BT, a major initiative, namely the BT 21st Century Network (BT21CN), a major project undertaken to renew BT's entire network/infrastructure, is important to illustrate the emergence of a platform-based approach by an incumbent telecom operator.

After the downturn of 2001/2002, large incumbent telecom operators in Europe (e.g., BT, DT, and FT) accumulated huge debts as a result of (a) their attempts to expand internationally and (b) their purchases of mobile spectrum licences.[17] Such high levels of debt combined with decreasing revenues from traditional voice services demanded new approaches for such firms to keep growing and sustaining a reasonable level of profitability. This undesirable situation called for change. The question was (and has been ever since) how to remain competitive in this turbulent environment (Fransman, 2007). The period from 2002 to 2005 could be seen as a period of significant uncertainty in the telecom industry in general, and for the incumbent telecom operators in particular.

The situation of BT, DT, and FT was in general very similar in terms of high levels of debt and market constraints. However, they took different approaches due to differences in their home markets and styles of management. Among the three operators, BT seems to have taken the most drastic decisions to recover from the unfavourable financial situation they were in regarding their debts and their ability to grow by increasing their income. BT sold its mobile business in 2001 while DT and FT retained theirs.[18] BT also decided to renew its infrastructure, establishing a

---

[17] An account of the telecom bust can be found in Fransman (2002). Regarding the level of debt, BT had £27.9 billion of debt at 31st March 2001 (BT, 2001:30, 31); Deutsche Telecom had €67 billion in 2001 (DT, 2001:U4); and FT debt was about €64.9 billion in 2001 (FT, 2001:88) and €68 billion by the end of 2002.

[18] mmO2, BT's mobile unit, was divested on 19th November 2001 (BT, 2002:9). T-Mobile is the mobile unit of DT (see, for example, Deutsche Telekomn's Annual Report 2001 (DT, 2001)). Orange is the mobile unit of FT (see, for example, France Telecom's Annual Report 2001 (FT, 2001)).

major programme called BT 21st Century Network (BT21CN)[19] while DT and FT agreed with the underlying changes that BT was propagating in the industry (through BT21CN), but proceeded at a slower pace.[20] BT21CN represents the transition of BT's traditional telecommunications network (based on several technologies such as Plesiochronous Digital Hierarchy (PDH), Synchronous Digital Hierarchy (SDH) and Asynchronous Transfer Mode (ATM)) to a new network based on the so-called IP/MPLS technology.[21] This change in the technology of the core network enables the incumbent telecom operators to be more flexible in the delivery of new services. Such flexibility in services is to a certain extent already enjoyed by Internet-based firms (such as Google, Yahoo, or Skype) as they have employed IP technology since their inception.

This new network based on the IP/MPLS technology is called NGN.[22] However, IP is not a new technology, since it has been deployed in the Internet sector for a long time (since the 1970's) and is widely used in, for example, local area networks (LANs). The novelty is that IP technology, in conjunction with MPLS, has been developing in terms of Quality of Service (QoS) in such a way that it has become possible to deploy it not only for data services, but also for the transmission of real-time services, including voice and video, with superior benefits compared to competing technologies, like ATM. In mid-2004, the ITU-T, the telecommunications standardisation body based in Geneva, created a focus group dedicated to standardising the architecture and protocols of the NGN.[23]

Along with the emergence of the IP technology as the basis of a robust platform for converged services (integrating voice, video, and data services in a single network),[24] the competitive environment of communications has been changing in the first decade of the 21st century, with fixed and mobile operators, Internet-related firms (such as Skype, Google, and Yahoo), cable TV operators, and smaller communications providers all competing for customers (and frequently the same

---

[19]BT issued a press release on 9th June 2004 announcing its plan to build BT 21CN.

[20]Interview with BT Senior General Manager, November 2005; interview with Deutsche Telecom Project Manager, November 2005; interview with FT Technical Manager, November 2005.

[21]It is not the aim of this research to elaborate on such technologies, but a historical perspective on network synchronisation involving PDH, SDH, and ATM can be found in Bregni (1998). More technical information about the application of MPLS in IP networks can be found in Awduche (1999).

[22]The expression NGN is now widely used by official institutions like the ITU-T and the OECD's Directorate for Science, Technology and Industry (see, for example, OECD (2005).

[23]More information is available on http://www.itu.int/ITU-T/ngn/fgngn/ (Accessed on 13 October 2008).

[24]Fransman (2002) provides an account of the influence of the IP and Internet on the telecommunications industry.

customers). The idea that voice, video, and data services could share the same IP, which is widely used in the public Internet, is driving convergence of services and contributed to the choice of the traditional telecommunications industry (represented by telecom incumbents such as BT, DT, and FT) to adopt the IP as "the" technology of choice to build their NGNs.[25] By 2005, major incumbent telecommunications operators had announced plans to migrate to the NGN, an all-IP platform that enables the delivery of a whole range of new multimedia services, besides the voice-only services.[26]

The situation presented in the first decade of the 21st century in the telecommunications industry is a result of various technologies being developed independently and reaching a point where they started to interact with each other and diffuse in their adjacent markets. In other words, different technological trajectories are overlapping at the services level. The fixed-line voice service is the most traditional one, with more than 100 years of history. More recently, in the 1990's, mobile communications has gained rapid acceptance in the market and began, during the 2000's, to take significant market from the fixed-line service. The cable TV industry, once related only to broadcasting TV (video) channels, is now offering broadband and voice service as well. Internet Service Provider firms, once perceived by firms and the general public as offering unreliable services, have been evolving and are now competing with telecom giants by offering voice and multimedia services. The emergence of Skype in 2003 and its subsequent purchase by e-Bay in 2005 raised the perception that low cost and good quality long distance voice calls are possible.[27] The Internet is usually regarded as having a marginal place in the market for robust telecommunications services. Nonetheless, it has had a major influence in the evolution of the telecom industry since the mid 1990s. As previously noted, the IP, supports the network infrastructure and the efforts to upgrade the speed of packet transfer referred to as broadband[28] and World Wide Web (web portal) which have been changing the perception of services by customers. Combining lower cost, convenience, and good enough quality

---

[25] An overview of NGN standards and architecture, including the role of IP can be found in Lee and Knight (2005), Carugi et al. (2005), and Knightson et al. (2005).

[26] See, for example, OECD (2005).

[27] An account of the initial impact of Skype and its purchase by e-Bay can be found in *Economist* (2005).

[28] The definition of broadband in this research follows the one found in OECD's report (see Paltridge (2001), where the downstream speed is a minimum 256 kbps for "always-on" service. The term broadband in this context is usually used to differentiate from the dial-up access with speed in the order of 10ths of kbps (e.g., 56 kbps). In 2009, it is common to find downstream speed of 2 Mbps or more.

for new interesting services, the Internet is changing sharply the way traditional telecommunication operators conduct their business and deal with innovation.[29]

The move to NGN represents the transformation of not only the infrastructure but also the business of incumbent telecom operators. The NGN provides more capabilities for them to deliver new services. It became ever more apparent to incumbent telecom operators that the aim of the new architecture being proposed is to decouple the infrastructure from the services layer.[30] This is a telecommunications industry version of the architecture already practiced in the computer industry: The decoupling of the network/infrastructure (hardware/operating system) from services (application software) means that every time a new application is introduced it is not necessary to change significantly the infrastructure (hardware/operating system). This reflects the concept of platform (and its reusability), where the network becomes the platform to deliver new services developed by the operator itself and in collaboration with other service providers. Incumbent telecom operators are now trying to implement this platform-based approach.

The platform architecture of BT21CN

Incumbent telecommunications operators, facing increasing competition and declining revenues in their traditional fixed-line voice services and further pressed by the downturn in 2001/2002, started to examine ways to change their "*modus operandi*" in order to remain competitive. The NGN was a response to this challenge at an industrial level. The evolution of the IP/MPLS technology seemed to happen at the right moment to be deployed at the core of the NGN.[31] The technology was available from several suppliers, although the standards were not yet mature. However, historically, BT was unaccustomed to waiting for standards to become mature in order to implement their network and services (Mansell, 1993).[32] There was (and continues to be) an execution risk, but the industry seemed to have reached an unprecedented agreement about the adoption of IP/MPLS.[33] The question was not whether to make the transition to NGN, but how to do so, and how quickly it could be done.[34] Incumbents like DT and FT in Europe,

---

[29] An initial discussion about the impact of the Internet on the traditional telecom operators can be found in Noam (1998).

[30] A discussion about the NGN architecture can be found in Knightson *et al.* (2005) and Lee and Knight (2005).

[31] Interview with BT Senior Manager, July 2005.

[32] Interview with BT Senior Manager, November 2005.

[33] Interview with AT&T Senior Manager, September 2005; interview with Telefónica Manager, October 2005; interview with DT Manager, March 2006; interview with BT Senior Manager, March 2006.

[34] Interview with BT Senior Manager, March 2006.

and NTT in Japan decided to make the transition at a slower pace compared to BT in the UK.[35] As explained in the previous section, BT set a time frame of five years to complete the migration, while others decided to migrate with no target date for completion, i.e., to revise implementation plans on the go.

BT's decision to migrate to NGN was organised as a project, which was called the BT 21st Century Network (BT21CN). As major undertakings, NGNs are designed to meet a customer's business and technical needs and they are also seen as a single and consistent network platform that enables new services at a lower cost.[36] BT21CN represents BT's effort to manage the transition to NGN and it can be seen, as a whole, as an integrated solution provided by eight preferred suppliers[37] with BT as customer.

BT21CN is set to deploy a network seen as a platform, and its major business and technical characteristics are identified in this section. Platforms are seen not only as technical, but also as business constructs to achieve strategic goals. As explained in the preceding section, the concept of network platform integrates the notion of: (i) product platform (Meyer, 2007; Meyer and Dalal, 2002), where the main characteristic is the re-use of subsystems or interfaces; and (ii) opening the interfaces to drive innovation in the industry (Gawer and Cusumano, 2002). Systems integration and project management are components of the capabilities that incumbent telecom operators need to develop in order to compete in the development of NGNs.

Telecom operators are users rather than producers or suppliers of complex products and systems (CoPS) (and will not have the advantages of said producers or suppliers, who have roadmaps for their products and systems, and who usually use product families to highlight their evolution or define different market niches). Thus, whenever a telecom operator needs to build a new network or upgrade it, the architecture and the way these systems are combined become fundamentally important. In many instances, the operator is concerned about building a coherent platform through architecture. The architecture is the "stable intermediate form" (cf. Simon (1996)) that is used by the operators and vendors in order to deal with

---

[35]As mentioned before, the target of DT, FT, and other incumbent operators would be between 2015 and 2020 for the completion of the transition to NGN (see footnote 8). Just for comparison, smaller operators such as C&W claimed that they would be making the transition to NGN in a few years (from interview with C&W Senior General Manager, March 2006. Another smaller operator, THUS, claimed that they had already made the transition, as they were founded using IP technology (from interview with THUS Senior Business Manager, October 2006).

[36]Interview with BT Senior Manager, September 2005.

[37]The eight preferred suppliers are: Alcatel, Lucent, Siemens, Cisco, Ciena, Ericsson, Huawei, and Fujitsu. Later in 2006 Alcatel acquired Lucent, forming Alcatel-Lucent. Siemens (the carrier business, which was dedicated to telecom operators, distinct from the enterprise business) made a joint venture with Nokia in 2007, establishing Nokia Siemens Networks.

complex choices and decisions. Large suppliers like Ericsson, Siemens, and Cisco possess their reference architectures that they try to sell on to telecom operators. Smaller telecom operators usually do not have enough power to ask the suppliers to change their architecture. However, large operators like BT usually do have such power.[38] BT has developed its own architecture for BT21CN and, while designing it, vendors were consulted and it became apparent that the technology required to implement the architecture was already available (Crane, 2005). With this architecture in mind, BT selected the current preferred suppliers. Potential suppliers needed to produce systems that were already compliant with the architecture or to have a clear migration path to reach such compliance. Of course, BT also required that the technological choice should be defended with a strong commercial proposal showing the whole-life cycle of the solution offered over ten years.[39]

Using the chosen architecture as a reference, BT communicated with and selected their suppliers. BT's chosen architecture divided the network into five major parts, and the suppliers were invited to submit proposals for each part.[40] During the tender process to select the vendors, BT disclosed the part of the architecture in which the potential supplier expressed interest, and reserved the role of integrating all the parts for itself. In large projects, the total systems integration is usually the responsibility of a prime integrator from the supplier side (Davies and Hobday, 2005). However, due to the scale and scope of the project, and the opportunities to gain knowledge that could be transferred to other areas, BT decided to take on the role of prime integrator itself. The architecture, in this sense, represented the tool through which BT intended to develop technological and system integration capabilities, and to understand whether the various offers from the suppliers would comply with their architecture. Hence, BT activities for BT21CN are largely concentrated on systems integration (Crane, 2005). Acquiring a competence in architectural design, BT learned how to integrate systems and equipment from various vendors. Interestingly, BT has taken the view that it will be able to sell this capability to other firms (telecom operators), and this is what they intended through internal initiatives targeting telecommunications operators in developing countries.[41]

The architecture of BT21CN is based on the principle of being "a single converged network carrying all services" and on "the idea of reusable service

---

[38]Interview with Alcatel Manager, May 2005.
[39]Interview with BT Senior Manager, March 2006.
[40]The five parts are: Access, metro, core, transmission networks, and i-node. The i-node is where the "intelligence" of the network resides (e.g., call control, network management, operation, and support systems). At least two suppliers were selected for each part, except for the i-node, according to the press release issued by BT on 28th April 2005.
[41]Interview with BT Senior Manager, March 2007.

components, so that a product designer can rapidly create and change the services BT provides" (Reeve *et al.*, 2005:13). The previous network was characterised by an "aggregation" of networks where, "each network [was] associated with a single service and support system (customer care and billing, for example) that have been developed for each network and service by its own internal software developers" (Reeve *et al.*, 2005:11). Thus, this "stove-pipe-" or "silo-" based approach became unsustainable for the requirement of faster service provision, and a platform-based approach was proposed.

From silos to platforms

The agreement in the industry over IP technology is an unprecedented and historical milestone in the telecommunications industry (Fransman, 2002).[42] The industry realised that IP was mature enough to transport, besides data, real-time voice, and video. The whole NGN can be considered as the "global platform" and behind it there is an architecture that was traditionally organised through technology-service silos. The silo structure means that the service is associated with the network (infrastructure). Thus establishing a new service means following a strategy of either constructing a new infrastructure, or modifying the existing infrastructure in significant ways. Both of these strategies are inadequate for meeting the aim of decreasing the time-to-market period for new services, and providing the flexibility and choice required by customers. Pressed by competition and declining revenues from its traditional fixed-line voice services, BT established a shorter time-to-market period for new services as a strategic objective. Thus, the IP technology provided the possibility of building an infrastructure based on a single networking technology for voice, data, and video services. It can also offer communications services that are decoupled from the infrastructure, so that changes in services do not necessarily imply changes in the infrastructure.

Another aspect of the platform-based approach was the initiative of BT to identify what they called "common capabilities," which were a set of functionalities that could be reused in several types of services.[43] "The aim was to produce a model where 80% of new services could be created from reusable elements"

---

[42] Also confirmed from interview with AT&T Senior Manager, September 2005; interview with Telefónica Manager, October 2005; interview with DT Manager, March 2006; interview with BT Senior Manager, March 2006.

[43] BT believed that a set of capabilities such as authentication, presence and location, storage, secure connections, directory and profile, customer portal, home, and office hubs, and so on could be identified and reused to mediate the underlying functions of the physical system and the solution/service demanded by the customer. For more information about these "common capabilities" see Levy (2005).

(Levy, 2005:49). This way, BT can develop services in a shorter time-to-market and hopefully at lower costs. This also had some impact on the organisational structure of BT, where the silo-based approach was detrimental to this new way of working, demanding a more horizontal, platform-based approach. The consequence of this is that "over time there has been a blurring of traditional customer segment boundaries, which has led to customer requirements for multiple services that transcend an organisation's internal partitions" (Levy, 2005:49). Meyer and DeTore (2001) pointed out that it may be difficult for firms to consider reusability as it may be difficult to make certain parts of the organisation to work together. However, the emergence of IP as the common protocol/technology for voice, data, and video services lowered the barriers for this to occur.

This adoption of the platform-based approach[44] has had a huge impact on the operational and support systems (OSSs) for the services offered by the telecom operators, and on the architecture of BT21CN.[45] OSSs were also based on silos, as each service tended to be implemented independently from the others. This creates major problems with interoperability as most of the systems use proprietary protocols.[46] Customer's requirements have evolved. They now possess different ways of requesting resources and services from different parts of the organisation, and this blurs the boundaries of the organisation's internal divisions. For example, in the past customers bought a voice service separate from a data service, and a back office system for each one, separately. Nowadays, it is possible to click for a voice application in a PC or laptop, which makes the distinction irrelevant (Levy, 2005).

Henderson and Clark (1990) argued that some firms have made innovations in the architecture of their products, i.e., in the way the components are interconnected, but with the same components. Their unit of analysis is the product. In BT21CN, innovation occurs at the network level, and the new services produced

---

[44]The platform-based approach was further emphasised in interviews with: IBM Technical Director, October 2005; Siemens Senior Business Development Manager, October 2005; Huawei Technical Director, March 2006. It was also advocated by incumbent telecom operators, such as BT (from interview with BT Senior General Manager, July 2005); Belgacom (from interview with Belgacom Senior Commercial Director, September 2005); Swisscom (from interview with Swisscom Senior Programme Manager, October 2005); DT (from interview with DT Technical Director, March 2006); FT (from interview with FT Senior General Manager, October 2006); Telefónica (from interview with Telefónica Managing Director, October 2006); Portugal Telecom (from interview with Portugal Telecom Senior Director, October 2006).

[45]Although BT made separate tender processes for the OSS and for BT21CN, they are interconnected. Some parts of the OSS (e.g., network management system) are directly related to the suppliers of BT21CN, but other parts (e.g., the billing system) are indirectly related. However, both OSS and BT21CN face the silo/platform issue.

[46]Interview with BT Senior Manager, March 2007.

by the network are uncertain to some extent. Taking the network as the unit of analysis (not a single product or discrete system), innovation occurs at both architecture and component levels. The NGN is not a "like-for-like" replacement of functionality, i.e., it does not simply replace the PSTN switch with an IP NGN router.[47] The architecture is modified in order to simplify it and achieve the reductions in operational expenditure that are expected after completing the network transformation. According to this new architecture, BT expects to reduce the number of network elements from 100.000 to 30.000.[48] The main technological component is also being changed. The IP router is more functional and allows a greater variety of network configurations. This gives the network designer more choices when designing the architecture and configuring the network with possible implications in the reduction of operational costs and more flexible services. Thus, both architecture and components[49] are changed.

The architecture and the platform-based approach are directly related to the characteristics of the physical infrastructure of BT21CN. BT is striving to develop capabilities to build this new infrastructure, creating new company-level capabilities, and leveraging existing ones. Such capabilities are also important when delivering integrated solutions to large customers through BT Global Services (BTGS).

Opening up the BT21CN platform

The idea of opening up the network as a platform for third-party firms to develop services, which was unpopular in the 1990s (Mansell, 1993), is becoming more and more accepted in the 2000s. This platform-based approach is highly influenced by the successful approaches taken in the IT industry, by Microsoft and Intel, for example (see Gawer and Cusumano, 2002).

One of the initiatives of BT to address the openness of it platform was Web21C.[50] This initiative is to expose capabilities (functionalities) through the publication of Application Program Interfaces (APIs) that developers can access and develop their own applications accessing BT network. BT makes available Software Development Kits (SDK) that allows others to develop new applications. This is already a practice common in the IT/software development industry, but is

---

[47]Interview with Juniper Technical Manager, March 2006; interview with Telefónica Senior Technical Manager, October 2006; interview with FT Senior Technical Manager, October 2006; interview with Cisco General Sales Manager, March 2007.
[48]Interview with BT Senior Technical Manager, March 2006.
[49]Components can be huge equipment or systems, many of them can be classified as CoPS as in Davies and Hobday (2005).
[50]The acquisition of Ribbit in 2008 overtook the Web21C initiative (Meyer, 2008).

a practice that BT needs to learn and find its optimum and effective point within its context. This is mostly what Internet-based firms like Amazon or Facebook are doing when they release their APIs, and is part of the platform idea of "unleashing innovation beyond the boundaries of the payroll, exposing capabilities for others to develop."[51] This is a fact in the open source community, where developers collaborate in the development of Linux without being paid for it. Focusing on people, the practical boundary is who is paying for the work of those people. Focusing on system, who is paying the cost of setting up and maintaining that system? Although the Web21C/Ribbit initiative is very immature at the moment, it proves the willingness of BT to play in this more uncertain market and to learn with it. To what extent the incumbent telecom operators should play in the domain of Google, Yahoo, and Microsoft is still an uncertainty, but BT and other incumbent telecom operators seem to be willing to do so, hoping to capitalise on it once the right opportunity emerges and as a protection against possible disruptive innovations which may prove fatal to their business in the future.

Evolving the platform architecture towards platform leadership

The huge traditional PSTN network was not anymore up to the task of competing with convergent services provided by Internet-based firms, cable TV, and satellite TV firms. Besides data, video is becoming an increasing component of the network traffic. The business model of selling a huge amount of voice services (the traditional business) was in significant decline. And the capability of BT for developing new products and services was far from the speed and effectiveness required by the competitive market.

Three contextual factors were identified as main contributors for the adoption of the platform-based approach by BT through BT21CN: (i) the emergence of the IP as a common technology to transport voice, video, and data; (ii) fierce market competition; (iii) BT's high debt and unfavourable positioning in the market.

As mentioned in this section before, IP won the battle against other technologies such as ATM to be the common technology to transport voice, data and video under the same infrastructure. This opened the possibility of BT and other telecom operators to simplify their network, designing a network architecture with a more stable common core, which is at the heart of the platform architecture. This unprecedented agreement in the telecom industry on the IP was not reached without struggle. Many other operators were already using it in large scale, most notably the new entrants, and the incumbent telecom operators were slower in adopting it. The massive adoption of IP, although done through an aggressive

---

[51]From a presentation by BT CTO Matt Bross at the Telemanagement Forum 2007, in Nice, France.

initiative such as BT21CN, was undertaken by BT after much of the doubts about the scalability of IP had dissipated, and after many smaller operators had already proved the feasibility of IP. One could argue that BT had in fact arrived late to the decision, despite the boldness of the decision.

With the convergence of the market for voice, video and data services after the Internet bubble burst in 2001/2002, market competition became fiercer. Not only the new entrants and other international incumbent telecom operators were competing for the customer, but also firms from other backgrounds, such as satellite TV (e.g., Sky), cable TV (NTL-Telewest) and Internet-based firms (such as Google, Skype) were changing the landscape of the telecommunications industry. Also BT had the disadvantage of having sold its mobile phone business operation (mmO2) in 2001 in order to deal with the high level of debt.

BT had an unfavourable position compared to other incumbent telecom operators such as FT and DT, who maintained their mobile business operations and were capitalising from the increasing market for mobile business. It is important to note that BT prepared itself for radical change, hiring external top managers (CEO and CTO) in the beginning of the 2000's, deciding not to promote internal staff. This may also have had a strong impact in the decision-making and realisation of BT21CN.

Although it can be argued that the platform-based approach was being practiced at various levels of BT, the emergence of this approach at the firm level as a whole was due to BT21CN which brought this approach to the core of the organisation. The platform architecture affected the organisational structure and processes of the firm as a whole.

As mentioned before about the platform architecture of BT21CN, the awareness of the architecture plays an important role in the development and construction of the platform. The architecture serves as a mental construction to visualise the whole of the network and highlights the advantages of the platform-based approach, contributing to the decision about the feasibility of deploying such platform. At the level of the architecture, the common feature continues to be to establish, maximise, and make use of the common part of the platform. This gives some stability to the architecture while providing the flexibility to change the end product or service according to customer and market needs. This is an empirical validation in the context of the service industry to the argument provided by Baldwin and Woodard (2009) of the platform architecture having a stable part and also a complementary part that is designed to vary. This equilibrium between having *at the same time* stable parts that can reduce costs and variable parts that can provide more flexibility to provide unique services and integrated solutions (and hence, increasing revenues) is what makes the platform-based approach attractive and powerful.

The stability/variability dichotomy calls for the reusability of components and subsystems within BT's processes. Such processes, components, and subsystems usually grow in a fragmentary way due to specific needs of different groups and systems along different times. This creates a level of complicatedness that makes process slow and ineffective for developing new products and services. The platform-based approach is an attempt to identify commonalities and address this fragmentation. This is done at the hardware level through the massive use of IP as the common technology to transport a high variety of services based on data, voice, and video. At the software level, BT tried to identify the so-called "common capabilities" (such as authentication, directory and profile, presence and location, etc.) containing functionalities that could be reused in the development of approximately 80% of new services (Levy, 2005). This addresses the internal issue of developing new services faster. On the other hand, BT is also addressing the openness of its platform allowing external actors to access its network/platform for developing new services and driving innovation in the industry. Although the idea of opening up some interfaces for external actors to develop new services using BT's platform sounds to be a good idea (which is being applied by other firms in the ICT sector such as Google and Apple), its implementation has been proving difficult for BT as an incumbent coming from monopolistic roots.

The internal approach of the platform architecture through reusability and maximising the commonalities of the infrastructure (hardware) and services (software) had a huge impact on BT's organisation structure and processes. The vertical silo-based structure and processes is replaced by a platform-based approach based on horizontal integration of processes. The impact is deemed to be positive in terms of reducing operational costs and making the processes more fluid across the organisation. This can create a virtuous self-reinforcing process that can make BT more cost effective and lean.

On the other hand, the external approach of the platform architecture entails the opening of the infrastructure for external actors to interact with it, developing new services and boosting the use of the network (and hopefully generating more revenues for BT). This is the more challenging side of the platform-based approach being deployed by BT. This requires a level of engagement of BT with external players that is unprecedented for incumbent telecom operators. This also requires a level of visibility, or platform leadership as Gawer and Cusumano (2002) put it, that is earned through well articulated interfaces and through global presence in the market. The evolution of BT platform depends on having the visibility and internationalisation co-evolving in an appropriate manner.

The BT21CN case focuses on the emergent side of platforms in a service industry. In this specific case, by the time BT decided to move on with the

BT21CN programme, there was not much agreement about the details of the standards to be applied in the construction of such platform, although there was a general agreement about the adoption of IP. The construction should be done while standards evolved.

Taking into account the evolutionary theory proposed by Gawer (2009a), where platforms may evolve from internal to supply chain and then to industry platforms, from this research, it seems that the platform developed by BT, represented by BT21CN, emerged as an industry platform. From the very beginning, the network (BT21CN) as a platform was not designed and manufactured internally, but with the support of the eight suppliers BT chose to build the platform. As an industry platform, BT21CN was designed to have interfaces that allow external firms and users to develop new services upon its platform. This apparent mismatch between evolutionary proposition by Gawer (2009a) and the empirical evidence provided by BT21CN can be explained by the "complexity" of the platform under analysis. The evolutionary theory proposed by Gawer (2009a) is based predominantly on mass-manufactured goods, such as automobiles, personal computers, microprocessors, and operating systems. In the same way that Davies and Hobday (2005) argued that the distinction between product and process innovation proposed by Utterback and Abernathy (1975) for mass produced goods does not happen for CoPS,[52] the evolutionary proposition by Gawer (2009a) may be suitable for mass-manufactured products, but less suitable for complex systems like the BT21CN. Thus, as a complex system, BT21CN as a platform has not started as an internal platform, but it was designed as an industry platform from the very beginning. It might be important to differentiate "product/service as platform" from "infrastructure as platform." In the first case, products like smartphones, tablets and operating systems are mass-produced instantiations of platforms, which offer a common "space" for provision and consumption of different types of applications and services. In the second case, the whole organisational structure and processes are redesigned in order to operate according to platform thinking. This is usually related to what is called in the industry as "business transformation."

Considering the internal and external of the platform-based approach, the internal approach seems more straightforward, whereas the external one seems more challenging. The internal approach, based on reusability of components and sub-systems, both at hardware and software levels, contribute to more efficient and effective processes in terms of time-to-market for new products and services, and to overall operational cost reductions. In practice, this is not an easy process to

---

[52]CoPS are defined as "high cost, engineering-intensive products, systems, networks, and constructs" (Hobday, 1998:690).

conduct, but BT seems to be taking advantage of the implementation of this internal approach. The success of the external approach, based on the openness of the platform to external actors, relies heavily on the global presence of the firms who become platform leaders (e.g., Intel, Microsoft, and Cisco as reported by Gawer and Cusumano (2002), and more recently, Apple and Amazon). This external approach seems to be suffering from some challenges inherent to the positioning and dynamics of market in which incumbent telecom operators work. One such challenge is the dynamics of globalisation of incumbent telecom operators, which seems to be less favourable than the motor car industry, or even for equipment suppliers in the telecommunications industry. Fransman (2002) suggests that the telecoms industry had not experienced significant global concentration of telecom operators as it happened for the equipment suppliers (e.g., Cisco) and for the motor car industry. And this situation seems to have persisted in the subsequent years when BT21CN was deployed. Also Gawer and Cusumano (2002) and Fransman (2002) report on the ambitions of NTT DoCoMo to establish a global platform, forging partnerships and relationships with third-party complementors. As with NTT, BT is having difficulties to establish a global platform, and to date no concentration of incumbent telecom operators has formed as it has happened for the motor car industry and for the telecom equipment suppliers. As the platform is the infrastructure, it is more difficult to internationalise compared to Internet-based firms (such as Google and Facebook, more based on software), and compared to other firms (such as Apple) whose product is platform-based (not its infrastructure). The limited success in the internationalisation process may also hamper the success of the platform-based approach in terms of platform leadership, i.e., in terms of creating a global platform for industry innovation. This case suggests that the success of the adoption of the platform architecture is highly dependent on the environmental dynamics here represented by the internationalisation process. It highlights the global presence as an added factor of environmental dynamics that influences the platform evolution, according to the framework offered by Tiwana *et al.* (2010).

Conclusion

This paper shows the important role of the IP as the common technology for the emergence of the platform architecture of BT21CN, providing its stable core at the infrastructure/hardware level. BT21CN was also designed having in mind a high degree of openness of its interfaces to external firms, increasing the variability of products and services that can be implemented using the same platform.

BT's platform-based approach, implemented through BT21CN, encompasses both the reusability of components and sub-systems and the openness to external actors in order to drive industry innovation *at the same time*. As this platform is constructed at a higher level of aggregation (i.e., at the organisational level), BT's platform is a complex system, emerging as an industry platform, not passing through the phase of internal platform. It is important to take into account the level of the complexity of the platform when considering the typology and the evolution of platforms. The pervasive use of the term "platform" requires the refinement of the typology of platforms in order for this concept to be useful in various circumstances, and lessons transferred between compatible types.

The platform architecture of BT21CN is composed by two complementary aspects: (i) the reuse of components (subsystems, interfaces, and processes) as a way to create new products and services; and (ii) as a platform for innovation, opening interfaces and allowing third-party firms to develop new services on top of the network. The reuse of components and sub-systems was illustrated by the identification of "common capabilities" by BT, they are reused to speed up the development of new services. From the business perspective, this internal platform-based approach has the aim of reducing the total operational costs and improving the time-to-market for new products and services. It seems that BT is taking advantage of this approach in the medium and long term. At the same time, BT is opening up its BT21CN platform and inviting external actors to develop new services using its network. Although this is becoming a usual practice in the ICT sector, it has been proving difficult for BT as an incumbent with monopolistic roots. BT's limited success to reach a more globalised market makes it difficult to establish BT21CN as a global platform. Without a more international and significant presence it makes difficult to bring third-party complementors which can confer market leadership to BT's platform. The platform-based approach differs from BT's previous strategies where technologies and services were treated in a more fragmented way and the implementation was less demanding in terms of time of deployment.

BT21CN represents a first-move for the incumbent telecommunications operators in the world, and this means that BT has experienced some of the transition problems earlier than other incumbent operators. BT21CN represents an instance of the network as a platform, a strategy which seems to be gaining momentum in the telecom industry, influenced by the convergence not only of technologies, but also of the practices and strategies of the IT industry. IP, as a common technology for voice, data, and video services, lowered the barriers for implementing a platform at BT's organisational level.

The outcome of BT21CN is a new infrastructure designed to be a global platform that challenges the way BT itself works. Changes reinforce and open the path for other changes some of which may be unforeseen. The platform-based approach being implemented at the infrastructure level forces the development of new capabilities, changing both the way BT innovates in services and the way BT relates with its customers and partners. This case shows that the success of the adoption of the platform-based approach is not dependent only on the adoption of the platform architecture, but also on the environmental dynamics represented by the internationalisation process. The appropriateness of the platform-based approach to BT is not clear yet. This raises the question of whether this is the right approach/strategy for BT and/or whether it is enough.

The platform resulted from BT21CN can be usefully viewed as a complex organisational, one-off platform. Its dynamics of innovation is different from the mass-manufactured platforms in terms of product and process innovation. BT21CN has not passed through the stage of internal platform. It was born as an industry platform. The evolutionary theory by Gawer (2009a) seems to be more applicable to mass-manufactured goods and services which go through an initial stage of internal product platform. This kind of categorisation and consideration may be important when considering the applicability of lessons learned from one innovation environment (e.g., mass produced products) to another (e.g., complex systems).

An important implication is for the strategy of other incumbent telecom operators which are adopting or planning to adopt the platform-based approach in full scale due to the limitation of the platform-based approach in terms of obtaining higher revenues from the openness of the platform to third-party complementors. Other incumbent telecom operators willing to implement the platform-based approach may benefit from the expected lower operational costs and decreased time-to-market for developing and launching new products and services. However, they may have difficulties in establishing its platform as a leading industry platform if their globalisation process has limited success, following a different pattern from equipment suppliers (such as Cisco), motor car industry, and Internet-based firms (such as Google, Facebook, and Amazon). In this case, a more balanced approach considering platform and product strategy may be more effective. Also, success in platform leadership seems to depend on the extent to which incumbent telecom operators can become software telcos, as the business model for platforms is much more dependent on software.

This research was exploratory (i.e., inductive and empirically-based) and highly dependent on interview data, and on the snowballing technique for gaining the perspectives of different interviewees. The interviews suggested other two key areas for further in-depth research. One area is the investigation of platform evolution in other types of service/infrastructural industries. A comparison with firms from other industries (like energy, transportation, and construction) would be useful including firms from the IT side (e.g., IBM) and from the professional service/consultancy side (e.g., Accenture). This would allow the investigation of whether the findings from telecom platforms also apply to other major sectors and incumbent firms. The second area is to investigate whether within telecom, the BT results have relevance to other cases (e.g., in DT, NTT, Telefonica, AT&T, Verizon, and FT) undergoing similar challenges/transitions. Those incumbent operators may be following different strategies and specific capabilities may be more or less relevant depending on the context in which they perform. In all the above, the view of customers/users of the platform may also be a potentially rich source of insight into the complexities of demand, and into the usefulness and limitations of the platform-based approach.

References

Aerts, ATM, JBM Goossenaerts, DK Hammer and JC Wortmann (2004). Architectures in context: On the evolution of business, application software, and ICT platform architectures. *Information and Management*, 41(6), 781–794.

Awduche, DO (1999). MPLS and traffic engineering in IP networks. *IEEE Communications Magazine*, 37(12), 42–47.

Baldwin, CY and CJ Woodard (2009). The architecture of platforms: A unified view. In *Platforms, Markets and Innovation*, A Gawer (Ed.), pp. 19–44. Cheltenham, UK: Edward Elgar.

Bregni, S. (1998). A historical perspective on telecommunications network synchronization. *IEEE Communications Magazine*, 36(6), 158–166.

BT (2001). BT Annual Report 2001. London.

BT (2002). BT Annual Report 2002. Available at http://www.btplc.com/report/index.shtml [Accessed 27 February 2009].

BT (2005). BT announces network transformation timetable. Press release issued on 9th June 9, 2004.

Carugi, M, B Hirschman and A Narita (2005). Introduction to the ITU-T NGN focus group release 1: Target environment, services, and capabilities. *IEEE Communications Magazine*, 43(10), 42–48.

Ciborra, CU (1996). The platform organization: Recombining strategies, structures, and surprises. *Organization Science*, 7(2), 103–118.

Crane, P. (2005). A new service infrastructure architecture. *BT Technology Journal*, 23(1), 15–27.

Davies, A and M Hobday (2005). *The Business of Projects: Managing Innovation in Complex Products and Systems*. Cambridge, UK: Cambridge University Press.

DT (2001). Deutsche Telekom Annual Report 2001. Bonn: Deutsche Telekom.

Economist (2005). The meaning of free speech. *The Economist*, September 17, 81–84.

Eisenhardt, KM (1989). Building theories from case study research. *Academy of Management Review*, 14(4), 532–550.

Flyvbjerg, B, N Bruzelius and W Rothengatter (2003). *Megaprojects and Risk: An Anatomy of Ambition*. Cambridge, UK: Cambridge University Press.

Fransman, M (2002). *Telecoms in the Internet Age: From Boom to Bust to...?* Oxford, UK: Oxford University Press.

Fransman, M (2007). *The New ICT Ecosystem: Implications for Europe*. Edinburgh: Kokoro.

FT (2001). France Telecom Annual Report 2001. Paris: France Telecom.

Gawer, A (2000). The organization of platform leadership: An empirical investigation of intel's management processes aimed at fostering complementary innovation by third parties. Unpublished PhD, Massachusetts Institute of Technology, Boston.

Gawer, A (2009a). Platform dynamics and strategies: From products to services. In *Platforms, Markets and Innovation*, A Gawer (ed.), pp. 45–76. Cheltenham, UK: Edward Elgar.

Gawer, A (ed.) (2009b). *Platforms, Markets and Innovation*. Cheltenham, UK: Edward Elgar Publishing Limited.

Gawer, A and MA Cusumano (2002). *Platform Leadership: How Intel, Microsoft, and Cisco Drive Industry Innovation*. Boston, MA: Harvard Business School Press.

Henderson, RM and KB Clark (1990). Architectural innovation: The reconfiguration of existing product technologies and the failure of established firms. *Administrative Science Quarterly*, 35(1), 9–30.

Hobday, M (1998). Product complexity, innovation and industrial organisation. *Research Policy*, 26(6), 689–710.

Kahn, BE (1998). Dynamic relationships with customers: High-variety strategies. *Journal of the Academy of Marketing Science*, 26(1), 45–53.

Knightson, K, N Morita and T Towle (2005). NGN architecture: Generic principles, functional architecture, and implementation. *IEEE Communications Magazine*, 43(10), 49–56.

Lee, C and D Knight (2005). Realization of the next generation network. *IEEE Communications Magazine*, 43(10), 34–41.

Levy, B (2005). The common capability approach to new service development. *BT Technology Journal*, 23(1), 48–54.

Mansell, R (1993). *The New Telecommunications: A Political Economy of Network Evolution*. London: Sage Publications.

Mansell, R and WE Steinmueller (2000). *Mobilizing the Information Society: Strategies for Growth and Opportunity*. Oxford: Oxford University Press.

Maylor, H and K Blackmon (2005). *Researching Business and Management*. New York, NY: Palgrave Macmillan.

Meyer, D (2008). BT guns for Android and Skype with Ribbit buy. *ZDNet UK*.

Meyer, MH (2007). *The Fast Path to Corporate Growth: Leveraging Knowledge and Technologies to New Market Applications*. Oxford, UK: Oxford University Press.

Meyer, MH (2008). Perspective: How honda innovates. *The Journal of Product Innovation Management*, 25(3), 261–271.

Meyer, MH and D Dalal (2002). Managing platform architectures and manufacturing processes for nonassembled products. *The Journal of Product Innovation Management*, 19(4), 277–293.

Meyer, MH and A DeTore (2001). Perspective: Creating a platform-based approach for developing new services. *The Journal of Product Innovation Management*, 18(3), 188–204.

Meyer, MH and AH Lehnerd (1997). *The Power of Product Platform*. New York, NY: The Free Press.

Meyer, MH and PC Mugge (2001). Make platform innovation drive enterprise growth. *Research-Technology Management*, 44(1), 25–39.

Muffato, M (1999). Introducing a platform strategy in product development. *International Journal of Production Economics*, 60–61, 145–153.

Muffato, M and M Roveda (2000). Developing product platforms: Analysis of the development process. *Technovation*, 20(11), 617–630.

Noam, EM (1998). The impact of the internet on traditional telecom operators. Paper presented at the Centro Study San Salvador, Venice, Italy.

OECD (2005). Next generation network development in OECD countries. Available at http://www.oecd.org/dataoecd/58/11/34696726.pdf [Accessed 1 November 2005].

Olleros, FX (2008). The lean core in digital platforms. *Technovation*, 28(5), 266–276.

Ottoson, S (2003). Dynamic product development of a new intranet platform. *Technovation*, 23(8), 669–678.

Oxford (1989). Oxford Advanced Learner's Dictionary. Oxford: Oxford University Press.

Paltridge, S (2001). *The Development of Broadband Access in OECD Countries*. Paris: OECD.

Reeve, MH, C Bilton, PE Holmes and M Bross (2005). Networks and systems for BT in the 21st century. *BT Technology Journal*, 23(1), 11–14.

Rochet, JC and J Tirole (2003). Platform competition in two-sided markets. *Journal of the European Economic Association*, 1(4), 990–1029.

Sawhney, MS (1998). Leveraged high-variety strategies: From portfolio thinking to platform thinking. *Journal of the Academy of Marketing Science*, 26(1), 54–61.

Schwartz, B (2004). *The Paradox of Choice: Why More is Less*. New York, NY: HarperCollins Publishers Inc.

Simon, H (1996). *The Sciences of the Artificial*. Cambridge, MA: MIT Press.

Standing, C and S Kiniti (2011). How can organisations use wikis for innovation? *Technovation*, 31(7), 287–295.

Suarez, FF and MA Cusumano (2009). The role of services in platform markets. In *Platforms, Markets and Innovation*, A Gawer (ed.), pp. 77–98. Cheltenham, UK: Edward Elgar.

Tatikonda, MV (1999). An empirical study of platform and derivative product development projects. *The Journal of Product Innovation Management*, 16(1), 3–26.

Tidd, J, J Bessant and K Pavitt (2005). *Managing Innovation: Integrating Technological, Market and Organizational Change*, 3rd Edition. Chichester, NJ: John Wiley & Sons Ltd.

Tiwana, A, B Konsynski and AA Bush (2010). Platform evolution: Coevolution of platform architecture, governance and environmental dynamics. *Information Systems Research*, 21(4), 675–687.

Utterback, JM and WJ Abernathy (1975). A dynamic model of product and process innovation. *OMEGA, The International Journal of Management Science*, 3(6), 639–656.

van de Paal, G and WE Steinmueller (1998). Multimedia platform technologies as a means of building consumer demand for data communication services. Working Paper No. 41. Mastricht: Merit.

Wheelwright, SC and K Clark (1992). *Revolutionizing Product Development: Quantum Leaps in Speed, Efficiency, and Quality*. New York, NY: The Free Press.

Chapter 4

Changing Innovation Roles of Foreign Subsidiaries from the Manufacturing Industry in China[*]

Wenqian Zhou, Vivek K. Velamuri and Tobias Dauth

Based on the observation of higher internationalisation of innovation activities of multinational companies towards emerging markets, this paper aims to provide insights on how foreign subsidiaries from the manufacturing industry are changing their innovation roles in China. Based on in-depth expert interviews conducted from the subsidiary perspective in China, this paper affirms that foreign subsidiaries are moving towards higher innovative activities. The study contributes to extant literature by revealing three characteristics, namely innovation capabilities, organisational structures, and interaction with the headquarters that differentiate subsidiaries' innovation roles with regard to their geography and magnitude of innovation. This study illustrates how these distinctive characteristics and their underlying elements advance as subsidiaries move towards greater innovation roles and discusses implications for managerial practice.

Keywords: Subsidiary; innovation; multinational companies; manufacturing industry; organisational change; China.

Introduction

China, the world's largest emerging economy, has become one of the most important investment destinations for multinational companies (MNCs) (Kinkel *et al.*, 2014; OECD, 2013). Even after the financial crisis in 2008, when global foreign direct investment (FDI) flows generally declined, China's FDI inflow has been constantly increasing and reached $129 bn in 2014, thus making it the largest FDI recipient in 2014 (United Nations, 2015; OECD, 2013). In addition to reallocating manufacturing sites and production lines, there is an increasing number of MNCs who are redistributing their product innovation activities to China.

---

[*]Originally published in *International Journal of Innovation Management*, 21(1), pp. 1–32.

While MNCs have been increasingly internationalising their innovation activities through overseas research and development (R&D) centres, the geography of activities until the late 1990s has largely been confined to the triad regions of North America, Western Europe and Japan (Bruche, 2009). Conducting innovation activities in China began to significantly increase towards the end of 1990 and after 2000. While China accounted for only 1–2% of global R&D sites between 1975 and 1995, it increased to 9% in 2004 with a projection for further ongoing growth (Doz et al., 2006). Following this trend, more than 1500 R&D centres of multinational firms have been reported to have been set up in China in 2015, increasing from less than 50 R&D units in early 2000 (Bruche, 2009; Jolly et al., 2015). At the same time, distribution of R&D sites to Western Europe and USA is gradually shrinking.

Several motivations have led to this development (Gassmann and Han, 2004). In part, this is a result of Chinese policy makers' sensitivity to transforming the country through technology (Stanley et al., 2013). The government has put a series of Five Year Plans in place to transition the country from "Made in China" to "Created in China". This includes activities such as an increase in spending on R&D and education (Veldhoen et al., 2012). Foreign companies benefit from these activities due to the abundance of qualified staff at still comparably low cost (Yip and McKern, 2014). Additionally, special economic and investment zones have been established that attract foreign investors by providing advanced infrastructure along with financial incentives. Another key reason for internationalising innovation activities to China is related to its market importance for leading global companies. For decades, the sheer size of China's market tempted MNCs to localise R&D activities in order to adapt their products to local cultural and market needs. This allowed tailored products to have shorter development timing and enhanced cooperation with local sourcing and manufacturing activities (Gassmann and Han, 2004).

However, despite the ongoing interest in subsidiary innovation and the emergence of China as a "new geography of innovation" (Bruche, 2009, p. 282) for MNCs, the research focus has mostly been on subsidiaries in mature economies like Europe or North America so far (Ghoshal and Bartlett, 1988; Ciabuschi et al., 2014; Nobel and Birkinshaw, 1998). Recent review papers call for further study of the phenomena of subsidiary innovation in emerging economies (Reilly and Sharkey Scott, 2014; Li et al., 2013), and specifically China (Kinkel et al., 2014) in which significant change of subsidiary innovation roles towards higher innovative activity is expected (Veldhoen et al., 2012). In order to enhance the understanding of the current and future role of subsidiaries in China, this study aims to answer two research questions from the subsidiary perspective in the manufacturing industry:

(1) What is the current innovation role of foreign subsidiaries in China and how is it changing?
(2) What are distinctive characteristics of the different innovation roles in China?

The paper is structured as follows: The section Theoretical Underpinnings gives insights from existing research linked to subsidiary roles as well as organizational innovation. The section Research Methodology and Research Sample describes the research methodology and the sample. The section Findings summarizes the findings with regards to subsidiary innovation roles in China and its distinctive characteristics. The last section Summary, Future Research Direction and Contribution contains a summary, further paths of research, contribution to literature, and managerial implications of the research are presented.

Theoretical Underpinnings

Subsidiary innovation roles in China

Until the late 1970s, international business literature has almost ignored the phenomena of subsidiary innovation. Traditionally, subsidiaries have been seen as inferior to their headquarters (Vernon, 1979). With the idea arising of MNCs being a network of subsidiaries which are led but not constrained by headquarters and the recognition of knowledge and resources spread within the network (Ghoshal and Bartlett, 1990), subsidiary roles have slowly been reconceptualised and repositioned as a contributor to MNCs competitiveness and innovation (Reilly and Sharkey Scott, 2014; Birkinshaw and Hood, 2001).

In the literature of MNC and subsidiary management, past research has tended to focus on the different strategic roles of subsidiaries according to their geography and the magnitude of their innovation activity (Ghoshal and Bartlett, 1988; Nobel and Birkinshaw, 1998). According to Nobel and Birkinshaw (1998), subsidiaries can take the role of local adaptors, international adaptors, and international creators. As a local adaptor, the local R&D role is to adapt innovations developed by the parent company. While as an international adaptor, the local R&D takes a more creative role and not only adapts parent company's innovations, but develops some innovations for the foreign market itself. Subsidiaries in the international creator role conduct long-term, globally relevant research and product development on their own suggesting a higher innovative role. Following this typology, Veldhoen *et al.* (2012) find that MNCs' innovation activities in China already go beyond local adaptations in a study about foreign and local innovation activities in

China. Around 40% of the surveyed MNCs are not only conducting R&D for the Chinese market anymore, but are developing products in China for foreign markets. This trend is expected to intensify: by 2022, the study shows that more than 60% will conduct R&D in China for global markets and intensify activities in applied research, fundamental research, and ideation adding to the expectation of China to become a global innovation hub (Veldhoen et al., 2012).

Beule (2012) distinguishes between quiescent, autonomous, confederate, and active subsidiaries in his study about subsidiary strategic evolution in China. Quiescent subsidiaries aim at providing cost effective ways of securing profits from the supply of goods of the parent company with limited creative autonomy to adapt products to local taste. Autonomous subsidiaries concentrate on the subsidiary's own market needs. They have fairly well-developed R&D facilities and carry out local adaptations. Confederate subsidiaries focus on cost-effective production and are largely dependent on the headquarters for strategic decisions and resources, whereas active subsidiaries receive or take responsibility for the creation, production, marketing, and further development of products and processes. Beule (2012) notes that in 1995, the majority of foreign subsidiaries in China were quiescent types. In the period between 1995 and 2005, he observes a shift of those subsidiaries towards more autonomous or confederate roles and eventually active subsidiaries as a result of increased affiliate capabilities, request for local autonomy, and corporate behest.

In a study about German factories' manufacturing and innovation strategy in China, Kinkel et al. (2014) cluster the subsidiaries into offshore, source, server, contributor, outpost, and lead factories according to a typology developed by Ferdows (1997). Offshore and source factories produce specific items at low cost while the strategic role of a source factory is broader than that of an offshore factory. For example, a source factory may select their own suppliers. Server factories supply specific national or regional markets, and they typically provide a way to overcome tariff barriers and reduce taxes and logistics costs. Contributor factories serve specific national or regional markets. Their responsibilities extend to product and process engineering, as well as to the development and choice of suppliers. The primary role of an outpost factory is to collect information. Such factories are placed in an area where advanced suppliers, competitors, research laboratories, or customers are located. Lead factories create new processes, products, and technologies for the entire company. They tap into local skills and technological resources and transform knowledge into useful products and processes. In the case of China, Kinkel et al. (2014) found that many subsidiaries have ample local competencies in areas like production, logistics, and procurement, but only basic competences for introducing innovations. This limits the responsibility of the local R&D units to less complex product adjustments of existing products. They note that the subsidiaries are gradually developing from server factories towards the strategic role of contributor factories, but have not achieved this stage yet. Given the current low state of innovation competencies and the strategic reasons related to know-how protection, the majority of German MNCs are far from establishing, and not intending to establish, lead factories with global responsibilities for new products, processes, and technologies in China (Kinkel et al., 2014).

Subsidiary role development into a more advanced position can be seen as an evolutionary process in which the subsidiary gains higher capabilities and business responsibilities. More mature subsidiaries are found to be able to contribute more creatively to technology generation within their MNE network having had time to evolve away from domestic orientation (Cantwell and Mudambi, 2005). Three main drivers have been proposed to initiate this change, namely headquarters, subsidiary, and environmental factors (Birkinshaw and Hood, 1998). The role of the subsidiary can be seen as a function of headquarters-determined factors such as corporate strategy, culture, and resource allocation to the subsidiary and the mandate it is given (Cavanagh and Freeman, 2012; Cantwell and Mudambi, 2005). At the same time, it is acknowledged that roles can be driven and "earned" by the subsidiary itself. Thus, even relatively weak subsidiaries can change their role and strategic position themselves through successful and significant entrepreneurial initiative (Schmid et al., 2014). The market environment, i.e., customers, suppliers, and governmental bodies, will in both cases influence headquarters' and subsidiaries' decisions regarding activities taken. In some cases, new laws and the provision of certain assets may be necessary before certain innovation can even be deployed (Teece, 2006). Various strategies may be applied to increase capabilities for subsidiary role development. For example, the investment into human resources, process development, and strategies for market and product development can each have a different effect on the emergence or adjustment and pay-off timing of the capability of the firm (Branzei and Vertinsky, 2006). However, while it is commonly suggested that greater investments into R&D facilitates innovation and enhances subsidiary innovation role, Mudambi and Swift (2014) suggest that over the long-term, successful firms need to transition between exploitation of existing capabilities and exploration of new capabilities and alternate between different strategies depending on the environmental situation to maximise firm performance and profit.

It is evident from the above that research on foreign subsidiary innovation roles is established but non-conclusive for the case of China. While some researchers see China as a "global innovation hub" (Veldhoen et al., 2012) in the near future with already relatively high innovation competences, others doubt this trend. In this study, we aim at providing some empirical insights to contribute to the understanding on the current subsidiary innovation role of foreign manufacturing subsidiaries and future direction of subsidiary evolution in China.

Characteristics of subsidiary innovation roles

While the characteristics of organisational innovation in general have been studied from various perspectives such as process, dynamic capabilities, or upper-echelon perspective (Crossan and Apaydin, 2010), the phenomena of subsidiary innovation has been studied predominantly from a resource-based and dynamic capabilities view. Here, the underlying idea is that innovation capabilities are a necessary prerequisite in order to become an innovator subsidiary (Pogrebnyakov and Kristensen, 2011). Subsidiaries with higher capabilities are having a greater business scope and responsibility, and thus are bound to have a greater innovation role (Birkinshaw and Hood, 1998). Innovation capability studies arise from the organisational capability theory which is closely aligned with the resource-based view (Börjesson *et al.*, 2014). In that, it is assumed that performance differences across firms arise due to firm-specific resources which cannot be easily imitated. Innovation capability can be viewed as a special tacit asset of a firm which includes a disperse set of characteristics like learning, R&D, manufacturing, marketing, organisational, resource exploiting, and strategic capability (Guan and Ma, 2003). The resource-based view on innovation capabilities is extended by the dynamic capabilities perspective. Teece *et al.* (1997) postulate that in order for the company to stay competitive, internal and external resources and competences need to be constantly renewed and reconfigured, especially in rapidly changing environments. Firms that possess this capability will continuously create new combinations of knowledge and transform this into innovative products, processes, and systems (Lawson and Samson, 2001). They have the ability to sense and seize opportunities and renew themselves by developing new marketable products and services. They can also make the correct decisions for additional investments to deliver value to customers and earn their cost of capital (Teece, 2006). Eisenhardt and Martin (2000) postulate that they do so via routines which encompass specific organisational and strategic processes like product innovation processes and alliancing with partners. While these processes tend to be stable with predictable outcomes in less dynamic markets, the processes are rather fragile and its outcome unpredictable in high-velocity markets.

To enhance dynamic innovation capabilities, a great deal of research acknowledges learning as a crucial aspect. Subsidiaries are recognised for their unique position in which they are able to draw knowledge for innovation from both the internal and external networks in which they are dually embedded

(Almeida and Phene, 2004). Tsai (2001), in a study of knowledge transfer in the intraorganisational network of MNCs, demonstrates the importance of gaining access to knowledge in order to increase innovation achievement. He points out that the centrality of a business unit will facilitate knowledge sharing, transfer, and information access to increase both innovation and business performance of the subsidiary. Figueiredo (2011) finds that subsidiaries who display world-leading innovative performance are likely to display a higher level of dual embeddedness. This suggests that subsidiaries with greater innovation roles are better connected in their networks.

An associated stream of research on MNC and subsidiary innovation relevant to the present study concerns the autonomy of subsidiaries. Subsidiary autonomy has been reported both conceptually and empirically as positively associated with subsidiary innovation in that it increases innovation intensity (Ciabushi and Martin, 2011). Autonomy has been defined as the responsibility or decision power that the subsidiary holds as opposed to the headquarters for strategic, functional, and operational areas (Collinson and Wang, 2012; Kawai and Strange, 2014). It includes decision freedom for activities such as local product development or product change, purchasing decisions for usage of local sources, and decisions on local market positioning (Collinson and Wang, 2012). Ciabushi and Martin (2011), who study the effect of autonomy and innovation development, see autonomy as a driver of subsidiary innovation in which more independent subsidiaries display a greater level of innovative performance. This positive relationship could possibly occur due to autonomy granting subsidiaries the freedom to appropriately aligning business strategies and local market conditions, encouraging subsidiaries to foster organisational learning and building capabilities, and providing incentives for the subsidiary to feel more responsible for the success of the firm (Kawai and Strange, 2014).

Another viewpoint that has been tackled in subsidiary innovation research as a characteristic of organisational innovation is the process theory. From a business process viewpoint on innovation, similar inputs transform to similar outputs via core processes (Crossan and Apaydin, 2010). In the context of organisational innovation, these core processes will transform ideas into commercial outputs (Hansen and Birkinshaw, 2007). Crossan and Apaydin (2010) see the process elements to include initiation, portfolio management, development and implementation, project management, and commercialisation. Gassmann and Han (2004), in a study about motivations and barriers of foreign R&D activities in China, note that a consistent management information system which includes a global standardised stage-gate process, but also local freedom in creative early innovation phases, is necessary for successful local innovation activities.

Pogrebnyakov and Kristensen (2011), in a study about innovation subsidiaries in China, acknowledge that subsidiaries in an innovator role possess the ability to develop and manage complex innovation processes. As such, they are able to orchestrate innovation processes across several innovation stages and even across subsidiaries.

To conclude the theoretical underpinnings, extant research indicates that structural changes in the global configuration of the MNC towards higher involvement in innovation activities in China are taking place. In that process, the upgrading of subsidiary characteristics will bring the subsidiary into a more innovative position. From the identified research following the classification model of Nobel and Birkinshaw (1998), it is suggested that subsidiaries in a greater innovation role conducting global development activities will display greater innovative capabilities, are better connected in their internal and external network, possess more decision freedom, and have better processes in place to steer innovations than subsidiaries in a lower innovation role conducting adaptation activities. Thus, the research frame, as depicted in Fig. 1, emerges.

In our study, we build on this research frame and explore the changing roles of subsidiaries in China and how their specific characteristics in different innovation roles concur with articulated theory.

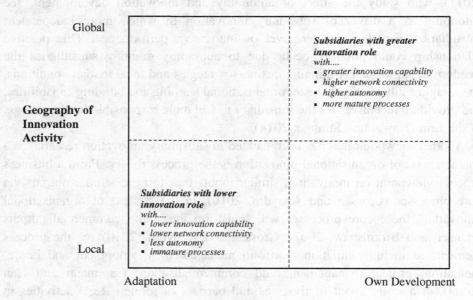

Fig. 1. Research frame: characteristics of subsidiaries' innovation roles.
Source: Based on Nobel & Birkinshaw (1998).

Research Methodology and Research Sample

Given the nature of the research questions, an exploratory study was conducted by in-depth interviews with subsidiary managers in China to obtain details about experiences in their subsidiary relevant to innovation activities.

The iterative data collection process took place between September and December 2014, and follow up questions were clarified in March 2016.

The database of the German Chambers of Commerce was accessed for initial sampling of foreign companies in China, followed by a snowball sampling strategy (Bryman and Bell, 2011). Purposive sampling was adopted to ensure that the subsidiaries were engaging in innovative activities and participants possessed specialist knowledge in the research field. In line with other innovation studies, this was ensured as only subsidiaries were contacted in which an R&D unit was established, thus indicating the strategic aim for conducting local product innovation activities (i.e., Fan, 2006; Li *et al.*, 2013). The subsidiaries all belong to an MNC of the manufacturing industry according to the International Standard Industrial Classification of All Economic Activities (United Nations, 2008). They are headquartered in Germany, France, or USA and belong to the Chemical, Machinery and Equipment, and Motor Vehicle Industries. These sectors account for in sum 15.7% of the foreign manufacturing industry and rank among the top 10 biggest sectors of the foreign manufacturing industry in China (Lemoine, 2000). A total of 14 experts involved in innovation activities from eight different subsidiaries agreed to be interviewed after guaranteed anonymity and a copy of the research findings (Table 1).

For three of the companies, China is among the top three of the most important markets. The subsidiaries in China have a workforce ranging from 120 employees to over 70,000 employees and sales of 15 million euros to 11,000 million euros in China. Therefore, we believe that the chosen subsidiaries do provide a typical and informative overview of the foreign manufacturing subsidiaries in China.

The interviews were based on a semi-structured interview guideline which guarantees a similar approach to all interviews, but also flexibility to the interview context. The questions have been derived based on the themes discussed in the theoretical underpinnings. For example, subsidiary innovation role, capabilities, and autonomy. The first level structure of this questionnaire consisted of guiding questions listed below, with several follow-up questions refining these:

(1) What is the current innovation role of your subsidiary in China?
(2) Has the innovation role changed in the past years, and if so, how?
(3) How do you expect the role to change in the future?
(4) How are the roles different, and what are their distinctive characteristics?

Table 1. Overview of researched subsidiaries and interviewees from the manufacturing industry in China.

Manufacturing industry classification	Subsidiary	Subsidiary innovation role	Interviewees
Chemicals and chemical products	Subsidiary A	Local/Regional development	Innovation Manager
	Subsidiary B	Local/Regional development	Marketing Manager
			R&D Manager
			Vice President supply chain (incl. Innovation Management)
	Subsidiary C	Local/Regional development	Vice President R&D
Machinery and equipment	Subsidiary D	Adaptation	General Manager
			Plant Director
	Subsidiary E	Adaptation	Head of purchasing
			Head of R&D
			Senior Production Manager
			Plant Director
Motor vehicles, trailers and semi-trailers	Subsidiary G	Adaptation	Innovation Manager
	Subsidiary H	Adaptation	General Manager
	Subsidiary I	Adaptation	Project Leader (responsible for setting up a new plant in China)

The interviewees were asked to provide an industry and company background and were asked to classify their subsidiary with regard to innovation magnitude and geography according to Fig. 1. They were further encouraged to provide related examples from their daily business.

Prior to conducting the interviews, one review session with academic mentors and two pilot interviews were conducted with professionals to scope the questions. An introduction to the topic as well as a shortened interview guideline was sent out prior to the interview in order to familiarise the participants with the subject. Eight interviews were carried out face to face, while the others were carried out over the phone lasting an average of one hour. Notes were taken during the interviews, and all but one interviewee gave permission for recording and transcription of the interview. Adding to this primary data, secondary data, such as press releases and internet research, were used to minimise bias of personal perspectives and enhance validity of the information through triangulation (Yin, 2014).

Following the logic of Alhojailan (2012), a thematic content analysis was used to evaluate the data. This process has been described as being particularly appropriate for analysis aiming at finding relationships between variables and comparing different sets of evidence that pertain to different situations in the same study (Alhojailan, 2012). We separated the data for subsidiaries which were classified by the interviewees as having an adaptation role or a local/regional own development role. For each data set, we followed a three-step approach as suggested by Miles and Huberman (1994) and Alhojailan (2012). We first reduced the data to sharpen and sort it in a way that conclusions could be drawn via open coding. In the second step, we displayed the data and organised it under first order concepts and further grouped it into second order themes. Finally, we drew conclusions out of the data by grouping the second order themes into core themes to frame the research. We used AtlasTi and Microsoft Excel for analyzing the data. To present our research findings, we followed an approach suggested by Gioia et al. (2012) to bring rigor to qualitative research by highlighting how data from open coding translates into the core themes that emerged. Figure 2 depicts this process for adaptation and local/regional own development subsidiaries.

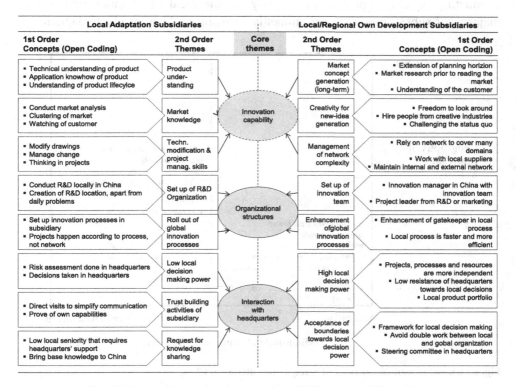

Fig. 2. Process of core theme generation of different subsidiary roles.

Findings

Having reviewed the transcripts and notes, the semi-structured interviews provided a wealth and depth of information addressing the research questions. The identification of core themes revealed several trends for the innovation roles of subsidiaries in China. First, we note that the prevailing subsidiary innovation role of subsidiaries in China is focusing on local adaptation activities, but affirming a change towards a greater innovation role as illustrated in the next section and depicted in Fig. 3. Second, we differentiate the innovation roles according to distinctive characteristics (Table 2). Quotations of the surveyed experts from the subsidiaries are used to illustrate major themes that arose.

Changing innovation roles in China

All respondents from the sample confirmed that the innovation role of their subsidiary in China has been strengthened over the past few years and will continue to intensify in the future. While subsidiaries from the Chemical Industry are already conducting development activities mostly for the local or regional market, the majority of subsidiaries are predominantly focusing on local product adaptations.

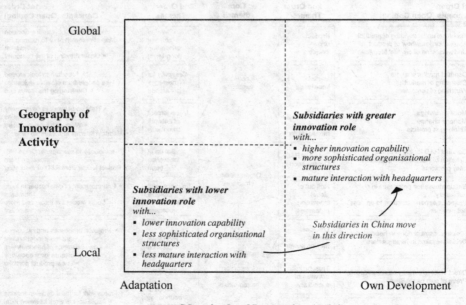

Fig. 3. Changing innovation roles of foreign subsidiaries in China.

Source: Based on Nobel & Birkinshaw (1998).

Table 2. Characteristics of subsidiary innovation roles in China.

Subsidiary innovation role		Local adaptation		Local/Regional own development
Distinctive characteristic	Most important element	Proof quotations	Most important element	Proof quotations
Innovation capabilities	Product understanding	"It is not enough to only understand one part of the product. We need to be able to have a complete overview of the product. It is also important that we have application knowhow, what our products are used for." (Subsidiary E, Head of R&D, Machinery Industry)	Market concept generation (long-term)	"It is our role to look long-term, spend time on conferences to come up with the big trends that will impact our business." (Subsidiary C, Vice President R&D, Chemical Industry)
	Market knowledge	"I am very much teaching my team to watch the market and the customer. (...) We have done a market analysis where we have clustered the market according to their accessibility for the public and to the industry. For example the tourist market is highly accessible to the public and to the industry, whereas the railway industry is highly accessible to the public but only limitedly accessible for new industries." (Subsidiary D, General Manager, Machinery Industry)	Creativity for new-idea generation	"So from the competitor understanding we are also much more competent than five years ago. From the raw material and suppliers' point of view, we were able to build a network. (...) Now we are in the region, we are able to work with suppliers in the region, build a network, and attend conferences where the knowledge can increase so much more than sitting in Europe and having some meeting minutes send by some European supplier." (Subsidiary B, R&D Manager, Chemical Industry)

(*Continued*)

Table 2. (*Continued*)

Distinctive characteristic	Subsidiary innovation role		Local adaptation		Local/Regional own development	
	Most important element	Proof quotations	Most important element	Proof quotations	Most important element	Proof quotations
	Technical modification & project management skills	"In the second phase, the R&D gains adaptation skills. They now understand interrelations. They can modify drawings. This requires construction skills and goes into change management, which is a higher art. I now need to drive change." (Subsidiary E, Plant Director, Machinery Industry)	Management of network complexity	"The employees in R&D need to be very much connected to their colleagues in other R&D centres. They need to be connected to people who are in America, in France, in Germany and in Japan." (Subsidiary C, Vice President R&D, Chemical Industry)		
Organisational structure	Set up of R&D organisation	"The R&D department started to be created. (...) Then the hiring process started and more and more new people came. At the end of the hiring process, there was the creation of the location, the R&D building, where all the people then moved. This is a big step to change the mind-set, so they are not stuck in the daily problems." (Subsidiary E, Senior Production Manager, Machinery Industry)	Set up of innovation team	"The project idea is coming from marketing or sales. There is the lab phase, pilot phase, production phase and market phase and marketing or sales are leading the overall process chain." (Subsidiary A, Innovation Manager, Chemical Industry)		

Changing Innovation Roles of Foreign Subsidiaries in China 111

	Roll out global innovation processes	"There are three huge innovation departments in Germany, and there are only two of us here right now in the innovation part which I lead. I hope that this will come into place in about one year time that the process will be here (...)." (Subsidiary G, Innovation Manager, Motor Vehicles Industry)	Enhancement of global innovation process	"We rolled out the global innovation process and developed it further. For example, we enhanced the gatekeepers. Now the general manager of China is in the gatekeeper team and is deciding for China. We have implemented a special process track which is much faster." (Subsidiary B, Vice President Supply Chain, Chemical Industry)
Interaction with headquarter	Low local decision making power	"The headquarters is supporting us by giving us people and money. But most importantly, they would be supporting us if they would be letting loose of some decision making and giving us opportunities." (Subsidiary H, General Manager, Motor Vehicles Industry)	High local decision making power	"It changes in that way that projects and processes and resources are more independent, and decisions are made within the local team." (Subsidiary B, Marketing Manager, Chemical Industry)

(Continued)

Table 2. (Continued)

Subsidiary innovation role		Local adaptation		Local/Regional own development
Distinctive characteristic	Most important element	Proof quotations	Most important element	Proof quotations
	Trust building activities of the subsidiary	"Direct visits in both directions — not the video conference, not the phone — but direct visits have helped to establish informal channels, which simplify the communication and establish trust." (Subsidiary D, Plant Director, Machinery Industry)	Acceptance of boundaries towards local decision power	"It will be a mistake to build up here an innovation hub in Asia for Asia and partially for the world which is completely independent from the rest, meaning the headquarters. This is not to control people here. This is first to avoid double work." (Subsidiary A, Innovation Manager, Chemical Industry)
	Request for knowledge sharing	"For the R&D center, we have many new projects coming, so for us, we have some challenges with the seniority of our staff. Overall, the work experience is only two years, which is quite short. So because of the low seniority we need a lot of support from expats from the headquarters." (Subsidiary H, General Manager, Motor Vehicles Industry)		

This has evolved from focusing on purely cost saving activities in the past. In subsidiary E, one manager describes this change as follows:

> "*The past role was mostly driven to be cost effective and to find cheap solutions. However, the local market is no longer only looking for cheap solutions. The solutions have to be new now and not only cost-efficient.*" (Subsidiary E, Senior Manager Production, Machinery Industry)

While the initial motivation to set up R&D in China has been cost driven in many cases, most subsidiaries' R&D departments today are to ensure that the products developed in the headquarters are adapted to the Chinese market on the basis of application engineering and variant development based on existing product platforms. This includes the adjustment of seat softness in the automotive industry due to a different perception of comfort and the adaptation of joysticks for the motor vehicle industry with additional protection due to higher pollution and dust in the Chinese environment. The surveyed experts clearly see market-driven reasons for the change in role, such as the special needs of the Chinese market requiring product modifications, increasing competition which requires speed in reaction, and partial Chinese governmental restrictions that require local production to access the market.

For the subsidiaries that are currently conducting local adaptation activities, most surveyed experts illustrate a strategic development towards a more innovative role. They emphasise on the strategic objective to intensify local and regional development activities in China in the upcoming 2–8 years. This includes the creation of their own products and platforms for Asia. For example, in the automotive industry, plans exist to develop China's own car concept in China for local customers' requirements such as the increased usage of the second seat row. While the ability to conduct local and regional development activities is the strategic goal, local adaptation activities are expected to co-exist. This is also confirmed by all interviewed experts from subsidiaries that are already conducting local and regional development activities.

This change in subsidiary role is especially driven through the initiative of the subsidiary. Building on the momentum of the strategic course for local innovation activities set by the headquarters, the subsidiaries actively engage in relationship management with the headquarters, highlight their performance, and strive for visibility. In subsidiary D, one manager mentions how he uses every opportunity during meetings or conferences with the headquarters to address the importance and potential of the Chinese market, the achievements of the local team, and further paths for growth in China and the region for the good of the overall company. This includes an increase in innovation costs for China and consideration of specific

Chinese customer needs in global development projects from the very beginning. To do this, active relationship management is necessary. One manager describes how he "hawks" ideas at the upper headquarters management and lobbies there for supporters. This facilitates further communication with the middle management in the headquarters, who oftentimes are described as being sceptic towards local innovation, requesting alignment and questioning local decisions. In this course, the local subsidiary will receive a bigger weight in pushing global decisions towards releasing more resources and responsibilities to the subsidiary to increase its innovation role.

At the same time, the subsidiary managers highlight how headquarters emphasise the strategic importance for local innovation activities in China by adjusting management structures to regional instead of formally functional management lines. This increases local responsibility and weight in the management team. This is oftentimes driven by the realisation that corporate functions lack the understanding of the local Chinese market and the change of the Chinese market from low cost to high-tech. This often requires their own developments for the region. Next to market reasons, governmental restrictions and regulations also lead to the necessity for local innovations. For example, subsidiary I has set up a R&D department in China in order to receive governmental subventions and licenses for sales. In subsidiary E, one of the interviewed managers mentions how Chinese importing restrictions of certain materials lead to the need of local product modifications.

However, while efforts are made on both sides to turn the subsidiary into a more innovative role, availability of innovative staff and cultural differences are inhibiting this change. For example, Chinese employees might understand organisational structures in a different way: While in Europe, many innovation projects are run in project organisations headed by a project leader. In subsidiary E, one of the interviewed managers laments that project support is only ensured in China if run by a department head. Also, global cost saving targets, in light of increasing competition, slow down the process of innovation role enhancement as less investments are made overall. Although China is still oftentimes granted exceptional R&D investments, the slowdown in Chinese economic growth might also inhibit required investments as headquarters are cautiously contemplating their resource allocation strategy.

Regarding future, long-term plans, the innovation direction for foreign subsidiaries that conduct development activities in China is more dispersed and dependent upon the size of the MNC. While smaller MNCs do not intend to conduct global innovation activities in China due to resource restrictions, larger MNCs are cautiously contemplating this idea for selected business areas.

The surveyed experts believe that China may develop into an important innovation location with regard to some selected technological segments. Focal points may be areas like energy and information technology, which are defined by the Chinese governmental research strategy. Other areas of interest arise due to the increasing importance and size of the Chinese market. In subsidiary A, the surveyed manager from the Chemical Industry notes that China is already by far the biggest market for tensid-based cleaning products. This trend could pave the way for establishing all global innovation activities for tensid-based cleaning products in China eventually. This trend is confirmed by another expert from subsidiary B who illustrates that the team overseeing the global R&D operations for this field will be based in Shanghai in the future. Increasing sectorial dominance of Chinese companies in the world as is seen in the telecommunications, mobile devices, or online service industries might further open up new paths for collaboration and global innovation possibilities. However, while ideas exist to move certain business units in the direction of global development, no grounded evidence was found in the expert interviews about a defined strategy for moving the subsidiary in China as a whole towards this role.

No data was found for subsidiaries conducting or planning to conduct global adaptation activities in China. While there are industries in which global adaptation activities are conducted in subsidiary locations (like IT), this does not seem to hold true for the subsidiaries of the manufacturing industry in China. This is possibly due to smaller volumes and thus less economies of scale in China compared to the main production location in headquarters, the limited knowledge about markets other than China, and long shipment lead times from China to the Western world.

The outlined change in foreign subsidiaries' innovation roles in China is depicted in Fig. 3. This shows the observed shift from a lower innovation role to a greater innovation role as subsidiaries move from local adaptation to local/regional own development activities.

Characteristics of different subsidiary innovation roles

In this section, we focus on summarising the different characteristics that local adaptation subsidiaries and local/regional development subsidiaries display. Through a thematic analysis of the interviews, the following characteristics have emerged as core themes for distinction with regard to the innovation role of the subsidiaries:

- **Innovation capabilities.** This refers to a set of tacit assets of an organisation that enables the organisation to introduce innovative outcomes.

- **Organisational structures.** This refers to organisational and procedural arrangements in the subsidiary to convert ideas into innovative outcomes.
- **Interaction with the headquarters.** This is derived from the interaction agreements between the subsidiary and headquarters.

Figure 3 presents the change in subsidiary roles in China, including the shift in subsidiary characteristics.

Table 2 presents details of the identified characteristics including underlying elements as well as proof quotations from the expert interviews that support the findings. Those characteristics are elaborated further in the next sections.

Characteristics of local adaptation subsidiaries

Innovation capabilities

The prevailing innovation role of subsidiaries in China is focusing on innovation activities in the area of local adaptations. This role has evolved from previously pure production-based subsidiaries which were concentrating mostly on optimisation activities and cost saving of existing products. While cost saving activities usually have a very narrow focus, and have a low level of complexity, conducting local adaptation requires a set of innovation capabilities: a sound product understanding about the product characteristics, market knowledge, technical modification, and project management skills are necessary elements for managing the change along the organisation. In subsidiary E, one manager describes the importance of product understanding as follows:

> "In order to make adaptations we needed to understand how the original products are, or were, working. This is the first step to go through: To understand and somehow assimilate the product from Western countries. As we understand this, we can start to make modifications." (Subsidiary E, Senior Manager Production, Machinery Industry)

Then, technical product knowledge needs to be paired with market knowledge about the customer base of the sales or marketing colleagues in order to generate ideas for product adaptations. A high collaboration between these market functions and R&D is necessary in order to drive the innovation in the correct direction. Furthermore, technical modification skills within the R&D and engineering teams are seen as elementary in order to change existing drawings, specifications, or formulas. For example, in the subsidiaries of the machinery industry, the surveyed experts name technical skills in mechanics and design, competences in testing and validation, understanding in electric, software, and control, as well as skills in using engineering systems as necessary.

Another capability that is seen as critical are project management skills to manage the product modification in the organisation. Especially in China, the subsidiaries mount a lack of good project managers who possess the ability to communicate clearly, motivate the team, track the tasks and milestones of the team members, and escalate across functions and hierarchies in order to steer a project in time, quality, and cost. In addition, these managers need to have an understanding of the business processes from neighbouring departments in order to manage the change in the organisation.

Organisational structures

The most prominent change when starting to conduct innovation activities in China is to establish R&D as its own organisational unit. While engineers have been set up at first for production support activities like testing of materials or validating products — thus sometimes functionally belonging to production or quality — they have now been gathered in a creativity centre, the R&D department. Next to 'classical' engineering work like testing and localisation support, they now have the clear task to watch the customer and the market for new possibilities of product adaptations apart from supporting former operational production problems.

In subsidiaries with sophisticated innovation management systems in the headquarters, oftentimes innovation managers or an innovation management unit are set up shortly after the creation of the R&D unit. Their task is to establish and develop the local innovation organisation, initiate and drive innovation projects, and provide consultancy and coaching. They serve as the interface for the global innovation teams and promote the idea of local innovation in China. However, this role is being perceived with some scepticism. Some interviewed experts express disbelief in the idea that innovation will centre on the role of one innovation manager; others express a lack of understanding of the concrete task of the innovation manager.

With regard to local innovation processes, global stage-gate processes are in the course of roll-out. These include typical gates with freezing points that split the innovation process into several phases including idea generation, idea elaboration, design, and implementation until start of production. While the set up of the local process is ongoing, many decisions are being done informally. In subsidiary G, the *status quo* is described as follows:

> "*Currently there is not one fixed process. We are in the midst of fixing our processes for our different departments. (...) It really depends on who is working on what, and figuring out the right*

> opportunities to talk to the management who is making the decisions." (*Subsidiary G, Innovation Manager, Motor Vehicle Industry*)

One difficulty in the roll-out of global innovation processes to China is the sustainability and application of the process within the local team. Though in some subsidiaries, the interviewed experts confirmed the establishment of a local innovation process, there is scepticism on the understanding among the employees. For example, in subsidiary E, one interviewed expert calls the provided trainings "*fake*" as they were too difficult for a newly established organisation in China with limited experience to understand and follow, consequently leading to confusion instead of application of the new process.

In summary, the organisational structures with regard to roles and processes in the local Chinese subsidiary can be described as follows: In subsidiaries, conducting local adaptation activities, R&D departments have been established as its own organisational unit with the clear task to conduct innovation. However, though the set up of local roles responsible for innovation and an innovation process are being rolled-out or are partially already established, the local organisation is yet to fully understand and apply it.

Interaction with headquarters

When questioned about interactions with the headquarters, the subsidiaries describe a high involvement of the headquarters regarding local decision making. This is especially true when local innovation organisations have been set up only recently. Processes are not defined yet and experience of the local team for conducting innovation projects is still low. In subsidiary E, the plant director describes the local innovation activities in China after being three years in operation as still "*submissive*" to the headquarters with a very strong leadership role in Europe. Though global innovation processes are partially rolled out locally, gate decisions and final releases are still made with high involvement of headquarters. Local portfolio decisions are often not tolerated or only with strict regulations.

While a frequent involvement of the headquarters in local decision making in adapting subsidiaries is prevailing, a change is noted by all subsidiaries. The local managers describe that they are working hard to build trust with the headquarters to receive greater freedom in decision making. This is oftentimes achieved through personal communication and networking. In subsidiary D, one of the managers describes that they tripled the travel budget for international innovation projects in order to foster networking between local and foreign colleagues. Combined with ongoing high performance of the local subsidiary, this fosters trust from the headquarters towards local decision making and capability:

> "The headquarters really trusts us and they support us. Of course a pre-condition is always that we meet our targets." (Subsidiary D, General Manager, Machinery Industry)

Another subject with regard to the interaction with the headquarters is the request for knowledge and information of the local subsidiary in order to become more innovative. As the Chinese innovation activities are usually still young in operation compared to the Western headquarters, a lot of effort is needed for them to catch up on the "base knowledge" like quality standards or engineering tools to build upon. Here, the headquarters are oftentimes still the primary knowledge pool that the local affiliates rely on to improve local capabilities. However, this relationship is often characterised by fear, like in subsidiary I:

> "There is a lot of fear to share too much information with China. A complete product design might disappear to the local competitor." (Subsidiary I, Project Leader, Motor vehicle industry)

To overcome this mistrust and hostility for knowledge sharing, a lot of sensibility and communication of the upper management team in the headquarters is needed. They need to foster willingness for the high-cost locations to share their knowledge with a low-cost location and emphasise the value added for the overall company performance.

In summary, the interaction with the headquarters from a subsidiary perspective which is conducting primarily local adaptation activities can be characterised by low local decision making power, initiatives of the local affiliate to build trust with the headquarters, and requests for knowledge sharing.

Characteristics of local/regional own development subsidiaries

Innovation capabilities

For conducting local or regional development activities on top of adaptations, the experts see following additional capability elements as critical for their subsidiaries: Generation of long-term market concepts, creativity for completely new idea generation, and management of internal and external network complexity.

The respondents describe development cycles as typically long-term; meaning planning horizons can last from 2 to 10 years with a high degree of uncertainty while product adaptation activities usually take less than one year. In order to pursue the correct development paths, it is important for the local organisation to understand market research, read the market, and anticipate customer needs well in advance. It is necessary to extract key information from the abundance of data and to draw the right conclusions out of it in order to generate long-term market concepts. At the same time, there needs to be a certain flexibility that market requirements might change along the development course and thus a tolerance towards change and failure.

Higher levels of creativity have been named as a necessary capability while moving up to do more innovative development activities. Usually this requires people with various experiences and backgrounds who have the ability to critically assess and challenge the *status quo* to come up with new ideas. These people are willing and able to try out new things independently and have the ability to find their own ways in a novel field. In subsidiary C, the Vice President heading the innovation team in China explains:

> *"It is not so easy for me to tell my staff what they need to work on exactly. I am telling them to work for a certain industry doing more or less that, find ideas and come back telling me what needs to be done. So in terms of creativity, in terms of independence it is a much bigger challenge."* (Subsidiary C, Vice President R&D, Chemical Industry)

The evidence from this study indicates that a higher level of complexity comes along with development activities. There are different partners involved, both internally and externally, which need to be managed. Internally, the team needs to be well connected within the R&D organisation of the MNC. This is because the development activities are typically broader in scope compared to adaptation activities. The local innovation teams are usually limited in size and resources covering a variety of markets and domains. Ergo, they need to draw knowledge from other sources. This requires high collaboration and communication with specialists around the world. For example, in subsidiary C, the local R&D specialists need to cover a range of domains like electronics, fluid dynamics, manufacturing, and simulation. The individual is not able to possess sufficient knowledge about all these fields, thus needs to be well connected to other specialists in headquarters or other subsidiaries. Externally, a network of suppliers and customers need to be managed. Though the interviewed experts acknowledge the importance of supplier involvement during innovation projects and high added value of the suppliers, they admit that the voice of the supplier is often not yet incorporated in China. Also, early involvement of customers and the set up of partnerships for common development projects can be improved to lever the advantages of dual embeddedness that subsidiaries are praised for (Figueiredo, 2011).

Organisational structures

In subsidiaries which are conducting local or regional development activities, both innovation roles and processes are at a higher maturity level compared to subsidiaries doing mostly local adaptations. The role of the innovation manager is defined and innovation teams are set up. Global innovation processes are rolled out locally and enhanced according to local needs. In subsidiary B, the regional R&D manager describes how since 2010 she has been responsible for the regional development activities in Asia with the clear task to develop the innovation pipeline adding track records of one successful innovation activity per year. Methods like competitor monitoring and trend analysis, together with market institutes, are well established among the local innovation organisation, and dedicated resources for innovation are available. Local innovation teams are set up in which the teams work cross functionally with R&D, marketing, and supply chain input:

> "We have established a team which is working very well together. Naturally, the influence of the country and the region has been increasing." (Subsidiary B, Vice President Supply Chain, Chemical Industry)

Additionally, the idea of reverse innovation activities is starting to spread. Local teams are starting to think about ways to expand their innovation activities to Western countries.

With regard to processes, local innovation processes are established and well understood among the organisation. Additionally, enhancement to the global stage-gate process is observed. For example, global processes tend to be rather rigid when meeting with management for decision making needs to be planned well ahead. In China, where local subsidiaries are of a smaller size and there is a high emphasis on speed to market, faster decision processes are established.

Interaction with the headquarters

For subsidiaries that are conducting local or regional development activities, a higher local decision freedom is observed. This is related to a higher level of trust of the global management towards the local activities. One manager from subsidiary B describes it the following way:

> "Previously the final decisions have been taken by the global team. Now the local team is making the final call. The global team is more informed about it." (Subsidiary B, Marketing Manager, Chemical Industry)

Local product portfolios for Asia are tolerated after alignment with the global team in which regional customer needs are considered.

Innovation processes are set up locally where gate decisions are taken by the local team. At the same time, there is a certain 'acceptance' of the local team towards boundaries and structures which are set within the MNC — i.e., monitoring of tasks and standardisation of processes to avoid double work. This is also described in subsidiaries A and B. Overall, as the local decision freedom is increasing, there seems to be a greater clarity as too which decisions can be made locally and which cannot. Involvement activities of the headquarters are more transparent and aligned, and decisions are less dependent on intrapersonal relationships and off-standard solutions. This is leading to a higher acceptance of the local team to follow global standards and decisions instead of constantly 'fighting' for more rights.

In summary, interaction with the headquarters can be characterised by high trust towards and decision freedom of the local team, as well as acceptance of boundaries of the subsidiary with regard to global standards and decisions.

Summary, Future Research Direction and Contribution

Based on the subsidiary typology developed by Nobel and Birkinshaw (1998) that differentiates subsidiary roles according to their magnitude and geography of innovation, we find that the prevailing role of the foreign subsidiaries from the manufacturing industry in China is to conduct local adaptation activities. At the same time, we demonstrate that there is a strong emphasis for changing this role towards higher innovative activities and conducting local and regional development, thus answering research question 1. With regard to research question 2, several characteristics have been identified that distinguish subsidiaries with low and higher innovation activities, namely innovation capabilities, organisational structures, and interaction with the headquarters. The study shows that subsidiaries conducting local adaptations and local/regional development activities display different underlying elements within those characteristics. The characteristics advance as the subsidiary's innovation role increases.

Future research direction

There are some limitations to the study which point towards future research. Though we carefully designed and documented the procedures taken in the research strategy for theory building and do believe we worked with typical subsidiaries from the manufacturing industry in China for generalisation purpose, data collection was primarily based on the viewpoints of the subsidiary and focusing on a narrow number of expert interviews per subsidiary. Consequently, in order to increase validity of the findings, it would be helpful to incorporate a larger sample of experts per subsidiary to ensure more robust results. Further, it would be interesting to account for headquarters' view on the strategic innovation directives for the subsidiary and provide a more abstract picture for the subsidiary network as a whole. This might shed light on reasons for resource allocation and subsidiary evolution like strategic moves to gain first mover advantage, specialisation of technologies in subsidiaries, relationship management with the local government, and utilisation of tax incentives when conducting local R&D and tariff barriers to import certain goods. It would also provide a headquarters perspective on the conflict between granting subsidiaries local autonomy for innovations and monitoring and restricting local decisions to manage a complex network of subsidiaries.

One interesting finding was that the chemical industry was conducting development activities in China already, whereas machinery and motor vehicle industries were largely focusing on local adaptation activities. This is possibly due to deeper technical knowledge, longer innovation cycles, and greater investments in local product development activities in the machinery and motor vehicle industry which limits the benefits of localisation of innovation. This is also possibly related to greater concerns for intellectual property protection. Thus, strategic emphasis is on conducting development activities centrally in the headquarters instead of locally in the subsidiaries. In any case, it would be interesting to explore reasons for industrial differences for subsidiary innovation in future research.

Contribution to theory

Regardless of the limitations, the study makes some important contributions to management literature. The emergence of foreign subsidiaries' innovation activities in China is a phenomenon that has rapidly increased since 2000 and is expected to carry on in the future. Past reviews have highlighted both the importance of subsidiary innovation activities and the existing knowledge gaps to understand this phenomenon, especially in emerging markets. With this contribution, the authors add to the understanding of changing innovation roles of foreign subsidiaries in China via empirically grounded data gathered from a subsidiary perspective. Though the prevailing innovation role of foreign subsidiaries from the manufacturing industry in China is local adaptation, the research affirms a strong emphasis to change this role towards higher innovative activities that include local and regional development. Yet, while for some selected business units, global innovation activities in China are in place or being contemplated, no evidence was found that the foreign manufacturing industry will transfer wide-ranging global innovation activities to China in the short-term. This affirms the findings of Kinkel *et al.* (2014) and partially contradicts the theoretical positions of Veldhoen *et al.* (2012).

The present study shows that subsidiary innovation is a different notion than that of independently owned firms because of the MNC context that has to be taken into account. Given that little is known about what occurs inside the subsidiary (Boojihawon *et al.*, 2007) and how its characteristics stimulate innovation in a nuanced way, the study determines clear differences between subsidiaries with low and high innovation roles. The study extends current research streams about subsidiary innovation which propose that the upgrading of subsidiary characteristics, like capability, network, autonomy, and process, will bring the subsidiary to a more innovative position. We built on these studies and consolidated the single aspects into an empirically grounded overview. The themes that we identified in our study converge into three main characteristics that describe subsidiaries with low and high innovation roles, namely innovation capabilities, organisational structures, and interaction with the headquarters. We confirm and show that subsidiaries in a greater innovation role display more advanced elements within those characteristics.

Managerial implication

The identified distinctive characteristics and their underlying elements may lever subsidiaries' innovation role; hence, this points towards some important implications for managerial practice. First, when setting up innovation activities in China, it is important to consider the local capability level with regard to product understanding, market knowledge, technical modification, and project management skills. Enough time should be allowed to build up base knowledge in the local team for using engineering software or working cross-functionally, which might be considered obvious in the headquarters. Only when a solid base has been built, enlargement of knowledge towards long-term planning, creativity, and network management may bring the subsidiary towards a higher innovative role.

Second, when copying and applying approved and existing organisational concepts of the headquarters to the local market, like setting up innovation teams and rolling out global processes, both the maturity stage of the local Chinese subsidiary, as well as cultural specifics need to be considered. For once, the knowledge level of the local team towards its functional role may be lower than in other subsidiaries in more advanced countries. This leads to the necessity of allowing enough time and resources to establish functional knowledge before introducing more advanced organisational concepts adding roles and process-thinking. Also, high sensitivity with regard to looking under the surface is needed. Though some roles and processes may have been officially established on paper, it may be necessary to check if they are actually understood and applied in practice before moving towards higher innovative activities. This oftentimes requires improvements to global processes and role enhancements.

Third, with regard to the interaction with the headquarters, trust between the subsidiary and headquarters is key to fostering local decision making authority. Headquarters' must also show a willingness to share knowledge in order for the subsidiary to move towards higher innovative activities. Apart from the networking activities often driven by the local team in China to build relationships and gain more local autonomy, it is especially important that there is support from the upper management in the headquarters. This helps mediate between the central middle management and the local team in order to increase the central team's willingness to release knowledge to a low-cost facility like China. This requires high sensitivity and communication in order to decrease mistrust which might otherwise paralyse the effort to conduct innovation activities in China. Once trust for knowledge-release from the headquarters towards the subsidiary is established, innovation capabilities and processes in the local organisation can be built up more rapidly and in accordance to global standards. The subsidiary may then eventually gain greater decision autonomy for innovation decisions reaching a state in which it both exerts its freedom and understands its boundaries for the good of the overall company.

References

Alhojailan, M (2012). Thematic analysis: A critical review of its process and evaluation. *West East Journal of Social Sciences*, 1(1), 39–47.

Almeida, P and A Phene (2004). Subsidiaries and knowledge creation: The influence of the MNC and host country on innovation. *Strategic Management Journal*, 25, 847–864.

Beule, FD (2012). A dynamic perspective on subsidiary strategy in China. *Science Technology Society*, 17(3), 365–383.

Birkinshaw, J and N Hood (1998). Multinational subsidiary evolution: Capability and charter change in foreign-owned subsidiary companies. *Academy of Management Review*, 23(4), 773–795.

Birkinshaw, J and N Hood (2001). Unleash innovation in foreign subsidiaries. *Harvard Business Review*, 79(3), 131–138.

Boojihawon, DK, P Dimitratos and S Young (2007). Characteristics and influences of multinational subsidiary entrepreneurial culture: The case of the advertising sector. *International Business Review*, 16, 549–572.

Börjesson, S, M Elmquist and S Hooge (2014). The challenges of innovation capability building: Learning from longitudinal studies of innovation efforts at Renault and Volvo cars. *Journal of Engineering and Technology Management*, 31, 120–140.

Branzei, O and I Vertinsky (2006). Strategic pathways to product innovation capabilities in SMEs. *Journal of Business Venturing*, 21(1), 75–105.

Bruche, G (2009). The emergence of China and India as new competitors in MNCs' innovation networks. *Competition & Change*, 13(3), 267–288.

Bryman, A and E Bell (2011). *Business Research Methods*, 3rd Edition. Oxford: Oxford University Press.

Cantwell, J and R Mudambi (2005). MNE competence-creating subsidiary mandates. *Strategic Management Journal*, 26(12), 1109–1128.

Cavanagh, A and S Freeman (2012). The development of subsidiary roles in the motor vehicle manufacturing industry. *International Business Review*, 21(4), 602–617.

Ciabushi, F and O Martin (2011). Effects of subsidiary autonomy on innovation development and transfer intensities. In *Entrepreneurship in the Global Firm (Progress in International Business Research, Volume 6)* AV Tulder, AT Tavares-Lehmann and R Van (eds.), pp. 251–273. Bingley: Emerald Group Publishing.

Ciabuschi, F, U Holm and OM Martín (2014). Dual embeddedness, influence and performance of innovating subsidiaries in the multinational corporation. *International Business Review*, 23(5), 897–909.

Collinson, SC and R Wang (2012). The evolution of innovation capability in multinational enterprise subsidiaries: Dual network embeddedness and the divergence of subsidiary specialisation in Taiwan. *Research Policy*, 41(9), 1501–1518.

Crossan, MM and M Apaydin (2010). A multi-dimensional framework of organizational innovation: A systematic review of the literature. *Journal of Management Studies*, 47(6), 1154–1191.

Doz, Y *et al.*, (2006). Innovation: Is global the way forward? A joint study by Booz & Company and INSEAD. *Insead and Booz Allen Hamilton*, pp. 1–13. Available at: http://www.boozallen.com/media/file/Innovation_Is_Global_The_Way_Forward_v2.pdf [Accessed June 21, 2014].

Eisenhardt, KM and JA Martin (2000). Dynamic capabilities: What are they? *Strategic Management Journal*, 21, 1105–1121.

Fan, P (2006). Catching up through developing innovation capability: Evidence from China's telecom-equipment industry. *Technovation*, 26(3), 359–368.

Ferdows, K (1997). Making the most of foreign factories. *Harvard Business Review*, 75(2), 73–88.

Figueiredo, P (2011). The role of dual embeddedness in the innovative performance of MNE subsidiaries: Evidence from Brazil. *Journal of Management Studies*, 48(2), 417–440.

Gassmann, O and Z Han (2004). Motivations and barriers of foreign R&D activities in China. *R&D Management*, 34(4), 423–437.

Ghoshal, S and CA Bartlett (1988). Creation, adoption and diffusion of innovations by subsidiaries of multinational corporations. *Journal of International Business Studies*, Fall, 365–385.

Ghoshal, S and C Bartlett (1990). The multinational corporation as an interorganizational network. *Academy of Management Review*, 15(4), 603–625.

Gioia, Da, KG Corley and AL Hamilton (2012). Seeking qualitative rigor in inductive research: Notes on the Gioia methodology. *Organizational Research Methods*, 16(1), 15–31.

Guan, J and N Ma (2003). Innovative capability and export performance of Chinese firms. *Technovation*, 23, 737–747.

Hansen, MT and J Birkinshaw (2007). The innovation value chain. *Harvard Business Review*, 85(6), 121–130.

Jolly, D, B McKern and G Yip (2015). The next innovation opportunity in China. *strategy + business*. Available at: http://www.strategy-business.com/article/00350?gko=e0f49 [Accessed March 27, 2016].

Kawai, N and R Strange (2014). Subsidiary autonomy and performance in Japanese multinationals in Europe. *International Business Review*, 23(3), 504–515.

Kinkel, S, O Kleine and J Diekmann (2014). Interlinkages and paths of German factories' manufacturing and R&D strategies in China. *Journal of Manufacturing Technology*, 25(2), 175–197.

Lawson, B and D Samson (2001). Developing innovation capability in organisations: A dynamic capabilities approach. *International Journal of Innovation Management*, 05(03), 377–400.

Lemoine, F (2000). FDI and the opening up of China's economy. *Centre d'Etudes Prospectives et d'Informations Internationales*, (11), 1–97. Available at: http://www.cepii.net/anglaisgraph/workpap/pdf/2000/wp00-11.pdf [Accessed February 26, 2016].

Li, X, J Wang and X Liu (2013). Can locally-recruited RD personnel significantly contribute to multinational subsidiary innovation in an emerging economy? *International Business Review*, 22(4), 639–651.

Miles, MB and AM Huberman (1994). *Qualitative Data Analysis. An Expanded Sourcebook*. London: SAGE Publications, Inc.

Mudambi, R and T Swift (2014). Knowing when to leap: Transitioning between exploitative and explorative R&D. *Strategic Management Journal*, 35, 126–145.

Nobel, R and J Birkinshaw (1998). Innovation in multinational corporations: Control and communication patterns in international R&D operations. *Strategic Management Journal*, 19, 479–496.

OECD (2013). FDI in figures. OECD, pp. 1–8. Available at: www.oecd.org/investment/statistics.htm [Accessed June 21, 2014].

Pogrebnyakov, N and JD Kristensen (2011). Building innovation subsidiaries in Emerging Markets: The experience of Novo Nordisk. *Research-Technology Management*, 54(4), 30–37.

Reilly, M and P Sharkey Scott (2014). Subsidiary driven innovation within shifting MNC structures: Identifying new challenges and research directions. *Technovation*, 34(3), 190–202.

Schmid, S, LR Dzedek and M Lehrer (2014). From rocking the boat to wagging the dog: A literature review of subsidiary initiative research and integrative framework. *Journal of International Management*, 20(2), 201–218.

Stanley, T, B Yang and R Ritacca (2013). *Innovated in China: New Frontier for Global R&D*. KPMG International Cooperative.

Teece, D, G Pisano and A Shuen (1997). Dynamic capabilities and strategic management. *Strategic Management Journal*, 18(7), 509–533.

Teece, DJ (2006). Reflections on "profiting from innovation." *Research Policy*, 35(8 SPEC. ISS.), 1131–1146.

Tsai, W (2001). Knowledge transfer in intraorganizational networks: Effects of network position and absorptive capacity on business unit innovation and performance. *The Academy of Management Journal*, 44(5), 996–1004.

United Nations (2008). International Standard Industrial Classification of All Economic Activities (ISIC), Rev. 4., pp. 1–291. Available at: http://unstats.un.org/unsd/cr/registry/regcst.asp?Cl=27 [Accessed December 12, 2014].

United Nations (2015). World Investment Report 2015, pp. 1–253. Available at: http://unctad.org/en/PublicationsLibrary/wir2015_en.pdf [Accessed March 3, 2016].

Veldhoen, S *et al.*, (2012). Innovation — China's next advantage? 2012 China innovation survey. A Benelux Chamber of Commerce, China Europe International Business School (CEIBS), Wenzhou Chamber of Commerce and Booz & Company Joint Report, pp. 1–12. Available at: http://www.strategyand.pwc.com/media/uploads/Strategyand_2012-China-Innovation-Survey.pdf [Accessed August 25, 2013].

Vernon, R (1979). The product cycle hypothesis in a new international environment. *Oxford Bulletin of Economics and Statistics*, 41(4), 255–267.

Yin, R (2014). *Case Study Research: Design And Methods*, 5th Edition. Los Angeles: SAGE Publications, Inc.

Yip, G and B McKern (2014). Innovation in emerging markets — the case of China. *International Journal of Emerging Markets*, 9(1), 2–10.

Chapter 5

Innovativeness and International Operations: Case of Russian R&D Companies*

Daria Podmetina, Maria Smirnova, Juha Väätänen and Marko Torkkeli

The number of Russian companies entering international markets has increased dramatically in the last 10 years. The development of innovative industries has intensified as well. Do innovations play significant role in internationalisation? Do innovators internationalize more actively? Does operating on international markets make companies more innovative? This paper studies innovations and internationalisation of companies in Russia, based on the survey of R&D-oriented companies located in the two most developed areas of Russia (St. Petersburg and Moscow). The study aims to identify the clusters of companies according to their exports and R&D expenditures, and fulfil in-depth analysis of innovations-related determinants that could explain the structure of the clusters. The main results of the study show the significant impact of innovation activities, competition and new product development on export intensity.

Keywords: Innovation; R&D; export intensity; internationalisation; Russia; competition.

Introduction

The opening of borders in the post-communist countries was highly anticipated by people and companies. However, together with obvious benefits of free trade, investments and economic development, there are a number of threats for domestic companies which have to be taken into consideration. Market liberalisation affected Russian companies with increased competition from imported goods, foreign direct investments (FDIs) and emerging of new effective companies. Companies had to learn to be competitive and find their own niche either on the domestic market or on the global one. In the communist system, domestic companies did not have to put much effort on the improving quality of products and services, personnel training, innovation and marketing research, because they were

---

*Originally published in *International Journal of Innovation Management*, 13(2), pp. 295–317.

enrolled in the direct sales, barter system, or had guaranteed governmental orders. The centralised research institutes were responsible for conducting research and providing technology development opportunities. The supply of technology was not often in balance with the demand from the enterprises. This not balanced connection between research institutions and companies still remains the weakest link in Russia (Krot, 2008). One of the most important factors that could contribute to the understanding of market players' changing behaviour is firm's attitude to innovations, technology development, technology transfer and commercialisation of innovations.

Globalisation has increased opportunities and pressures for domestic firms in emerging economies to innovate and to improve their competitive position (Gorodnichenko et al., 2008). Exporting allows firms in developing countries to enlarge their markets and to benefit from economies of scale (World Bank, 2001). Besides, export and import operations proved to be effective channels of technology transfer between countries (Pack, 1993). Some companies in the developing countries establish R&D centres or acquire companies from developed countries in order to obtain skills and knowledge (Bell and Pavitt, 1993). Russian economy is currently highly dependent on export of natural resources, such as oil and gas. Last eight years, Russian gross domestic product (GDP) has been growing more than 5% annually, thanks to high oil and gas prices on world markets. FDIs and exports are important cooperation channels for developing countries and the rest of the world.

Despite the substantial science base and the strong technology education, education innovation activity has been modest in Russia; only about 1.4% of GDP is spend on R&D and about 60% of R&D is publicly financed (Rosstat, 2007). The business sector is minor actor in R&D because only about 10% of industrial enterprises reported technological innovations in 2007, while the average in the European Union is 50% (Rosstat, 2007; OECD, 2005). The amount of R&D personnel in Russia is relatively high: about 1.3% of total labor force, compared with less than 2% in OECD countries. In theory, this should positively influence level of innovation capacity of companies in Russia, but only half of R&D personnel work as researchers, which means that the share of support personnel is extremely high.

Many researchers claim there is interdependency between innovation, competition and decision to internationalise (Sutton, 2007; Gorodnichenko, 2008; Wakelin, 1998; and others). And there is even more research support on the fact that the internationalised companies tend to transfer their experience from the international operations into increased innovativeness on the domestic market (Molero, 1998; Filipescu, 2007; Castelliani and Zanfei, 2006; and others). Authors agree with the statement that "these two features (internationalisation and innovation process) reinforce each other to the extent that today's economic analysis has to consider both of them simultaneously when trying to account for new dynamic of the firms operating at the international level" (Molero, 1998). The innovation is a wide concept, and it is not guaranteed that innovation successfully commercialized in one country would be as successful in another country, and vice versa.

Despite the natural resource export orientation is the growth driver of Russian economy. However, the development of knowledge intensive sector has been remarkable in the last five years. The export intensity of R&D oriented companies could depend on numerous factors, such as innovative and absorptive capacities, competition, industry, size of the company, level of regional development, etc.

In this paper, we aim to study how Russian companies decide on innovation-internationalisation challenge, how innovativeness is reflected on export intensity, and how competition matters in this context. We aim to find dependencies in Innovation–Export phenomena and try to track the interconnections between them.

The objective is to define set of variables, influencing innovation and internationalisation and test them on the data of over 170 Russian companies, collected in year 2008. The sample consists of companies, active in innovations or representing an industry with high innovation intensity (Frascati manual, 1993; Oslo manual, 2007).

The paper is structured as follows. Section 1 is the introduction; it describes the research gap of the study between innovation literature and international business literature and sets the objectives for research. Section 2 reviews the literature and studies the main theories on innovations and internationalisation, focusing on the main influencing factors, and formulates hypotheses and sets of variables. Section 3 describe the data and the methodology of the research. Section 4 presents the results of analysis of the data. Section 5 concludes the results of the study and sets the perspectives for future research.

Literature Review and Hypotheses Development

The existence of a strong relationship between internationalisation and innovation is obvious for technology oriented companies, when international technology transfer is a form of export per se (Robinson, 1988). The understanding of innovation has expanded from pure product and process innovations to organizational and even marketing innovations (Oslo manual, 2007). Moreover globalization pushes companies to enter foreign markets and acquire specific knowledge in order to implement technology and business innovations. Thus, "innovation has moved from an international reality dominated by the idea of technology transfer, where agents develop knowledge and transfer it to other countries, to a much more complicated situation where, although, that reality has not disappeared, there are also new ways of developing innovation in which the international ambit also affects the creation of knowledge stage and which multinational companies acquire new protagonism" (Molero, 2008).

There is substantial research evidence on the dual relationship between innovation and internationalisation. Filipescu (2007) tested empirically the prediction of product-cycle models of international trade which shows that innovation drives exports of firms in industrialized countries.

There are many factors influencing the dual relationship between innovation and internationalisation: firm's heterogeneity and internationalisation modes, relationship between (economic and innovative) performances and a further mode of internationalisation (Castellani et al., 2007), influence of innovation characteristics on firm's behaviour and relationship between trade and innovation on firm level (Wakelin, 1998), size of the company, innovativeness and export (Wakelin, 1998), and influence of the firm's technological capacity on both its decision to export and its export intensity (Lopez Rodriguez and Garcia Rodriguez, 2005). Some factors could be classified as domestic, exporting, controlling non-manufacturing activities abroad and manufacturing abroad (Castellani et al., 2007), or exporter versus non-exporters (Filipescu, 2007; Wakelin, 1998), non-exporting, low exporting, high exporting (Lachenmaier and Wossmann, 2006). Based on these theories we form our first hypothesis:

H1: *There is a relationship between innovation and internationalisation.*

The successful innovations also depend on macro-economic conditions, for example, the amount of effective demand within the national economy (Geroski and Walters, 1995), and the accessibility of foreign markets (Hughes, 1986). The behaviour of innovating firms depends on their own decision-making, but "is shaped by institutions that constitute constraints and incentives for innovations, such as laws, health regulations, subsidies, taxes, public expenditures, etc. Additionally, micro-economic conditions (e.g., market conditions, competition, and price setting) and macro-economic conditions (e.g., wealth, inflation, openness) will influence the decisions about innovation taken by firms" (Faber and Hasen, 2004). The development of a market economy in Russia has to be based on networks of innovative companies utilizing FDIs (Dyker, 2004).

Cooperation with foreign partners

The import of technology was complemented by a huge effort to develop local capabilities in the developing countries in East Asia: the technology cooperation with foreign partners covers not only "the acquisition of competencies for operating and maintaining, but also the acquisition of various combinations of design, engineering and project management skills". Companies invest in postgraduate education and training in the developed countries for their personnel to get enrolled into the informal international networks (Bell and Pavitt, 1993).

Dyker (2004) studies the process of development and dissemination of technology in Russia through the cooperation between Russian organizations and foreign firms. It is important to understand that FDI in Russia facilitates the technology transfer from abroad. The interesting point is that success of privatization in Russia can be estimated by "the diversity of enterprise forms, sizes, and strategies which is essential for knowledge diffusion and generation".

Co-operation and licensing deals with partners from developed economies are ways to speed up the innovation development process in Russia. But for Russian companies and research institutes, it is difficult to find partners when Russian scientists are not educated to prepare business plans or create new ventures. Venture capital industry in Russia is mainly foreign-owned, but on the other hand, foreign direct investments in R&D are quite modest. Probably the highest foreign R&D investment occurs in the ICT sector. Sun Microsystems, Motorola, Microsoft and Intel have R&D or dedicated development centres with more than 200 workers in St. Petersburg or Moscow (OECD, 2005).

Entry mode and export intensity

Innovations can also be factors, facilitating entry to international markets (Basile, 2001; van Dijk, 2002). Internationalization itself can be regarded as an innovation for the firm, whereas knowledge is a vital source. Innovations and R&D play important roles in overcoming barriers to internationalisation, but being conditional on having entered export market, R&D does not increase export intensity level when such R&D is treated as endogenous (Harris and Li, 2008). The customer orientation component of market orientation influences the innovative capability of the firms (Akman and Yilmaz, 2008).

Entry to the foreign market can play "a more creative role" serving as an instrument for introduction and diffusion of innovations. Geroski considers two types of entry, "imitative" and "innovative", which are different by nature, employing different management and control mechanisms and having different barriers and effects. However, innovative entry is rather seldom observed (10% of all new firms are innovative entrants). "Entry often plays a major role early in the life of most products...in early stages, outsiders are the source of most innovations and use these as a vehicle of entry...there is a shift from product towards process innovation..." (Geroski, 1991).

The effect of R&D and innovation on export is industry- and country-dependant. Both positive and negative effects could be found in the literature. There is a positive effect of R&D on exports for large samples of Brazilian and German firms, respectively (Willmore, 1992; Wagner, 2001). And there is a negative effect of R&D for Indian engineering firms (Lall, 1981). We can also suggest that non-exporting and low exporting strategies are prevalent among non-innovating firms. Innovators showed export share at 12.6% higher than non-innovators (Lachenmaier and Wossmann, 2006).

Innovating and non-innovating firms behave differently in terms of the probability of export and the level of export. Thus, the innovative capacity fundamentally changes the behaviour of the firm. Large innovating firms do more exports. Small innovating firms are more domestic (Wakelin, 1998). Based on these theories we propose to test relationship between innovations and export on the survey data (Hypothesis 1.1).

H1.1: *There is positive relationship between innovations and international operations. Companies with higher R&D expenditures have higher export intensity.*

Competition

Since the work of Joseph Schumpeter in 1934, innovation has been recognised as a tool for "competition and dynamic efficiency of markets" (Mansury and Love, 2008). The monopolistic firms were considered to be more willing to finance their R&D (less competition, more innovations). From the other perspective, implementing innovations is significant driver for increasing competitiveness of domestic companies both on the home and foreign markets.

Many researchers (Aghion *et al.*, 2005) hypothesized that there is inverted U-shaped relationship between intensity of competition and extent of innovation. However, research related to transition economies has proved that competition has a negative effect on innovation. No support was found for the inverted-U effect of competition on innovation (Gorodnichenko *et al.*, 2008). However, transitional countries should be the biggest beneficiaries of globalization, especially from the transfer of capabilities of FDIs (Sutton, 2007), because competition caused by foreign companies should strengthen domestic companies. We aim to test the effect of competition on innovation and suppose it is negative rather than positive (Hypothesis 2).

H2: *Competition has rather negative effect on innovations.*

As a factor of competition, innovation contributes to explaining heterogeneity in export behaviour (Basile, 2001). Technological resources can generate a double competitive advantage for a firm, in lowering costs by creating new and more efficient production processes, and in differentiation by means of product innovations (Lopez Rodriguez and Garcia Rodriguez, 2005). Mansury and Love (2008) state that "it is fair to say that theoretical support for the proposition that competition is good for innovation exists, but that it is yet far from conclusive". We suggest that effect of competition on non-exporting companies is more pronounced (Hypothesis 2.1.)

H2.1: *Competition has stronger effect on non-exporting companies and companies with low export intensity.*

Economic performance and productivity

Numerous academic studies have found a positive relationship between innovation and firm performance in manufacturing (Crépon et al., 1998; Mairesse and Mohnen, 2003). However, "reflecting the lack of maturity of the analysis of service sector innovation, studies of the relationship between innovation and business performance in the service sector are still relatively rare" (Mansury and Love, 2008). More productive firms are more likely to be engaged into internationalisation activities, and firms with high engagement in foreign activities also exhibit better economic and innovative performances (Castellani et al., 2007). The importance of the effect of the product markets structure on innovation activity and effect of innovation on productivity growth has been studied also by Geroski (1994).

Pianta and Vaona (2007) tested relationship of labour productivity levels and the diffusion of innovations in firms and proved that industries with a good export performance have to rely in all countries on improvements in both products and processes. "Advanced economies, such as the European ones, can expand their foreign markets only through a strategy of technological competitiveness where innovation plays a key role" (Pianta and Vaona, 2007). Considering the above mentioned academic studies, we aim to test relationship between labour productivity and export of the companies in our sample. The proposition is that companies with higher productivity are more export oriented (Hypothesis 3).

H3: *The more productive companies are more export oriented.*

Size

While traditional "Schumpeterian approach" states that small firms are not strong in introducing innovations and increasing productivity, recent research has found that small firms are not any weaker in innovation performance. They spend less on R&D than large firms, but they outperform large firms when considering innovation counts (Pianta and Vaona, 2007). Small and medium-sized enterprises report that the most important factors hampering their innovative activity include under-developed infrastructure in the area of technology commercialization, incomplete legislation, and lack of financing (OECD, 2005). Both the size of the firm, and FDI and capital employed play an important role in export intensity (Jauhari, 2007).

Often, for both developing and industrialised countries, an inverted U-shaped relationship between size and export propensity has been found (e.g., Wagner, 1995; Kumar and Siddharthan, 1994). Another explanation for the non-linear relation between exports and size is pointed out by Wakelin (1998). "Although size is an advantage in exporting, this may not apply to very large firms which can be more orientated towards the domestic market due to, for example a domestic monopoly giving them no incentive to export" (Wakelin, 1998). Hypothesis 4 is formulated in order to specify the effect of company's size of the export orientation of innovative companies.

H4: *The larger innovative companies are more export oriented, the small innovative companies are more domestic oriented.*

Patents

Product innovations, patents and process innovations positively and significantly affect both the decision to export and the export intensity. Technological capacity of the firm is the key factor in its international competitiveness, providing it with greater capacity to enter and sell products in foreign markets. R&D spending has a positive effect on export intensity (Lopez Rodriguez and Garcia Rodriguez, 2005). Faber and Hesen (2004) tested the relationships among R&D and other innovation activities, patents granted and sales of product innovations, and proved that patents do depend on sales of product innovations. The attitude of firms towards patenting reflects their orientation on innovation. This orientation has not only a positive effect on the number of patents granted as reported before, but also on the number of successfully introduced product innovations (Porter, 1990). We suggest that the effect of new product development and patents on the export intensity is positive (Hypothesis 5).

H5: *Product innovation and patents have positive effect on export intensity.*

Data and Methodology

The study is based on the survey of 176 R&D oriented Russian companies conducted in early 2008. The sample was drawn on companies, active in innovations or representing an industry with high innovation intensity (Frascati manual, 1993; Oslo manual, 2007). Thus, the sample was based on expecting the firms to be innovation-oriented and emphasizing R&D as a source of their long-term competitive advantage. Innovativeness indicators, such as R&D expenditure, new product development, and patenting activity, are used to evaluate the innovative capacity on the firm level.

There are significant difficulties in obtaining data in Russia due to low willingness of firms to disclose information, higher opportunism and strict knowledge-protection policies, in particular in the innovation-active industries. The procedure of data collection had to be made with guarantees of confidentiality of all the data gathered and limited opportunities to present the details of the companies taken part in the study in reports and further publications. The data gathering was conducted as follows. At the first stage of collecting the information, the interviewer approached the companies by the phone and allocated the qualified respondent. Usually the respondent represented the top management body. Then, the interviewer offered him/her to answer the questions. The response rate equalled 17%.

An important advantage of our study is that we have combined data on R&D expenditures (officially reported) and data on innovation activities and patents, reported by companies in out interviews. This approach allows avoiding the common method bias. Concerns about the common method use arise when both dependent and independent variables are measured by the same key informant (Luo et al., 2006; Podsakoff et al., 2003).

Most studies use mainly patents data and R&D expenditures, which is problematic. Patents have several weakness because they measure inventions rather than innovations, they are very industry-, country- and process-dependant, and companies often use other methods to protect their inventions. Using R&D expenditures can also been problematic, because not all innovations are generated by R&D expenditures. R&D does not necessary lead to innovation, and formal R&D measures are biased against small firms.

To achieve sample results, a number of industries and regions were included in the sample. The survey was conducted in Saint-Petersburg and Moscow — the Russian regions with the highest impact of FDIs and highest innovation sector development (Väätänen et al., 2007; Torkkeli et al., 2009). The industrial composition of the sample is as follows (Table 1): the largest number are service companies (27.8%), followed by machine building (22.7%), ICT (14.2%), electronics (14.2%), energy, oil and gas industry (7.4%), and construction (6.3%).

However, when analysing the share of sales in certain industries, machine building is the leader (30%), followed by ICT (24%), energy, oil and gas industry

Table 1. Industries and R&D expenditures.

	Share, %	Sales, %	R&D exp./sales, %
Services	27.8	9.5	1.6
Machine building	22.7	29.8	3.3
ICT	14.2	24.3	0.6
Electronics	14.2	10.0	5.6
Energy, oil and gas	7.4	19.7	2.0
Others	7.4	4.4	1.1
Construction	6.3	2.3	0.6
All	100.0	100.0	2.3

Table 2. Industries and export.

	Exporters/total, %	Exporters, % in industry	Exports/sales, %
Machine building	13.1	57.5	20.3
Services	10.2	36.7	10.3
ICT	9.1	64.0	11.9
Electronics	7.4	52.0	39.7
Energy, oil and gas	3.4	46.2	10.9
Others	2.3	30.8	20.7
All	45.5	45.5	17.0

(20%), electronics (10%) and services (9.5%). The industrial composition of the sample indicates companies' R&D orientation. The average share of R&D expenditure of sales is 2.3% when including all companies (and 6.5%, when including only companies with R&D expenditures). The highest share of R&D is in electronics (5.6%) and machine building (3.3%).

Enterprises are classified as exporting and non-exporting in order to analyse link between export and innovations in Russian companies. The share of exporting companies is high — 45.5% (Table 2). By the number of companies, the most export intensive is ICT sector, followed by machine building, electronics, and services. By the share of exports in the total sales, the leading industries are electronics, machine building and ICT.

The average sales per exporting company is slightly lower (59.9 million euros) than non-exporting company — 60.3 million euros (Appendix A). Productivity (sales per employee) of exporting companies is higher — 17.5 thousands euros per employee against 15.8 thousands euros per employee for non-exporting companies. There are 15% of foreign companies among exporters, and 10.4% among non-exporters.

If considering R&D companies separately (Appendix B), the most R&D intensive industries are machine building and ICT. However, the share of R&D expenditures of total sales is highest in electronics, machine building and energy, oil and gas sectors. The share of exporters is higher for R&D companies than for other companies (21.7% against 17%).

The key method of the study is to link the firm's innovativeness and the level of internationalisation — measure through export activities (Table 2). The interaction effects are tested separately for each dependent variable by applying the methods corresponding with the level of measurement (cross-tabulation, T-test for independent samples, ANOVA, linear regression analysis and GLM univariate test).

Dependent variables

As our *dependent variables*, we measure export activity of firm i as (a) whether firm i exported in a given year t ($EXPORT_D$), (b) a volume of export by a given firm ($EXPORT$), and (c) export as a share of sales of firm i in a given year t ($EXPORT_S$). See Table 3 for definitions of variables used in the study to explore the data and test the hypotheses formulated in the previous part of the paper.

Table 3. Definition of variables.

Variable	Description
EXPORT	Export of a firm in year t
$EXPORT_D$	Dummy variable equals 1 if company i exports in year t
$EXPORT_S$	Export as a share of sales
R&D	R&D expenditure of a firm in year t
$R\&D_D$	Dummy variable equals 1 if company i has R&D expenditure in year t
$R\&D_S$	R&D expenditure as a share of sales
$R\&D_{RDE}$	Ratio between the R&D expenditure of firm i and the number of R&D employees
COMPETITION	Importance of competition from imports in the market for the main product line/service in the domestic market
PRODUCTIVITY	Labour productivity, euro/person
$PRODUCTIVITY_{RD}$	Labour productivity of R&D employees, euro/person
SIZE	Size of firm in terms of a number of employees (small and medium sized — less than 200 employees, large — more than 200 employees)
$EMP_{R\&D}$	Number of R&D employees in the firm i
NPD	Number of technologically new or significantly improved products introduced by firm i during the last three years
PATENTS	Number of patents that the firm i has applied for the last three years
$PATENTS_E$	Ratio between the number of patents the firm i has applied for over the last three years and the number of employees
$PATENTS_{RDE}$	Ratio between the number of patents the firm i has applied for over the last three years and the number of R&D employees

Independent variables

Our key *independent variables* (Table 3) are linked to the field of innovation activities of the firms in the sample, and cover R&D expenditures of the firm (*R&D, R&D$_D$* and *R&D$_S$*), number of technologically new or significantly modified products introduced (*NPD*), labor productivity (*PRODUCTIVITY* and *PRODUCTIVITY$_{RD}$*), and number of patents (*PATENTS, PATENTS$_E$* and *PATENTS$_{RDE}$*).

We also consider the role of the competition from the side of the imports on the key product/service line in domestic market for the firms in our sample (*COMPETITION*). We also analyze the role of the size of the firm by proposing the variable *SIZE* that is based on splitting the sample into sub samples of small, medium and large firms.

Results

The study aims to identify the clusters of companies according to their exports and R&D expenditures and fulfil in-depth analysis of innovations-related determinants that could explain the structure of the clusters. The distribution of the firms in the sample according to our key variables — *EXPORT$_D$* and *R&D$_D$* is presented in Table 4. This distribution allows splitting the sample into four clusters:

Cluster 1: Non-exporting innovators [*R&D, but no export (29.0%)*]
Cluster 2: Non-innovating exporters [*export, but no R&D (14.0%)*]
Cluster 3: Non-exporting non-innovators [*no export, no R&D (25.6%)*]
Cluster 4: Exporting innovators [*both export and R&D (31.3%)*]

When describing the clusters, we see that there is no significant difference in total sales of the firms. The relationship between clusters and sales of the companies is presented in Appendix C.

Table 4. Export and R&D expenditures.

	No R&D expenses	R&D expenses	Total
No export	Cluster 3	Cluster 1	96
	45 (25.6%)	51 (29.0%)	
Export	Cluster 2	Cluster 4	80
	25 (14.0%)	55 (31.3%)	
Total	70	106	176

Note: Pearson chi square = 4.447 (0.035).

Hypotheses testing

While testing research hypotheses, the main emphasis was set on understanding the mechanisms, underlying both exporting and innovation decisions. There can be a link between firm's innovation activities and its internationalisation (see literature review in Section 2). There is no clear research evidence, in particular when considering Russia. The research question thus stays open and requires further investigation. Distribution of sample firms across the clusters assumes plurality of motivations underlying both internationalisation and innovation decisions. We limit our research to a number of key variables linked to firm's innovation activities.

Testing relationship between innovation and internationalisation

The detailed distribution of firms when analyzing both R&D expenditures and export as a share of sales are presented in Table 5. Table 5 shows that there is statistically significant relationship between export and innovation activities by the firms in a sample. The four clusters were defined and described earlier: R&D, but no export (29.0%), Export, but no R&D (14.0%), No export, no R&D (25.6%), and Both export and R&D (31.3%). There is a correlation between the R&D expenditure ($R\&D$) and export ($EXPORT$). When looking in depth of the export-innovation relationship on example of the sample firms, we may find a statistically significant relationship on the level of export and R&D expenditures as share of

Table 5. Export ($EXPORT_S$) — R&D expenditure ($R\&D_S$) relationship.

Export as % of sales		R&D expenditures as % of sales				
		0	Less than 5%	5–9%	More than 10%	Total
0	Count	45	25	11	15	96
	% of Total	25.6%	14.2%	6.3%	8.5%	54.5%
From 1% to 25%	Count	11	4	5	1	21
	% of Total	6.3%	2.3%	2.8%	0.6%	11.9%
From 26% to 50%	Count	10	17	7	2	36
	% of Total	5.7%	9.7%	4.0%	1.1%	20.5%
From 51% to 75%	Count	2	2	3	3	10
	% of Total	1.1%	1.1%	1.7%	1.7%	5.7%
From 76% to 100%	Count	2	8	2	1	13
	% of Total	1.1%	4.5%	1.1%	0.6%	7.4%
Total	Count	70	56	28	22	176
	% of Total	39.8%	31.8%	15.9%	12.5%	100.0%

Note: Pearson Chi-square $= 24.490$ ($p = 0.017$).

Table 6. R&D operations.

	Export	No export
Share companies with R&D	69%	53%
R&D expenditure/sales, %	3.7	3.8
R&D expend./R&D pers.	4,565.9	3,541.7
R&D expenditures. TOP4, %		
Internal R&D	19.9	22.6
Machinery & equipment	21.8	18.0
Acquisition of external knowledge	15.3	14.8
Acquisition of external R&D	10.6	12.3

sales (export and R&D intensity). In total, 60% of firms ($n = 106$) have R&D expenditures, while only 45% ($n = 80$) are exporting.

The existing relationship between exporting and innovativeness, revealed by our data, is not easy to explain. We need to address the variables that were selected as independent ones, to try to explain the selection mode of the firms in a sample between the spending on R&D and making decision to go international, since in many cases this seems to be a matter of compromise.

When comparing R&D expenditure, there are no significant differences between exporting and non-exporting companies (Table 6). The share of R&D of the total sales is 3.7% for exporting companies and 3.8% for non-exporting companies. Similarly, there are surprisingly few differences between exporting and non-exporting companies in the structure of R&D spending. Exporting companies spend 3.0% more on acquisition of machinery and equipment and 0.5% more on acquisition of external knowledge. Non-exporting companies spend 2.7% more on internal R&D and 1.7% more on acquisition of external R&D.

Testing the role of competition in domestic market

Competition is one of the factors that we could use as an explanation for both driving the innovation and export activities. We measure competition as perceived importance of competition from the side of import. But at the same time, there are some significant differences in perception of the level of competition by firms from different clusters (Table 7). For the defined clusters, we can see statistically significant differences in terms of perceived competition.

The highest level of competition is perceived by exporting firms with R&D expenditure, while the lowest level of competition is perceived by non-exporting firms with R&D activities. The explanation for this difference is not in the influence on the level of R&D expenditure. We did not find significant results by running the regression with *COMPETITION* as independent variable, influencing R&D, while the impact of *COMPETITION* on $EXPORT_S$ was significant (Appendix D). Thus, the stronger the competition in the home market, the more the firm will be inclined to export. The same is not true for the level of the R&D expenditure of firms.

There is a statistically significant relationship between the level of competition as perceived by the given firm, and the cluster the firm belongs to (Table 8). As we see from the regression results above, these differences are largely explained by the driving power of competition that is influencing the export activity of the firm. Due to our cluster approach, we see that the most firms, perceiving the competition as a serious factor, are combining the R&D and exporting activities.

Table 7. Competition perceived by clusters [mean (std. deviation)].

	No R&D expenses	R&D expenses
No export	2.63 (1.41)	2.42 (1.26)
Export	3.08 (1.25)	3.41 (1.29)

Note: $F = 6.832$ (0.000).

Table 8. Distribution of firms across the sample.

How important is competition from imports in the market (from "not important" to "extremely important")		R&D and Export — Cluster number of case				
		Cluster 1	Cluster 2	Cluster 3	Cluster 4	
		R&D, no export	Export, no R&D	No export, no R&D	Export + R&D	Total
Not important	Number of firms	15	3	12	4	34
	Percentage	8.5%	1.7%	6.8%	2.3%	19.3%
Slightly important	Number of firms	14	6	11	9	40
	Percentage	8.0%	3.4%	6.3%	5.1%	22.7%
Fairly important	Number of firms	9	3	7	12	31
	Percentage	5.1%	1.7%	4.0%	6.8%	17.6%
Very important	Number of firms	9	10	7	11	37
	Percentage	5.1%	5.7%	4.0%	6.3%	21.0%
Extremely important	Number of firms	3	2	6	13	24
	Percentage	1.7%	1.1%	3.4%	7.4%	13.6%
These products cannot be imported	Number of firms	1	1	2	6	10
	Percentage	0.6%	0.6%	1.1%	3.4%	5.7%
Total	Number of firms	51	25	45	55	176
	Percentage	29.0%	14.2%	25.6%	31.3%	100.0%

Note: Pearson Chi-square $= 27.663$ (0.024).

Testing the role of productivity in influencing firm's export activity

When testing the research hypothesis we measure both overall labor productivity and productivity of R&D employees. This hypothesis is partly supported, since there is a significant positive impact from the side of the productivity of the R&D employees, while the overall labor productivity has a significant, but negative influence on the export of a given firm (Appendix E).

Testing the size effect on both innovations and internationalisation of the firm

The size of the firm may have a crucial role in firm's innovating and exporting activities. The relationship between the size of the firm and the cluster the firm belongs to was proved to be insignificant. We applied t-test for independent samples across a number of variables that could help us explain the differences in innovative and exporting activities according to the size (Table 9). Indeed, smaller

Table 9. Results of testing the difference between small/medium sized and large companies.

		All sample	
Size of the firm by the number of employees		Mean	Std. deviation
EXPORT	SME	1,151,832.791***	1,334,983.2590
	LSE	39,874,664.903***	46,900,918.6148
PRODUCTIVITY	SME	37,539.195***	16,220.1458
	LSE	31,176.803***	13,174.8124
R&D	SME	202,230.859***	229,884.5340
	LSE	3,640,291.565***	5,873,241.4924
$R\&D_S$	SME	0.0897**	0.09570
	LSE	0.0484**	0.04624
$R\&D_{EMP}$	SME	15.81***	20.951
	LSE	573.69***	1,312.426
$R\&D_{RDE}$	SME	24,775.0618	51,062.74785
	LSE	40,295.8503	79,133.44190
PATENTS	SME	19.07	24.935
	LSE	14.00	9.266
$PATENTS_E$	SME	0.40018***	0.741290
	LSE	0.01808***	0.028480
$PATENTS_{RDE}$	SME	3.6409	8.77868
	LSE	0.2107	0.29686
NPD	SME	6.6500	3.85630
	LSE	6.1429	2.82468

Notes: ***$p < 0.001$, **$p < 0.01$.

Table 10. New product development (NPD).

	Export	No export
New product introduced in the last 3 years	28.8%	26.0%
New product developed by:		
Own company	91.3%	88.0%
In cooperation with others	8.7%	12.0%
Turnover, 2006, distributed %		
New product	21.1%	17.2%
Significantly improved	44.4%	43.0%
Unchanged	34.5%	39.8%
Average duration of NPD from idea to market (months)	13.86	13.56

firms are more limited in terms of R&D expenditures, have less employees, and less exports. But they out-perform the larger firms in terms of higher labor productivity, higher share of R&D spending as percentage of sales and higher number of patents per employee and per R&D employee. These innovation activities, though, do not lead directly to higher export sales.

Testing the role of product innovation and patents on export intensity

The basic indicators of new product development (NPD) are presented in Table 10. In the last three years, 28.8% of exporting companies have introduced new products compared with 26.0% of non-exporting companies.

NPs were mainly developed by own company — 91.3% of exporters and 88.0% of non-exporters. Non-exporters are more likely to co-operate with external partners in the product development phase. The sales mix also does not show large differences between exporters and non-exporters.

The final regression model tests the hypothesis on the role of product innovation (*NPD*) measured as a number of new technological products or significantly modified products introduced by the firm over the last three years and number of patents (*PATENTS*) in enforcing the exporting activities of the firm (Appendix F). Following the results of the previous analysis we also have included into model R&D expenditures of the firm (*R&D* and $R\&D_S$).

The regression test shows that export is influenced by the number of new products (*NPD*) and total R&D expenditure (*R&D*). From the previous test, we learned that R&D is significantly higher by the larger firms (Table 9). The number of patents has no direct influence on export activity that may be explained by some limitations of our study (industries selected, stage of internationalisation and innovation activities of the firms in the sample). Development and introduction of new technological products or significantly modified products seems to be one of the drivers underpinning higher internationalisation activities in Russian firms, while correlation with the level of R&D expenses was already supported by previous tests. It is interesting that the share of the R&D expenditures has no significant effect on export, while the share of the R&D is slightly (insignificant) higher in smaller and medium sized firms. Nevertheless, this factor has not proved to be significant determinant in export development and is one more explanation for easier internationalisation of larger companies.

Conclusion

The research paper studied the innovations and internationalisation of R&D oriented companies in Russia. The main results of the study show the impact of innovation activities, competition and new product development on export intensity.

Test results show that there is a statistically significant relationship between export and innovation activities by the firms in the sample. There is a correlation between the R&D expenditure (*R&D*) and export (*EXPORT*). When looking in-depth at the export-innovation relationship of the companies, we can detect a statistically significant relationship on the level of export and R&D expenditures as share of sales (export and R&D intensity). This relationship is again statistically significant ($p = 0.017$), and implies that there is a strong link between innovativeness and export behavior of the firm.

We measure competition as perceived importance of competition from the side of import. But at the same time, there are some significant differences in perception of the level of competition by firms from different clusters. For the clusters defined, there are statistically significant differences in terms of competition perceived. The highest level of competition is perceived in case of exporting firms with R&D expenditure, while as the lowest level of competition is perceived by non-exporting firms with R&D activities.

There is a statistically significant relationship between the level of competition as perceived by the given firm, and the cluster the firm belongs to. Results of regression analysis show that these differences are largely explained by the driving power of competition that is influencing the export activity of the firm. Due to our cluster approach, we see that the most firms, perceiving the competition as a serious factor, are combining the R&D and exporting activities.

When testing the research hypothesis, we measured both overall labor productivity and productivity of R&D employees. This hypothesis was partly supported, since there was a significant positive impact from the side of the productivity of the R&D employees, while the overall labor productivity had a significant, but negative influence on the export of a given firm.

The size of the firm may have a crucial role in firm's innovating and exporting activities. The relationship between the size of the firm and the cluster the firm belongs to was insignificant. T-test was applied for independent samples across a number of variables that could help explain the differences in innovative and exporting activities according to the size.

Indeed, smaller firms are more limited in terms of R&D expenditures, and have less employees and less export. But they out-perform the larger firms in terms of higher labor productivity, higher share of R&D spending as percentage of sales and higher number of patents per employee and per R&D employee. These innovation activities, though, do not lead directly to higher exports. However, when analyzing the innovators only, the exports by large innovators seem to be higher than those of whole sample.

The final regression model results revealed that exports are influenced by the number of new products (NPD) and total R&D expenditure ($R\&D$), while number of patents and share of R&D expenses had an insignificant relationship. The results of the study are subject to limitations due to the cross-sectional nature of the survey, selection of pro-innovation oriented sectors and limited number of regions presented in the study.

Further research of the relationship between exporting and innovativeness should provide interesting perspectives, especially to increase understanding of the mechanism underlying both exporting and innovations decisions. We aim to analyze traditional inward and outward internationalisation modes, together with level of cooperation with foreign partners, acquisition of foreign technology, outsourcing modes, and others in order to estimate the influence of internationalisation on companies' innovation abilities and on their innovation output.

Appendices

Appendix A. Financial indicators.

	Export	No export
Percentage of total	45.3%	54.7%
Sales/company, mln €	59.9	60.3
Employees/company	3,431	3,805
Productivity (Sales/employees)	17,468	15,853
Percentage of Foreign companies	15.0%	10.4%

Appendix B. Industries and R&D companies.

	Share of R&D exp, %	R&D/sales, %	Export, %	Number
Electronics	10.8	8.5	60.2	11
Energy, oil and gas	13.4	4.9	14.7	5
Machine building	32.9	4.8	21.9	16
Services	6.4	4.0	2.7	8
Construction	1.1	2.1	0.0	0
Others	4.4	1.8	29.0	2
ICT	31.0	0.7	14.8	11
All	100.0	3.7	21.7	55

Appendix C. Cluster — Sales relationship (€ mln).

	No R&D expenses	R&D expenses
No export	55.43* (108.28**)	64.63 (157.03)
Export	64.66 (142.28)	57.78 (102.7)

Note: $F = 0.058$ ($p = 0.982$), *Mean, **Std. deviation.

Appendix D. Impact of competition on the share of export in sales.

	Dependent variable — $EXPORT_S$		
$R^2 = 0.072$	T-value	B-coefficient	Sig.
COMPETITION	−3.672	0.268	0.000

Appendix E. Influence from productivity and productivity of R&D employees on export.

	Dependent variable — EXPORT		
$R^2 = 0.107$	T-value	B-coefficient	Sig.
Labour productivity, euro/person	−2.173	−0.239	0.033
Labour productivity (R&D employees), euros/person	2.522	0.278	0.014

Appendix F. Linear regression model.

	Dependent variable — EXPORT		
$R^2 = 0.800$	T-value	B-coefficient	Sig.
NPD	2.592	0.477	0.041
$R\&D_S$	−1.491	−0.273	0.186
R&D	4.007	0.749	0.007
PATENTS	0.061	0.012	0.953

References

Aghion, P, N Bloom, R Blundell, R Griffith and P Howitt (2005). Competition and innovation: An inverted U relationship. *Quarterly Journal of Economics*, 202, 701–728.

Akman, G and C Yilmaz (2008). Innovative capability, innovative strategy and market orientation: An empirical analysis in Turkish software industry. *International Journal of Innovation Management*, 12(1), 69–111.

Basile, R (2001). Export behaviour of Italian manufacturing firms over the nineties: The role of innovation. *Research Policy, Elsevier*, 308, 1185–1201.

Bell, RM and K Pavitt (1993). Technological accumulation and industrial growth: Contrasts between developed and developing countries. *Industrial and Corporate Change*, 2(2).

Castellani, D and A Zanfei (2007). Internationalisation, innovation and productivity: How do firms differ in Italy? *The World Economy*, Special Issue on "Exports and Growth".

Crépon, B, E Duguet and J Mairesse (1998). Research, innovation and productivity: An econometric analysis at the firm level. *Economics of Innovation and New Technology*, 7, 115–158.

Dyker, D (2006). *Closing the EU East-West Productivity Gap*. Imperial College Press.

Faber J and AB Hesen (2004). Innovation capabilities of European nations: Cross-sectional analysis of patents and sales of product innovations. *Research Policy*, 33, 193–207.

Filipescu, D (2007). Innovation and internationalization. A focus on the Spanish exporting firms. Research work, Doctoral programme: Creation, strategy and management of the firm, Universitat Autonoma de Barcelona, Business Economics Department.

Frascati, M (1993). The measurement of scientific and technological activities: Proposed standard practice for surveys of research and experimental.
Geroski, PA and CF Walters (1995). Innovative activity over the business cycle. *Economic Journal*, 105, 916–928.
Geroski, PA (1991). *Market Dynamics and Entry*. Cambridge, MA: Blackwell.
Geroski, P (1994). *Market Structure, Corporate Performance and Innovative Activity*. Oxford University Press, Oxford.
Gorodnichenko, Y, J Svejnar and K Terrell (2008). Globalization and innovation in emerging markets. *IZA Working Paper* No. 3299, Available at SSRN.
Harris, R and QC Li (2008). Exporting, R&D, and absorptive capacity in UK establishments. *Oxford Economic Papers*, 28 March, Oxford University Press.
Hughes, K (1986). *Export and Technology*. Cambridge: Cambridge University Press.
Jauhari, V (2007). Analyzing export intensity of the selected electronics firms in India. *International Journal of Innovation Management*, 11(3), 379–396.
Krot, P (2008). The Russian innovation system — An internaitonal perspective. *Research Report 206*, Lappeenranta University of Technology, Finland.
Kumar, N and NS Siddharthan (1994). Technology, firm size and export behaviour in developing countries: The case of Indian entreprises. *Journal of Development Studies*, 31.
Lachenmaier, S and L Wossmann (2006). Does innovation cause exports? Evidence from exogenous innovation impulses and obstacles using German micro data. *Oxford Economic Papers*, 58, 317–350.
Lopez Rodriguez, J and R Garcia Rodriguez (2005). Technology and export behaviour: A resource-based view approach. *Internaitonal Business Review*, 14, 539–557.
Luo, X, RJ Slotegraaf and X Pan (2006). Cross-functional "coopetition": The simultaneous role of cooperation and competition within firms. *Journal of Marketing*, 70, 67–80.
Mansury, MA and JH Love (2008). Innovation, productivity and growth in US business services: A firm-level analysis. *Technovation*, 28, 52–62.
Molero, J (1998). Patterns of internationalization of Spanish innovatory firms. *Research Policy*, 27, 541–558.
Molero, J (2008). The challenges of the internationalization of innovation for science and technology policies. *ICEI Paper, Instituo Compluense de Estidios Internacionales*.
Oslo Manual (2007). *Guidelines for Collecting and Interpreting Innovation Data*, 3rd ed.
Pack, H (1993). Technology gaps between industrial and developing countries: Are there dividends for latecomers?. In World Bank (ed.), *Proceedings of the World Bank Annual Conference on Development Economics 1992*, Washington DC.
Pianta, M and A Vaona (2007). Innovation and productivity in European Industries. *Economics of Innovation and New Technology*, 16(7), 485–499.
Podsakoff, PM, SB Mackenzie and J-Y Lee (2003). Common method biases in behavioural research: A critical review of the literature and recommended remedies. *Journal of Applied Psychology*, 88(5), 879–903.

Robinson, R (1988). *The International Transfer of Technology: Theory, Issues, and Practice*. Cambridge, Massachusetts: Ballinger Publishing Company.

Porter, ME (1990). *The Competitive Advantage of Nations*. London: Macmillan.

Rosstat (2007). Data from Official Russian Statistical Agency (1998–2008).

Schumpeter, JA (1934). *Theory of Economic Development*. Cambridge: Harvard University Press.

Sutton, J (2007). Quality, trade and the moving window: The globalization process. *The Economic Journal*, 117(524), F469–F498.

Torkkeli, M, D Podmetina and J Väätänen (2009). Knowledge absorption in emerging economy — Role of foreign investments and trade flows in Russia. *International Journal of Business Excellence*, 2(3/4).

van Dijk, M (2002). The determinants of export performance in developing countries: The case of Indonesian manufacturing. *Eindhoven Centre for Innovation Studies, The Netherlands, Working Paper*.

Väätänen, J and D Podmetina (2007). International technology transfer in the Russian economy — The effect of foreign direct investment spillovers. *Global Business and Finance Review*, 12(3), Special issue.

Wagner, JW (1995). Review of beyond the bilingual classroom: Literacy acquisition among Peruvian amazon communities, by Barbara Trudell. *Notes on Literacy*, 21(2), 56–59.

Wakelin, K (1998). Innovation and export behaviour at the firm level. *Research Policy*, 26, 829–841.

Willmore, L (1992). Transnationals and foreign trade: Evidence from Brazil. *Journal of Development Studies*, January, 314–335.

Chapter 6

Don't Get Caught on the Wrong Foot: A Resource-Based Perspective on Imitation Threats in Innovation Partnerships[*]

Foege J. Nils, Erk P. Piening and Torsten-Oliver Salge

Innovation partnerships can be a double-edged sword. While they are important vehicles for learning and value creation, such partnerships also increase a firm's vulnerability to unintended knowledge leakage and imitation by others. In this study, we go beyond previous research by studying the imitation threats induced by innovation partnership portfolios rather than individual alliances. Drawing on the resource-based view, we develop and test a model that links salient structural attributes of partnership portfolios and distinct forms of imitation. Results from our analysis of 803 German manufacturing firms support our prediction that a firm's probability of being imitated increases with the partnership variety of its portfolio. We also find that firms can mitigate this threat by carefully selecting innovation partners and using appropriation mechanisms.

Keywords: Imitation; innovation partnership; partnership portfolios; open innovation; resource-based view.

Introduction

Especially firms operating in technology-driven industries see themselves confronted with increasing technological complexity, shorter product life cycles, and rapidly changing customer demands. As part of their strategic response, many firms have sought to establish collaborative relationships with a wide array of innovation partners (Hoffmann, 2007; Lavie, 2006, 2007; Parmigiani and Rivera-Santos, 2011; Wassmer, 2010; Wind and Mahajan, 1997). Given the prevalence of this phenomenon today, the literature has gradually moved from its traditional focus on single dyadic alliances towards a more holistic analysis of a firm's partnership portfolio (Sivakumar *et al.*, 2011; Wuyts *et al.*, 2004). Drawing on

---

[*]Originally published in *International Journal of Innovation Management*, 21(3), pp. 1–42.

partly diverging conceptual foundations and terminologies, research on strategic alliances (Jiang *et al.*, 2010; Lavie and Miller, 2008), and open innovation (Chesbrough, 2003; Laursen and Salter, 2006) has begun to shed light on the emergence, configuration, management, and performance implications of such partnership portfolios.

There is now mounting evidence that maintaining broad partnership portfolios composed of diverse members (e.g., suppliers, customers, research institutions) can indeed boost a focal firm's, i.e., the portfolio holder's, innovative performance, as it not only broadens access to technology and market knowledge, but also enables risk and cost sharing among partners (Hooley *et al.*, 2005; Jiang *et al.*, 2010; Laursen and Salter, 2006; Leiponen and Helfat, 2010). That said, managing diverse alliances simultaneously is costly and challenging in many ways. Value appropriation concerns are a particularly critical issue that arises from innovation partnerships (Giarratana and Mariani, 2014; Laursen and Salter, 2006; Lavie, 2007; Manzini and Lazzarotti, 2016). It is precisely by opening up the innovation process, i.e., by rendering organisational boundaries permeable, that the focal firm risks exposing otherwise secret knowledge to its innovation partners or third parties (Dahlander and Gann, 2010; Martinez-Noya *et al.*, 2013). Although some attention has been devoted to the issue of knowledge leakage and interorganisational imitation in dyadic alliances (Kale *et al.*, 2002; McEvily *et al.*, 2004; Oxley and Sampson, 2004), studies examining the specific, presumably more pronounced imitation threats emanating from multifaceted, geographically dispersed innovation partnership portfolios are still missing.

In the present study, we therefore draw on the resource-based view (RBV) of the firm to develop a model that specifies the relationship between the partnership variety of a focal firm's innovation partnership portfolio and its risk of being affected by imitation, defined as the unsolicited use of technical inventions, products and business models, brand names, and designs by others. We test our model using comprehensive data from 803 German manufacturing firms, thereby contributing to the literatures on innovation partnerships and value appropriation in three important ways.

First, we shed new theoretical and empirical light on the conditions under which a collaborative approach to innovation is associated with detrimental effects stemming from imitation. Extending previous research on the imitation risks associated with dyadic alliances, we specifically elucidate the potential downside of maintaining a broad portfolio of functionally, geographically, and temporally diverse innovation partnerships. As most firms tend to be involved in multiple collaborative arrangements at the same time, focusing on the partnership portfolio as the unit of analysis is likely to provide a more realistic appreciation of the imitation threats firms are exposed to. Indeed, for various reasons (e.g., difficulties to identify opportunistic behaviour due to a higher degree of complexity), the imitation threat emanating from a portfolio of multiple, diverse partnerships is expected to be greater than the sum of the risks of individual alliances (Li et al., 2012).

Second and related, we provide new insights into how the focal firm's decisions concerning the configuration and governance of its partnership portfolio affect its risk of being imitated. Research on the contingency factors that enable firms to manage the trade-off between capturing the benefits, yet avoiding the costs of multiple innovation partnerships is still scant (Hsieh and Tidd, 2012). Against this backdrop, we explore the role of partner type (e.g., customer or supplier), partner location (e.g., domestic or international), and innovation phase the partnership is focusing upon (e.g., R&D or commercialisation stage). We also examine the moderating role of firms' intellectual property (IP) protection strategy and internal R&D activities as two potentially important isolating mechanisms safeguarding the focal firm from partnership-induced imitation. Knowing with whom, where, and when to collaborate in order to minimise imitation threats and avoid getting caught on the wrong foot is not only theoretically meaningful, but also of great practical importance.

Finally, our study provides a test of key assumptions of the RBV of the firm as an increasingly prevalent conceptual platform for research on technology and innovation management. RBV theorising, however, has remained ambiguous regarding whether innovation partnerships facilitate resource imitation given the greater permeability of organisational boundaries or, conversely, impede imitation attempts by means of greater causal ambiguity induced by collaborative innovation processes (Ketchen et al., 2007; Kozlenkova et al., 2013). Our study helps elucidate this persisting tension by identifying those factors that shape the direction and intensity of the link between innovation partnerships and the threat of imitation.

Theory and Hypotheses

A resource-based perspective on innovation partnership portfolios

According to the RBV, firms can be conceptualised as bundles of productive resources, that is, tangible and intangible assets (e.g., machinery, human capital, organisational structures), which are semi-permanently tied to the firm (Wernerfelt, 1984). Assuming that resources are heterogeneously distributed across firms and imperfectly mobile, proponents of the RBV have emphasised that valuable, rare and inimitable resources are the major source of interfirm performance differences (Barney, 1991; Peteraf, 1993). Possessing a favourable resource endowment, however, does not guarantee that a firm achieves a competitive advantage, much less that this advantage can be sustained over time (Newbert, 2007; Sirmon *et al.*, 2007). Importantly, organisations need to possess the ability to effectively exploit their resources and protect them from imitation in order to generate and appropriate value (Barney and Hesterly, 2012). In line with this reasoning, resource-based theorising suggests that a firm's ability to manage interorganisational collaborations is contingent on its knowledge, processes, and supporting structures (Dyer and Nobeoka, 2002; Kale *et al.*, 2002). In particular, the ability to harvest the benefits of innovation partnership is seen as a function of the firm's absorptive capacity (Lavie and Miller, 2008; Vasudeva and Anand, 2011), that is, "the ability of a firm to recognise the value of new, external information, assimilate it, and apply it to commercial ends" (Cohen and Levinthal, 1990, p. 128).

Innovation partnerships and interfirm relationships more generally can both act as a valuable, rare, and hard to imitate resource *per se* (Barney, 2014; Dyer and Singh, 1998) and fuel the continuous renewal of the organisational resource base (Hoffmann, 2007; Lavie, 2007). Yet at the same time, collaborating firms also face value appropriation challenges (Alexy *et al.*, 2013; Giarratana and Mariani, 2014; Katila *et al.*, 2008; Lawson *et al.*, 2012). In this context, Alvarez and Barney (2004, p. 625) argue that the "twin tasks of gaining access to those resources to generate rents associated with a market opportunity, and doing so in a way that enables this economic actor to appropriate at least some of the rents that were generated can become quite complicated". As innovation partnerships entail a risk of involuntary knowledge spillovers, collaborating firms need to consider how to prevent the imitation of technologies, products, and business models by their partners and their partners' partners (Kozlenkova *et al.*, 2013; Martinez-Noya *et al.*, 2013; Ritala and Hurmelinna-Laukkanen, 2013). Within resource-based theorising, inimitability is seen as a critical attribute of resources, without which a firm cannot sustain a competitive advantage stemming from innovation (Liebeskind, 1996; Polidoro and Toh, 2011). Given the theoretical and practical importance of this issue, scholars have devoted considerable attention to identifying the factors that make resources inimitable (McEvily and Chakravarthy, 2002; Reed and DeFillippi, 1990), or what Rumelt (1984) refers to as isolating mechanisms. These isolating mechanisms include legal property rights (e.g., patents and registered designs), secrecy, causal ambiguity, and first-mover or lead time advantages (Barney, 1991; Dierickx and Cool, 1989; Mahoney and Pandian, 1992; Peteraf, 1993).

Innovation partnerships and imitation threats

A collaborative approach to innovation enhances the potential for problem solving and helps to identify and assimilate essential knowledge to innovate on domestic and global markets (Chesbrough *et al.*, 2006; Lewin *et al.*, 2009; Sivakumar *et al.*, 2011). For a number of reasons, though, innovation partnerships are also likely to increase a focal firm's vulnerability to violations of its IP (Almirall and Casadesus-Masanell, 2010; Dahlander and Gann, 2010; Schmiele, 2013). First, jointly generated IP is not only known to, but also often claimed by the involved parties, causing legal conflicts over property rights especially in absence of a clear IP strategy among partners (Bercovitz and Feldman, 2007; Grimpe and Kaiser, 2010). This can be particularly harmful, if the focal firm operates on international markets, where litigations can be costly, lengthy, and uncertain with regard to their outcome (Schmiele, 2013). Second, collaborative innovation provides partners with insights into a focal firm's innovation capabilities (Wuyts *et al.*, 2004), thereby enabling them to replicate superior innovative performance (Ketchen *et al.*, 2007; Martinez-Noya *et al.*, 2013). De Rond and Bouchikhi (2004), for instance, report how a previously collaborative innovation partnership between a pharmaceutical giant and a small biotech company turned competitive, when the smaller partner decided to market and sell to third parties chemical structures similar to those jointly developed as part of the partnership. When collaborating with a broad portfolio of partners, firms face the dilemma of being required to share their knowledge with these partners in order to maximise value creation, while protecting its knowledge from undesirable spillovers to maximise value appropriation (Grimpe and Kaiser, 2010; Li *et al.*, 2012; Martinez-Noya *et al.*, 2013). Innovation collaboration may also indirectly increase imitation threats, when innovation partners do not act opportunistically themselves, but function as a channel for third parties to obtain and infringe critical knowledge. As a case in point, competitors may gain access to valuable technological knowledge of the focal firm by collaborating with its suppliers or associated scientific institutions rather than with the focal firm itself (Dyer and Nobeoka, 2002).

The risk of imitation becomes particularly evident when viewing innovation partnerships through the theoretical lens of the RBV. Accordingly, it can be expected that the extent to which a firm's resources and capabilities are readily observable decreases the causal ambiguity surrounding their relationship with performance, and thus the challenges and costs of imitation (Liebeskind, 1996; Reed and DeFillippi, 1990). As stated by Barney (2014, p. 26), "It is in the self-interest of firms to keep information about the emergence of many of their resources and capabilities in-house, to reduce the threat of imitation." Empirical evidence corroborates this argument. For example, Ethiraj and Zhu (2008) findings suggest that the amount of available information determines whether competitors can effectively (e.g., at lower cost and risk) imitate innovations. In this regard, Kogut and Zander (1992, p. 394) use the "*poker hand*" metaphor, according to which imitation will rapidly ensue, once a firm has revealed its cards — i.e., its knowledge. Being provided voluntarily with information by the focal firm and obtaining insights into its operations, opportunistic partners are well positioned to understand the value of the focal firm's resources and how they are best used (Li *et al.*, 2008; Martinez-Noya *et al.*, 2013). This may trigger the desire among partners to imitate in the first place. In particular, intense innovation partnerships can provide imitators with the rare opportunity to observe a firm's competencies in use. This is highly problematic, as the inability to do so is widely regarded as deterring imitation (Barney, 2014), because outside actors cannot fully understand how and why certain resources contribute to firm performance without considering the idiosyncratic context in which they are put to use (King, 2007; Polidoro and Toh, 2011). Collaboration enhances an imitator's ability to capture the focal firm's tacit, sticky knowledge embedded in its operating routines and managerial practices, because the transmission of this kind of knowledge requires social interaction (Li *et al.*, 2008). From this perspective, keeping innovation partners at arm's length and reducing interaction appears necessary in order to protect IP (Roy and Sivakumar, 2011).

While the arguments provided thus far similarly apply to dyadic alliances, simultaneously engaging in collaborative arrangements with various domestic and international innovation partners has specific implications for understanding imitation threats. We expect the focal firm to become more vulnerable to imitation the greater the variety of its partnership portfolio (i.e., the number of the different partner types with which the firm collaborates in different phases of the innovation process), if nothing else because more potential imitators have more opportunities to get access to the firm's proprietary knowledge. In particular, a greater number and diversity of innovation partners amplify the degree of managerial complexity and impede the monitoring of potentially opportunistic partners (De Leeuw *et al.*, 2014; Li *et al.*, 2012; Oxley and Sampson, 2004). Managing a partnership portfolio composed of multiple partner types such as suppliers, customers, and research institutions involves different exchange relationships and requires specific contractual rules as well as IP protection mechanisms (Laursen and Salter, 2006). This is especially the case when simultaneously engaging in collaboration during various phases (e.g., idea generation, R&D, commercialization) of the innovation process, each of which poses specific challenges regarding the protection of IP (Manzini and Lazzarotti, 2016). Moreover, there might be a trade-off between collaboration breadth and depth in that firms collaborating with a fewer number of different partners will be more likely to be able to build trusting relationships with their partners than those with multiple partners (Meuleman et al., 2010). This is consequential, as trust is widely considered an effective safeguard against opportunism in alliances (Dyer and Singh, 1998). In light of these arguments, we propose:

H1: *The greater the variety of a focal firm's innovation partnership portfolio, the greater will be its risk of being affected by imitation.*

Partner type, partner location, and innovation phase as contingency factors

To shed further light on how the structural configuration of a partnership portfolio influences imitation threats, we go beyond the aggregate effect of partnership variety and explore the specific effects of *partner type*, *partner location*, and *innovation phase* on imitation.

Partner type

Any innovation partnership with competitors, suppliers, customers, and research institutions entails a certain risk of opportunistic behaviour and knowledge leakage (Bercovitz and Feldman, 2007; Hernandez et al., 2015). Nevertheless, a focal firm's risk of being imitated is likely to vary based on the specific types of innovation partners being included into its portfolio (Diestre and Rajagopalan, 2012; Nieto and Santamaría, 2007). In particular, we expect that, given their strategic position towards the focal firm, certain partners will be more inclined than others to misuse these relationships for accessing and exploiting the focal firm's proprietary knowledge for their own advantage (Emden et al., 2006; Ritala and Hurmelinna-Laukkanen, 2013; Sivakumar et al., 2011).

According to this reasoning, scientific innovation partnerships, that is, collaborative arrangements with universities and research institutes, are likely to be the partner type of choice for firms seeking to minimise appropriability concerns. Scientific institutions tend to have fewer incentives to act opportunistically and also typically lack complementary assets required to commercialise IP generated as an outcome of joint research themselves (Bercovitz and Feldman, 2007; Martinez-Noya et al., 2013). That said, scientists and practitioners might still have conflicting interests with regards to IP. While managers are interested to keep IP proprietary and secret in an attempt to ensure value appropriation, publishing technological advancements is critical for academics to enhance their scientific reputation (Bercovitz and Feldman, 2007).

Collaborative research endeavours with customers and suppliers, or what has been referred to as vertical partnerships, are deemed less hazardous for unintended knowledge leakage and imitation than horizontal partnerships with competitors, which are widely regarded as particularly risky (Bercovitz and Feldman, 2007; Miotti and Sachwald, 2003; Nieto and Santamaría, 2007; Oxley and Sampson, 2004; Sivakumar *et al.*, 2011). It should be noted, however, that collaborative arrangements with suppliers are by no means immune to imitation. In particular, a supplier might turn into a future competitor through forward integration or serve as a conduit through which technological knowledge of the focal firm leaks to third parties that are not only customers of the supplier, but also competitors of the focal firm (Martinez-Noya *et al.*, 2013). Indeed, the same may be assumed for corporate customers and, to a lesser extent, research institutions.

However, current competitors engaged in innovation partnerships with the focal firm do not only have the strongest incentives and complementary assets to exploit proprietary knowledge of the focal firm, but are also likely to possess the absorptive capacity necessary to identify, assimilate, and utilise this knowledge. This assumption can be made as competitors serve similar markets and hence tend to possess comparable knowledge bases, which enable imitating firms to understand the value and applicability of their partner's technological competencies (Diestre and Rajagopalan, 2012; Ritala and Hurmelinna-Laukkanen, 2013; Yang *et al.*, 2010). A high degree of knowledge overlap between innovation partners increases the partner-specific absorptive capacity (Dyer and Singh, 1998), which facilitates interfirm knowledge transfer regardless of occurring voluntarily or involuntarily (Emden *et al.*, 2006). Based on the arguments presented, we expect that:

H2: *Horizontal innovation partnerships will be associated with a greater risk of being affected by imitation for the focal firm than vertical and scientific innovation partnerships.*

Partner location

The issue of geographic location has received considerable attention in various fields such as research on strategic alliances (Phene and Tallman, 2014; Sivakumar *et al.*, 2011), innovative search (Ahuja and Katila, 2004), and headquarter-subsidiary relationships (Goerzen and Beamish, 2003). Conceptualising distance in multidimensional terms (e.g., geographical, cultural, and institutional distance), this body of work highlights both the potential benefits (e.g., access to complementary knowledge) and challenges (e.g., coordination, control, and knowledge transfer) associated with distal exchange relationships (Lavie and Miller, 2008; Zaheer and Hernandez, 2011). In an effort to extend these studies mainly concerned with the general performance implications of distance, in the present research, we explore how the location of innovation partners in a focal firm's partnership portfolio affects its imitation threat. We do so by comparing the risks of collaborating with domestic (i.e., located in Germany), continental (i.e., located in Europe), and international (i.e., located elsewhere) innovation partners.

There are, however, conflicting arguments concerning the relationship between partner distance and the threat of imitation. On the one hand, it is argued that proximity to an innovation partner fosters not only intended, but also unintended knowledge spillover (Giarratana and Mariani, 2014; Schmiele, 2013). In particular, proximity in terms of geography, culture, language, institutional conditions, and organisational knowledge has been associated with improved interaction and knowledge sharing in innovation partnerships by facilitating face-to-face communication between actors and the development of trusting relationships (Boschma, 2005; Ben Letaifa and Rabeau, 2013). Although, good interorganisational and interpersonal relationships are often regarded as an effective safeguard against opportunistic behaviour (Dyer and Singh, 1998; Lavie and Miller, 2008), other scholars have argued that familiarity increases the focal firm's vulnerability to unintended knowledge leakage (Boschma, 2005; Li et al., 2008). In line with resource-based theorising, frequent personal interaction can be seen as a prerequisite for understanding and exploiting tacit, causally ambiguous knowledge (Kogut and Zander, 1992; Oxley and Sampson, 2004; Wuyts et al., 2004; Zaheer and Hernandez, 2011). As such, proximal innovation partners are best equipped to imitate even complex technical inventions, products, and business models that would otherwise be very difficult to replicate (Li et al., 2008). This is especially true, as firms may underestimate the risk of opportunism in alliances with seemingly well-known partners (Boschma, 2005), and thus miss the opportunity to employ adequate protection mechanisms. The ability of proximal innovation partners — especially domestic ones — to assimilate and exploit the focal firm's IP can also be traced back to their greater absorptive capacity compared to foreign partners, which stems from having a similar cultural, institutional, and cognitive background (Giarratana and Mariani, 2014; Lavie and Miller, 2008). Finally, proximity makes it easier for opportunistic partners to hire away key employees of the focal firm who may provide access to otherwise secret knowledge.

On the other hand, despite distance-induced difficulties to obtain insight into the focal firm's operations and lower levels of absorptive capacity (Lavie and Miller, 2008), there are reasons to suggest that distal innovation partnerships can be risky in their own right. First, foreign innovation partners might have stronger incentives to imitate the focal firm. Indeed, imitation is a prevalent strategy used by foreign firms to mitigate the liabilities of foreignness, that is, the costs associated with the international expansion of operations (Salomon and Wu, 2012; Zaheer, 1995). Second, research suggests that geographic, cultural, institutional, economic, and linguistic distance between collaboration partners can lead to conflicts, mistrust, and lack of commitment. Such interorganisational tensions provide a fertile ground for the misappropriation of the focal firm's IP (Katsikeas et al., 2009; Robson et al., 2008). Finally, the difficulty of monitoring violations of property rights and protecting them is likely to increase with distance. In particular, formal governance mechanisms tend to be less effective in preventing imitative behaviour by foreign as opposed to domestic partners. Litigations in a foreign country can be costly, time consuming, and uncertain in terms of their expected outcomes due to oftentimes weaker appropriability regimes abroad (James et al., 2013; Schmiele, 2013; Zaheer, 1995). Especially emerging countries often lack legal mechanisms to prosecute and penalise imitators, rendering the enforcement of IP rights challenging at best (Keupp et al., 2010). This governance vacuum in global partnerships is also likely to increase the chances of co-created IP being contested and commercially exploited or shared with third parties in particular in international markets given the ambiguities of delineating, evaluating and enforcing boundary-crossing ownership claims (Hernandez et al., 2015). Initial evidence supports these arguments in that offshoring R&D to distant countries increases the risk of infringement of valuable IP by host country competitors (Schmiele, 2013).

Bridging the two positions outlined above, we argue that both too much and too little distance to its innovation partners increases a focal firm's risk of being affected by imitation. This implies that domestic and international innovation partnerships are associated with a particularly high imitation threat. Continental partnerships characterised by a moderate degree of geographic, cultural, and institutional distance, in contrast, are expected to be less risky. Our reasoning is broadly consistent with Lavie and Miller (2008) finding that a moderate degree of alliance portfolio internationalisation yields the highest level of firm performance. Since differences to continental, in this case European, innovation partners are noticeable but not excessive, the focal firm can establish high quality exchange relationships with geography keeping its partners at arm's length. At the same time, chances are high that any unauthorised exploitation of IP by firms operating on the same continent will be detected and sanctioned legally or socially, thereby further discouraging imitative behaviour. Taken together, we expect:

H3: *Continental innovation partnerships will be associated with a lower risk of being affected by imitation for the focal firm than domestic and international innovation partnerships.*

Innovation phase
Finally, differences in the structural characteristics of collaborative activities carried out in early stages (e.g., idea generation, R&D), mid stages (e.g., design, prototyping), and later stages (e.g., commercialisation) of the innovation process

may hold implications for the resulting risk of imitation (Hussinger, 2006; Paasi et al., 2010). Only few empirical studies have explicitly examined this issue in previous research, with Manzini and Lazzarotti's (2016) case study being a notable exception. They shed light on the specific imitation risks and IP management challenges that firms are facing in different phases of the innovation process. In particular, their study suggests that imitation risks are most pronounced in early stages of innovation processes. In line with this observation and supported by a number of theoretical reasons, we expect that a partnership portfolio featuring a high proportion of early-stage innovation partnerships relative to those in mid and late stages of the innovation process will expose the focal firm to particularly high imitation risks.

Early-stage partnerships pertaining to joint ideation and R&D activities have been characterised as involving close interaction and extensive exchange of core technological knowledge between partners in order to develop new ideas and getting them to work (Laursen and Salter, 2006; Li et al., 2008). While intensive interaction, for example, between engineers of innovation partners allows for the transfer of tacit knowledge and may foster creative problem-solving, it also increases the likelihood of unintended knowledge leakage (Li et al., 2012; Roy and Sivakumar, 2011). It is hence through close collaboration that front-end intangible resources become vulnerable to outside imitation (Kozlenkova et al., 2013).

In contrast, the interaction in later stages of the innovation process can be expected to be less intense and restricted to solving highly specific problems related to prototyping or commercialising already existing products or services. In these stages, firms often engage in innovation partnerships to gain access to complementary assets such as distribution and marketing capabilities (Harhoff et al., 2003; Kozlenkova et al., 2013). As such, the focal firm can pursue selective revealing strategies more effectively by providing partners with information relevant to solve the specific problem at hand rather than with the core technological knowledge (Alexy et al., 2013; Harhoff et al., 2003). Without this knowledge about the technological principles underpinning a new product, service, or business model, imitating firms will find it difficult to succeed in their replication efforts, for example, trough reverse engineering (McEvily et al., 2004). Moreover, the pre-existence of a product or process technology in later stages of the innovation process enables the focal firm to delineate its IP rights more clearly from the outset (Pisano, 1989).

Such delineation attempts are likely to be notably more difficult in early-stage partnerships involving joint idea generation and concept development activities (Li et al., 2008). These innovation activities are characterised by a higher degree of ambiguity given the lack of information and the considerable uncertainty about available courses of action and possible innovation outcomes (Carson *et al.*, 2006; Manzini and Lazzarotti, 2016). It will then be most difficult — if not practically impossible — to disentangle the respective contributions of each partner and allocate IP rights accordingly (Bercovitz and Feldman, 2007). This might well provide a fertile breeding ground for future IP disputes. Such ambiguity also increases the likelihood of opportunistic behaviour by partners, as acts of opportunism are less likely to be detected when perceptions of partner behaviour are ambiguous (Carson *et al.*, 2006; Oxley and Sampson, 2004). In line with the arguments presented above, Katila *et al.* (2008) found that entrepreneurial firms are more likely to take the risk of engaging in corporate investment relationships in later stages of technology ventures, as it is easier to protect more mature technologies. In light of these arguments, we hypothesise:

H4: *Early-stage innovation partnerships will be associated with a greater risk of being affected by imitation for the focal firm than mid-stage and late-stage innovation partnerships.*

IP protection and internal R&D as isolating mechanisms

Resource-based theory suggests that the strength of isolating mechanisms protecting resources from imitation is critical for appropriating the rents of collaborative innovation activities (Lawson *et al.*, 2012; Wang *et al.*, 2009). This is even more important for firms with a broad and diverse portfolio of innovation partnerships. As the risk of knowledge leakage is likely to increase with the number of innovation partners, firms engaged in multilateral alliances have been found to devote particular attention to knowledge protection by means of governance structures (Li *et al.*, 2012). In the present study, we explore whether both legal and strategic isolating mechanisms, as represented by a focal firm's formal IP protection strategy and its internal R&D activities, can mitigate partnership-induced imitation threats. As previous evidence on this issue is limited to dyadic alliances, it remains to be seen whether these protection mechanisms are also effective in governing partnership portfolios featuring heterogeneous actors and exchange relationships.

IP protection

First, formal IP protection mechanisms such as patents, trademarks, and copyrights have long been acknowledged as important imitation barriers within resource-based theorising (Mahoney and Pandian, 1992; Peteraf, 1993; Rumelt, 1984). By establishing property rights, firms can, at least temporarily, protect knowledge residing in new products or processes from imitation and preserve their rent streams (Lawson *et al.*, 2012; Teece, 1986; Thomä and Bizer, 2013). For example, empirical evidence suggests that obtaining a patent increases the returns of an innovation by around 47%, since it provides a temporary monopoly over the knowledge contained within the innovation (Jensen *et al.*, 2011). In the context of innovation partnerships, formal IP protection is not only assumed to prevent collaboration partners from using a focal firm's core knowledge in their own operations, but also to facilitate controlled knowledge transfer between firms by defining clear property rights (Chesbrough *et al.*, 2006; Huang *et al.*, 2013; Ritala and Hurmelinna-Laukkanen, 2013). Yet, at the same time, it has been argued that patents and other property rights are often narrowly defined, costly and time-consuming to enforce, and provide only limited protection, especially in countries with weak appropriability regimes (Somaya, 2012). It is even possible that legal property rights foster imitation instead of preventing it as, for example, patents require the codification and formal disclosure of some of a focal firm's knowledge (Hurmelinna *et al.*, 2007).

However, we expect that the safeguarding effect of formal IP protection mechanisms will outweigh the potential disadvantages stemming from disclosure. Infringing a focal firm's IP rights or legally inventing around them by creating substitute technologies can be costly, time consuming, and risky (Polidoro and Toh, 2011; Reitzig *et al.*, 2007). Alleged infringers often face substantial litigation costs and may also suffer reputational damage that restrains other firms from collaborating with them in the future. Taken together, we argue that imitation barriers raised by patents and other IP rights will lower the occurrence of partnership-induced imitation, as others will perceive imitation as a less promising strategic alternative (Polidoro and Toh, 2011). Thus:

H5: *A focal firm's IP protection strategy will moderate the positive association between the variety of its innovation partnership portfolio and its risk of being affected by imitation, in such a way that this relationship will be weaker the stronger the IP protection.*

R&D intensity

Internal R&D activities constitute the second, at first sight less obvious, isolating mechanism deterring imitation of firms pursuing a portfolio approach to innovation partnerships examined in this study. In the innovation literature, in-house R&D is mainly viewed as a complementary activity to external knowledge sourcing (Cassiman and Veugelers, 2006; Ritala and Hurmelinna-Laukkanen, 2013). In line with the notion of absorptive capacity, firms need to develop prior related knowledge to be able to identify, assimilate, and exploit knowledge from external sources (Cohen and Levinthal, 1990). Yet, by engaging in R&D activities, firms can also create effective barriers to imitation (Cassiman and Veugelers, 2002; James et al., 2013). Knowledge generated through extensive in-house R&D is often firm-specific, complex, and tacit in nature, and thus causally ambiguous (i.e., it is unclear how and why certain knowledge resources contribute to competitive advantage) and difficult to imitate (Helfat, 1994; Kozlenkova et al., 2013; Reed and DeFillippi, 1990).

Firms that seek to imitate innovations based on knowledge with these characteristics must not only gain access to the knowledge itself, but also need to understand and replicate the organisational routines through which such knowledge has been generated and can be exploited (Wang et al., 2009; Yang et al., 2010). It is especially, the context-dependent nature of the R&D process and the important tacit element it involves that act as potentially powerful barriers to imitation (Helfat, 1994). This is consistent with the argument that the ease with which innovation partners can imitate a focal firm's resource stock is related to the

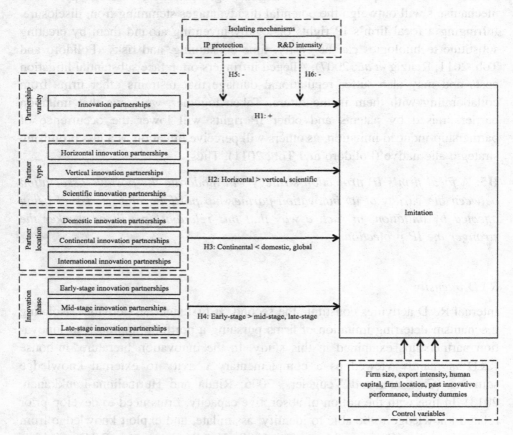

Fig. 1. Conceptual model on innovation partnerships and imitation.

characteristics of the process through which this resource stock has been accumulated (Dierickx and Cool, 1989). Particularly, strong imitation barriers can be expected to arise from combining internal and external knowledge resources during the innovation process (Grimpe and Kaiser, 2010; Sirmon et al., 2007). The causal mechanisms underlying the emergence of such new knowledge combinations are highly complex and causally ambiguous (Ketchen et al., 2007). Amongst others, it has been argued that such R&D-induced complexity and ambiguity overwhelm managerial cognitive capacities for imitation (Ethiraj et al., 2008), thus buffering the focal firm from imitation threats. Lead time advantages, i.e., advantages stemming from early timing of developing and introducing new products or processes, are another mechanism through with a focal firm with a high commitment to R&D can deter imitation (James et al., 2013). The preemptive access to scarce assets, learning-curve effects, and switching costs are assumed to make the imitation of first-mover firms costly and time-consuming for others (Lawson et al., 2012; Lieberman and Montgomery, 1988). In sum, we propose that:

H6: *A focal firm's internal R&D intensity will moderate the positive association between the variety of its innovation partnership portfolio and its risk of being affected by imitation, in such a way that this relationship will be weaker the higher the R&D intensity.*

Figure 1 summarises our conceptual model.

Methods

Research design and sample

This study draws on data from two consecutive waves of the Mannheim Innovation Panel (MIP) collected in 2007 and 2008. The MIP is commissioned by the German Federal Ministry of Education and Research and carried out by the Centre for European Economic Research (ZEW). It follows the methodology for large-scale firm-level innovation surveys outlined in the Oslo Manual (OECD, 2005). The MIP has been designed to collect information on the innovation activities of German firms operating in 22 different industry sectors. In 2008 (2007), 6,684 (4,914) of the 18,109 (25,862) firms surveyed returned usable questionnaires, yielding a response rate of 36.9 (19.0)%. Non-response analyses using telephone interviews were conducted and provided no evidence of any non-response bias that might be a source of concern for our study (Klingebiel and Rammer, 2014).

Our main analyses are based on a subsample of German manufacturing firms. Often competing on the basis of technology leadership, these firms are not only

Table 1. Sample description.

Industry membership		Innovation partnership type[a]	
Food	5.85%	Innovation partnerships	83.81%
Textiles	5.48%	Horizontal innovation partnerships	10.34%
Paper	11.21%	Vertical innovation partnerships	80.45%
Chemicals	7.85%	Scientific innovation partnerships	39.10%
Plastics	6.48%	Domestic innovation partnerships	82.57%
Glass	5.73%	Continental innovation partnerships	48.82%
Metal	16.56%	International innovation partnerships	28.64%
Engineering	12.45%	Early-stage innovation partnerships	77.58%
eTechnology	9.71%	Mid-stage innovation partnerships	70.63%
Medical instruments	10.96%	Late-stage innovation partnerships	85.18%
Vehicle construction	4.11%		
Furniture	3.61%		
Firm size		*Imitation*[b,c]	
<50 Employees	48.07%	In general	37.90%
50–249 Employees	35.24%	At home	14.96%
>249 Employees	16.69%	From abroad	15.69%
Breakdown by continent		*Breakdown by country (top 5)*[b]	
1. Europe[d]	51.42%	1. Germany	42.51%
2. Asia	44.94%	2. China	33.20%
3. North-America	3.64%	3. USA	3.64%
4. Oceania, South-America	0.00%	4. India, Turkey	3.24%
		5. Taiwan	2.83%

Notes: 803 total observations; [a]share of firms with at least one specific innovation partnership, [b]does not add up to 100% since multiple answers were possible, [c]247 firms reported to have been affected by at least one instance of IP imitation between 2005 and 2007, [d]including imitation in Germany.

particularly likely to engage in collaborative innovation activities, but also face the highest risk of being imitated by others given their strong manufacturing and technological capabilities (Li *et al.*, 2008). To test our hypotheses, we matched data from the 2007 and 2008 MIP waves, yielding a final sample of 803 manufacturing firms that provided usable responses to both waves. Table 1 describes the composition of our final sample of 803 German manufacturing firms. About 83.31% of these firms are small- and medium-sized enterprises (SME) with up to 250 employees — a stylised fact that reflects the importance of SMEs in the German economy. A total of 83.81% cooperated with at least one innovation partner, with the average firm engaging in 5.00 distinct forms of innovation partnerships as described by unique combinations between partner type (e.g., customers, suppliers) and innovation phase (e.g., idea generation, design). About 82.57% engaged in domestic innovation partnerships, while 48.82% cooperated with innovation partners from Europe and 28.64% with partners located elsewhere abroad. Taking a closer look at partner type variety, 80.45% collaborated with suppliers or customers (vertical innovation partnerships) and 39.10% drew on knowledge from universities and other research institutions (scientific innovation partnerships). However, only 10.34% partnered with competitors (horizontal innovation partnerships). Innovation partnerships occurred across all three stages of the innovation process, with early- (77.58%) and late-stage partnerships (85.18%) being notably more frequent than mid-stage partnerships (70.63%). Table 1 also reveals that imitation was a highly pervasive problem that affected 37.90% of all firms in our sample during the period from 2005 to 2007, with 42.51% of imitators originating from within Germany, followed by China (33.2%), and the USA (3.64%).

Measures

Dependent variables

In absence of a standard metric, we develop our measure for imitation by drawing on four binary indicators. These capture whether a focal firm was affected by the imitation of its (1) technical inventions, (2) products and business models, (3) brands and descriptions, or (4) designs between 2005 and 2007. Based on these self-reports of experienced IP infringement, we compute an overall imitation score as the count of the distinct types of imitation the focal firm was affected by. Accordingly, our measure ranges from 0 (no imitation) to 4 (all four types of imitation) and reflects the extent to which the focal firm is affected by imitation.

Independent variables

Partnership variety. We measure the variety of a focal firm's innovation partnership portfolio as the number of distinct partnerships it was engaged in between 2005 and 2007. For this purpose, we count for each firm the number of unique binary "partner type"-"innovation stage" combinations. We consider the six innovation partner types (i) business customers, (ii) consumers, (iii) material suppliers, (iv) service providers, (v) competitors as well as (vi) universities and research institutes along with the five innovation stages (i) idea generation, (ii) R&D, (iii) design, (iv) testing and production preparation, and (v) implementation and market introduction. A firm's overall partnership variety can consequently range from 0 (no partnership type used) to 30 (innovation partnerships with all six partner types across all five innovation stages). To examine the role of partner type (Hypothesis 2), partner location (Hypothesis 3), and innovation phase (Hypothesis 4), we construct three sub-indices for each.

As for *partner type*, we develop separate indices for horizontal innovation partnerships (number of innovation stages with at least one active innovation partnership with a competitor, range from 0 to 5), vertical innovation partnerships (number of innovation stages with at least one active innovation partnership with business customers, consumers, material suppliers, or service providers, range from 0 to 20), and scientific innovation partnerships (number of innovation stages with at least one active innovation partnership with a university or research institute, range from 0 to 5).

As for *partner location*, the sub-index for domestic innovation partnerships is computed as the count of unique combinations among the six innovation partner types introduced above and three partner location types within Germany. These are (i) local partnerships (i.e., within 20 km from the focal firm), (ii) regional partnerships (i.e., between 20 km and 100 km from the focal firm), and (iii) national partnerships (i.e., more than 100 km from the focal firm). The resulting index hence ranges from 0 (no domestic innovation partnerships) to 18 (innovation partnerships with all six partner types in all three German locations). Similarly, the indicator for continental (international) innovation partnerships emerges as the count of unique combinations among the same six innovation partner types and the partner location within Europe (outside of Europe). This index is thus limited to the interval between 0 (no continental (international) innovation partnerships) and 6 (innovation partnerships with all six partner types in European (non-European) countries).

Finally for *innovation phase*, we calculate indices for early-stage innovation partnerships (number of innovation partner types in the idea generation and R&D stages, range from 0 to 12), mid-stage partnerships (number of innovation partner types in the design and testing stages, range from 0 to 12)[1] and late-stage partnerships (number of innovation partner types in the implementation and market introduction stage, range from 0 to 6). All innovation partnership variables were finally standardized for formal testing of Hypotheses 2–4.

Moderating variables

Our first moderating variable IP protection is binary and captures whether the focal firm relies on (i) patents, (ii) registered designs, (iii) design patents,

---

[1]We consider early-stages (i.e., idea generation and R&D) to comprise efforts related to developing the critical technical characteristics of a product, whereas efforts in mid-stages (design and prototyping) comprise the aesthetic design and prototyping important for successful production.

(iv) trademarks, or (v) copyrights for protecting its IP rights. Following prior research (Laursen and Salter, 2006; Leiponen and Helfat, 2010), we calculate a focal firm's R&D intensity as the ratio of its annual R&D expenditures to its annual sales.

Control variables

We include a number of control variables widely used in the innovation literature to account for possible confounding factors (Laursen and Salter, 2006; Leiponen and Helfat, 2010). First, we control for firm size measured as the natural logarithm of the total number of employees, as larger firms are more visible and hence more likely to become the target of imitation. Second, we follow (Cassiman and Veugelers, 2006) and include export intensity to account for inter-firm differences in internationalisation and hence global visibility to potential imitators. Third, we use the percentage of employees holding a university degree as a proxy for a firm's human capital. Fourth, we employ a full set of industry dummies to control for potential inter-industry differences in firm performance and exposure to imitation.

Analysis

The dependent variable employed for testing Hypotheses 1–6 (i.e., imitation) is a count variable with non-negative integer values assumed to follow a Poisson distribution. Its standard deviation (0.92) exceeds its mean value (0.50), indicating overdispersion. This overdispersion, however, is accounted for by excess zeros (Greene, 2011), with 37.90% of all observations having a non-zero value for imitation. A significant Vuong statistic ($z = 4.41; p < 0.01$) points to the superiority of a Zero-Inflated Poisson model correcting for overdispersion due to excess zeros (Vuong, 1989). The first step of this two-step approach involves estimating the probability of zero-observations by means of a Logit model. As German manufacturing firms are less likely to be at the centre of imitators' attention if they have been performing poorly with respect to their innovation activities in the past three years (Ethiraj and Zhu, 2008), or are located in the still deprived Eastern part of Germany (Grimpe and Kaiser, 2010), we use past innovative performance and a firm location dummy (0 for West and 1 for East Germany) in addition to the constant as our zero-inflation parameters. To measure a focal firm's past innovative performance, we draw on data from the 2007 MIP wave and capture the 2006 revenue share from new and significantly improved products launched between 2004 and 2006. As part of the second step, a standard Poisson model with robust standard errors is employed to explain inter-firm differences in the exposure to imitation.

Results

Descriptive results

Table 2 depicts the descriptive statistics and pairwise correlations for the full sample of 803 firms. The table shows that a focal firm's imitation probability increases most notably with the variety of its continental, vertical, and early-stage innovation partnerships as well as with its use of IP protection mechanisms. It is also worth noting the mean export intensity of 29% among our sample firms, which is a feature that is not only characteristic of German manufacturing, but also most strongly correlated with the risk of imitation.

Regression results

Table 3 presents the results from Zero-inflated Poisson regression analyses explaining inter-firm differences in imitation. Our base model (Model 1) only contains the zero-inflation parameters in step 1 and the full set of control variables in step 2. In line with our expectations, the selection equation of the base model reveals firms with low past innovative performance located in East Germany to be less likely to be affected by imitation. Our key variable sets of interest are then introduced sequentially in Models 2–6.

Consistent with our arguments, the relationship between the variety of a focal firm's innovation partnership portfolio and its exposure to imitation is positive and statically significant in Model 2. Hypothesis 1 is hence supported. The incidence rate ratio (IRR) not reported here indicates that a focal firm's risk of being infringed increases by a factor of 1.26 (26.6%) for every one standard-deviation increase in its partnership portfolio variety.

Hypothesis 2 suggested that a focal firm would be exposed to greater imitation risks when engaging in horizontal as opposed to vertical and scientific innovation partnerships. The estimates presented in Model 3, however, indicate that vertical innovation partnerships are associated with the greatest imitation threat, significantly exceeding the risk induced by both horizontal (Chi-squared $= 9.30$; $p < 0.01$) and scientific innovation partnerships (Chi-squared $= 5.10$; $p < 0.05$). Indeed, we find the effect of horizontal and scientific innovation partnerships to remain statistically insignificant, thus offering no support for Hypothesis 2.

In Hypothesis 3, we proposed continental innovation partnerships to be associated with a lower imitation threat than their domestic or international counterparts. Contrary to our theoretical arguments, Model 4 reveals that only continental innovation partnerships exhibit a statistically significant imitation-enhancing effect. The difference in coefficient estimates, however, fails to reach statistical

Table 2. Descriptive statistics and correlations.

Variables	1	2	3	4	5	6	7	8	9	10	11	12	13	14	15	16	17	18
1. Imitation	1																	
2. Innovation partnerships	0.25*	1																
3. Horizontal innovation partnerships	0.01	0.28*	1															
4. Vertical innovation partnerships	0.24*	0.95*	0.14*	1														
5. Scientific innovation partnerships	0.18*	0.58*	0.12*	0.34*	1													
6. Domestic innovation partnerships	0.15*	0.60*	0.20*	0.56*	0.36*	1												
7. Continental innovation partnerships	0.21*	0.47*	0.16*	0.44*	0.30*	0.45*	1											
8. International innovation partnerships	0.19*	0.38*	0.13*	0.35*	0.28*	0.39*	0.63*	1										
9. Early-stage innovation partnerships	0.27*	0.89*	0.25*	0.83*	0.59*	0.59*	0.46*	0.37*	1									
10. Mid-stage innovation partnerships	0.17*	0.88*	0.24*	0.84*	0.49*	0.47*	0.38*	0.30*	0.62*	1								
11. Late-stage innovation partnerships	0.19*	0.72*	0.22*	0.73*	0.29*	0.40*	0.33*	0.28*	0.48*	0.58*	1							
12. IP protection	0.32*	0.36*	0.03*	0.32*	0.33*	0.31*	0.33*	0.29*	0.36*	0.27*	0.27*	1						
13. R&D intensity	0.11*	0.29*	0.10*	0.20*	0.38*	0.25*	0.27*	0.34*	0.30*	0.22*	0.20*	0.26*	1					

(*Continued*)

Table 2. (*Continued*)

Variables	1	2	3	4	5	6	7	8	9	10	11	12	13	14	15	16	17	18
14. Past innovative performance	0.20*	0.25*	0.10*	0.19*	0.26*	0.20*	0.24*	0.25*	0.26*	0.18*	0.16*	0.29*	0.38*	1				
15. Firm size	0.07*	0.08	−0.01	0.06	0.09	0.03	0.01	0.02	0.10*	0.04	0.02	0.07	0.04	0.02	1			
16. Export intensity	0.21*	0.23*	0.01*	0.18*	0.28*	0.13*	0.38*	0.38*	0.24*	0.18*	0.11*	0.35*	0.24*	0.24	0.10*	1		
17. Human capital	0.15*	0.24*	0.05*	0.17*	0.31*	0.20*	0.19*	0.26*	0.24*	0.16*	0.19*	0.26*	0.46*	0.33	0.05	0.21*	1	
18. Firm location	−0.17*	0.00	0.01	−0.01	0.05	0.02	−0.08	−0.08	−0.04	0.03	0.06	−0.12*	0.05	−0.03	−0.06	−0.18*	0.13*	1
Mean	0.50	5.00	0.14	4.10	0.75	3.74	1.14	0.52	2.55	1.76	0.70	0.59	0.02	1.22	243	0.29	3.10	0.34
Std. dev.	0.92	3.91	0.51	3.28	1.11	3.13	1.48	0.98	2.07	1.65	0.86	0.49	0.04	1.94	2011	0.28	2.12	0.47
Min	0	0	0	0	0	0	0	0	0	0	0	0	0	0	2	0	0	0
Max	4	24	5	17	5	18	6	6	10	10	6	1	0.15	8	55562	0.85	8	1

Notes: 803 total observations *p* < 0.01.

Table 3. Zero-inflated Poisson regression analyses explaining imitation.

Variable	Model 1	Model 2	Model 3	Model 4	Model 5	Model 6
Step 2: Poisson regression[b]						
Control variables						
Constant	0.294	−0.237	−0.260	−0.238	−0.272	−0.199
	(0.257)	(0.240)	(0.244)	(0.232)	(0.256)	(0.240)
Firm size[a]	0.077	−0.082	−0.077	−0.039	−0.074	−0.079
	(0.057)	(0.066)	(0.065)	(0.063)	(0.066)	(0.068)
Export intensity	0.226	0.237	0.245	0.111	0.269	0.236
	(0.248)	(0.257)	(0.256)	(0.254)	(0.257)	(0.257)
Human capital	0.020	−0.010	−0.011	−0.004	−0.020	−0.014
	(0.034)	(0.034)	(0.034)	(0.034)	(0.035)	(0.034)
Industries	included	included	included	included	included	included
Main effects						
IP protection[a]		0.601***	0.596***	0.636***	0.592***	0.580***
		(0.110)	(0.111)	(0.110)	(0.110)	(0.108)
R&D intensity[a]		−0.114*	−0.115*	−0.126*	−0.120*	−0.072
		(0.065)	(0.067)	(0.066)	(0.063)	(0.064)
Innovation partnerships[a]		0.231***				0.364***
		(0.061)				(0.091)
Horizontal innovation partnerships[a]			−0.035			
			(0.062)			
Vertical innovation partnerships[a]			0.213***			
			(0.054)			
Scientific innovation partnerships[a]			0.041			
			(0.052)			
Domestic innovation partnerships[a]				0.015		
				(0.059)		
Continental innovation partnerships[a]				0.153**		
				(0.073)		
International innovation partnerships[a]				−0.038		
				(0.074)		
Early-stage innovation partnerships[a]					0.214***	
					(0.064)	
Mid-stage innovation partnerships[a]					−0.045	
					(0.070)	
Late-stage innovation partnerships[a]					0.105**	
					(0.046)	
Moderating effects						
Innovation partnerships × IP protection						−0.178*
						(0.107)
Innovation partnerships × R&D intensity						−0.094*
						(0.049)
Step 1: Zero-inflation logit regression[b]						
Constant	0.245	−0.284	−0.264	−0.152	−0.272	−0.281
	(0.198)	(0.253)	(0.253)	(0.224)	(0.256)	(0.240)

(*Continued*)

Table 3. (*Continued*)

Variable	Model 1	Model 2	Model 3	Model 4	Model 5	Model 6
Past innovative performance	−0.320*** (0.101)	−0.249** (0.106)	−0.266** (0.116)	−0.268** (0.116)	−0.275** (0.120)	−0.251** (0.109)
Firm location	1.082*** (0.245)	1.146*** (0.284)	1.123*** (0.289)	1.036*** (0.276)	1.129*** (0.289)	1.123*** (0.282)
Total observations	803	803	803	803	803	803
Zero observations	579	579	579	579	579	579
Nonzero observations	224	224	224	224	224	224
Maximum likelihood R^2	0.114	0.184	0.185	0.177	0.188	0.191
Chi-squared	41.124***	120.803***	127.32***	112.932***	136.127***	124.5***

Notes: 803 total observations; zero-inflated Poisson model with "innovative performance" and "firm location" as zero-inflation parameters; robust standard errors in parantheses; [a]standardized measures; [b]step 1: logit regression estimating probability of not being imitated; step 2: Poisson regression estimating the extent of imitation experienced.
***p < 0.01, **p < 0.05, *p < 0.10.

significance with regards to both domestic (Chi-squared = 1.57; $p > 0.1$) and international innovation partnerships (Chi-squared = 2.43; $p > 0.1$). Hypothesis 3 is thus not supported.

According to Hypothesis 4, we expected a focal firm's risk of being infringed to be higher for early-stage than for mid- and late-stage innovation partnerships. In line with this expectation, the coefficient of early-stage partnerships in Model 5 is positive and statistically significant, indicating that early-stage innovation partnerships notably increase the imitation threat a focal firm is exposed to. In relative terms, early-stage innovation partnerships carry a significantly greater imitation risk than mid-stage innovation partnerships (Chi-squared = 5.41; $p < 0.05$). Interestingly, we also detect an imitation-enhancing effect of late-stage innovation partnerships. Although the coefficient estimate is notably smaller relative to early-stage innovation partnerships, the difference fails to achieve statistical significance (Chi-squared = 2.12; $p > 0.1$). Hypothesis 4 is thus partially supported.

Figure 2 illustrates how the imitation threat a focal firm is exposed to varies with the functional (i.e., partner type), geographical (i.e., partner location), and temporal (i.e., innovation phase) configuration of the innovation partnership portfolio.

As for the moderating role of IP protection postulated in Hypothesis 5, our analyses reveal a weakly significant negative interaction effect between IP protection and innovation partnerships (Model 6). This corroborates our theoretical expectation that IP protection has the potential to shield the focal firm against

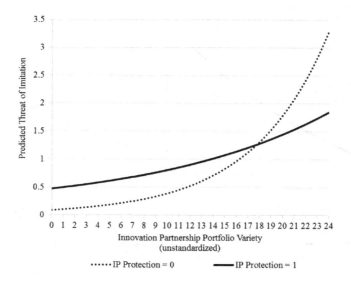

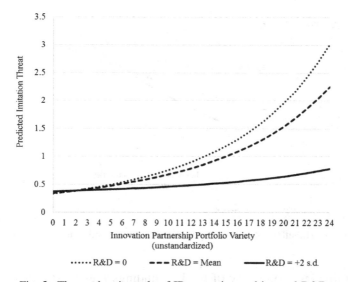

Fig. 2. The moderating role of IP protection and internal R&D.

partnership-induced imitation. This offers support for Hypothesis 5. Similarly, in support for Hypothesis 6, Model 6 reveals a marginally significant negative interaction effect of R&D intensity on the link between innovation partnerships and imitation. Figure 3 illustrates the practical effectiveness of both moderators in buffering the focal firm from collaboration-induced imitation especially at high levels of partnership portfolio variety.

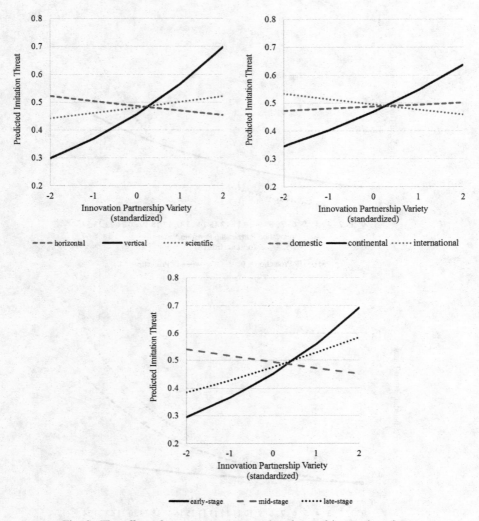

Fig. 3. The effect of partner type, partner location and innovation phase.

Table 4 provides a summary of the key findings presented thus far.

Post hoc analyses

We performed several robustness checks to examine the sensitivity of our main results presented above to changes in estimation procedure, construct measurement, and sample selection. We also conducted additional analyses to illuminate questions that emerged during our analyses and sought to rule out alternative explanations.

Table 4. Summary of results.

	Hypothesis	Result	Conclusion
H1	Innovation partnerships	supported	The imitation probability a focal firm is exposed to tends to increase with the variety of its innovation partnerships.
H2	Horizontal > vertical and scientific innovation partnerships	not supported	The size of the general imitation-inducing effect of innovation partnerships is largest for vertical innovation partnerships, followed by horizontal and scientific innovation partnerships.
H3	Continental > domestic and international innovation partnerships	not supported	Only continental innovation partnerships exhibit a statistically significant imitation-inducing effect. That said, the differences in effect sizes relative to both domestic and global innovation partnerships fail to reach statistical significance.
H4	Early- > mid- and late-stage innovation partnerships	partially supported	The size of the general imitation-inducing effect of innovation partnerships is larger for early-stage than for mid-stage, but not for late-stage innovation partnerships.
H5	IP protection × innovation partnerships	supported	The size of the general imitation-inducing effect of innovation partnerships is smaller for firms with than without formal IP protection mechanisms.
H6	R&D intensity × innovation partnerships	supported	The size of the general imitation-inducing effect of innovation partnerships is smaller for firms with than without strong internal R&D activities.

Notes: All conclusions are based on the average effects observed in our study of German manufacturing firms.

As for the *robustness checks* performed, we found fully consistent results when using alternative estimation techniques for count data including standard Poisson, Negative Binomial and Zero-Inflated Negative Binomial regression. Similarly, consistent, though less efficient estimates emerged when operationalising imitation as a dichotomous variable and replicating all main analyses by means of probit models. Results also remained stable when excluding all non-collaborating manufacturing firms (16.19% of the sample) or all non-innovating firms (41.34% of the sample) from our analyses. Moreover, we introduced and tested an alternative measure for partnership variety. We constructed it as the number of relevant

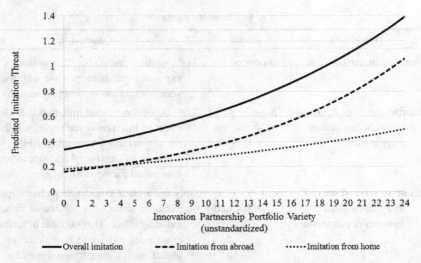

Fig. 4. The main effect of innovation partnership variety on imitation.

innovation partnerships types used or not used in the innovation process (i.e., (i) business customers, (ii) consumers, (iii) material suppliers, (iv) service providers, (v) competitors as well as (vi) universities and research institutes). Accordingly, this alternative ranges from 0 (no innovation partnerships) to 6 (innovation partnerships with all six types). Our results remained fully robust when using this variable in our estimations. The same applies when using an ordered instead of a binary measure of IP protection.

As for the *additional analyses* conducted, we first sought to explore the role of imitators' geographical origin. For this purpose, we created two count variables capturing the extent of IP infringement a focal firm experienced at home (i.e., from imitators located in Germany) and from abroad (i.e., from imitators located outside of Germany). Several noteworthy findings emerged from these more granular analyses. Interestingly, IRRs revealed a one standard-deviation increase in innovation partnerships to be associated with a 17.91% increase in the imitation risk experienced at home and a 35.59% increase in the imitation risk from abroad. As illustrated in Fig. 4, innovation partnerships induce first and foremost imitation threats from abroad and to a lesser extent from home.

Second, we examined possible differences between illegal imitation (i.e., imitation of a technical invention, product, brand, or design that was legally protected at the time of imitation) and legal imitation (i.e., imitation of a technical invention, product, brand, or design without any legal protection at the time of imitation). Interestingly, we found the variety of a focal firm's innovation partnership portfolio and its constitutive functional, geographical, and temporal dimensions to be associated with a greater threat of *illegal* imitation, though not of *legal* imitation. This reinforces the adverse nature of partnership-induced imitation that German manufacturing firms are exposed to. Third, we examined the extent to which the results obtained for all manufacturing firms in our sample equally hold for firms operating either in high-tech or in low-tech industries. Despite comparable results with regards to Hypotheses 1–4, three notable differences emerged: These pertain to (i) the higher exposure of high-tech firms to imitation risks fueled by scientific innovation partnerships, (ii) the greater effectiveness of formal IP protection as an isolating mechanism for high-tech firms, and (iii) the greater effectiveness of R&D as an isolating mechanism for their low-tech counterparts. Fourth, we found smaller firms to be more vulnerable to partnership-induced imitation probability than larger firms, a finding that might be explained by their limited opportunities to engage in selective revealing and IP enforcement activities (Ketchen et al., 2007). Last but not least, we sought to unveil the effect of being imitated on the financial performance of the focal firm. For this purpose, we matched data from the 2009 wave of the MIP and captured a focal firm's return on sales in 2008, measured on an ordinal scale from 1 ($<0\%$) to 7 ($>15\%$). This allows for a temporal sequencing of imitation and performance. Financial performance is an ordinal variable measured on an unequally distributed interval scale. An ordered probit regression model is hence most appropriate (Greene, 2011). As depicted in Table A.1, the coefficient estimate of imitation is negative and statistically significant. This supports the traditional assumption of resource-based scholars that falling victim to imitation tends to come at the cost of tangible decreases in financial performance (Ordanini et al., 2008).

Finally, we sought to rule out *alternative explanations* for the main findings presented in our study. In particular, our main argument rests on the assumption that a broader innovation partnership portfolio leads to the focal firm being more exposed to imitation risks. Although, panel or experimental data appear needed for a conclusive assessment, we examined this critical issue of causality conceptually, quantitatively, and qualitatively. Conceptually, the threat of positive reverse causality, i.e., of greater imitation incidents leading to an even broader innovation partnership portfolio, does not appear plausible apart from a theoretically possible tit-for-tat pattern leading firms that have fallen victim to imitation to expand their partnership portfolio in a retaliatory effort to engage in imitative behaviour themselves. Imitation incidents triggering a sequential closure of the innovation process, in contrast, is a reaction consistent with behavioural arguments suggesting that organisations adapt their search rules, when the current search mode yields unsatisfactory outcomes (Cyert and March, 1963). In presence of such negative reverse causality, we would tend to under- rather than overestimate the true effect of the variety of a focal firm's innovation partnership portfolio on the imitation threat it is exposed to. In addition to reverse causality, our findings might be compromised by unobserved heterogeneity. Perhaps most intuitively, a focal firm's reputation for technology and innovation leadership could affect its attractiveness as both an innovation partner and an imitation target leading to potentially spurious estimates. We took several measures to address this issue quantitatively. First, all our models contained a focal firm's past innovative performance as a proxy for its reputation for technology and innovation leadership. Second, we conducted a sub-sample analysis, as part of which we examined the main effect separately for firms with below- and above-median levels of innovative performance. The imitation-enhancing effect of partnership variety emerged in both sub-samples and thus appears to affect firms with low and high levels of innovative performance. Third, as both reverse causality and unobserved heterogeneity might give rise to endogeneity bias, we instrumented partnership variety within a two-stage least-squares (2SLS) framework (Garriga *et al.*, 2013). Importantly, valid instruments needed to be correlated with the partnership variety though not with the threat of imitation. These conditions were satisfied for external R&D (dummy variable to indicate whether the focal firm issued R&D contracts to third parties) and a dummy variable that indicates whether the focal firm had introduced administrative innovations for organising external relations with other firms of public institutions during the period from 2004 to 2006. 2SLS analyses using these two instruments enabled us to replicate our main effect of partnership variety on the threat of imitation at the same level of statistical significance. The size of our main effect declined only moderately from 0.231 ($p < 0.01$) in the original model to 0.168 ($p < 0.01$) in the instrumented model. Overall, our conceptual arguments and supplementary quantitative analyses render alternative explanations for our main conclusion that broad innovation partnership portfolios increase the threat of imitation a focal firm is exposed to less probable. To corroborate this claim also qualitatively, we conducted five interviews with R&D professionals routinely involved in innovation partnerships. Interviewees highlighted that being embedded in a broad portfolio of innovation partnerships will often come at the cost of a greater exposure to imitation risks. The primary challenge hence consists in sharing sufficient knowledge to make the diverse set of partnerships work, while at the same time protecting parts of the knowledge base to contain the threat of imitation. The following quote is particularly insightful in that it illustrates both the substantial imitation threat of vertical partnerships and the buffering role of IP protection.

"A client in Asia received from us a solution to a problem on which they had requested assistance. The solution included a proprietary element of ours, which emerged at a later date as their proprietary technology. [...] We immediately engaged prominent attorneys in that country to represent us despite the history of litigation by foreigners in that environment working against us, and fairly quickly resolved the dispute through a proper licensing agreement and reimbursement for the substantial legal costs. That same client has been back now twice for assistance in resolving technology problems, in a relationship that operates on a more congenial basis now."

Discussion and Conclusion

The purpose of this study was to examine, both theoretically and empirically, the underexplored value appropriation challenges in innovation partnership portfolios. Drawing on the RBV, we developed and tested a conceptual model linking salient structural attributes of a focal firm's innovation partnership portfolio and the imitation threat it is exposed to. Two main findings emerged after testing the proposed model with data from German manufacturing firms. First and most importantly, having a broad portfolio of innovation partnerships indeed increases the risk of falling victim to illegal imitation. Importantly, this effect persists even after controlling for past innovative performance and instrumenting the partnership variety to account for its potential endogeneity. The magnitude of the imitation threat depends on the specific configuration of the innovation partnership portfolio along the three dimensions partner type (i.e., highest for suppliers and customers), partner location (i.e., highest for continental partners), and innovation phase (highest for early-stage partnerships). Second, firms can partially mitigate partnership-induced imitation threats by employing formal IP protection and internal R&D as isolating mechanisms. These findings have several important implications for research and theory, which we discuss below.

Implications for research and theory

First, our findings provide novel insights into the challenges of imitation that occur in complex innovation partnership portfolios and go above and beyond those commonly observed in dyadic alliances (Kale *et al.*, 2002; McEvily *et al.*, 2004; Oxley and Sampson, 2004). Consistent with the idea that firms engaged in collaborative innovation face a trade-off between maximising the incoming knowledge flows and minimising unintended spillovers (Alexy *et al.*, 2013; Boudreau, 2010), we found that the broader the variety of a focal firm's innovation partnership portfolio, the higher the likelihood of being imitated. This supports the proposition that portfolio diversity in terms of partners' functional role, geographic origin, and temporal involvement increases not only complexity, but also the risk of goal conflict and opportunistic behaviour (Cui and O'Connor, 2012; Duysters and Lokshin, 2011). The imitation threat induced by a diverse partnership portfolio will hence typically exceed the sum of the individual risks associated with each alliance (Li *et al.*, 2012). Partnership-induced imitation that increases with the variety of the partnership portfolio might therefore act as one of the causal mechanisms underpinning the frequently observed decreasing marginal returns to both alliance portfolio diversity (Cui and O'Connor, 2012; Jiang *et al.*, 2010; Lavie and Miller, 2008; Sivakumar *et al.*, 2011) and open innovation (Laursen and Salter, 2006; Salge *et al.*, 2013).

Second and related, we revealed how the specific configuration of a focal firm's partnership portfolio affects the imitation risk it is exposed to. More specifically, our study complements extant alliance portfolio research by illuminating the role of specific partner characteristics — most notably partner type, partner location, and innovation phase — in shaping value appropriation rather than value creation (Cui and O'Connor, 2012; Jiang *et al.*, 2010; Lavie, 2007; Lavie and Miller, 2008). Perhaps most importantly, our findings indicate that the collaboration-induced imitation threat a focal firm is exposed to tends to be more pronounced the larger the share of vertical, continental, and early-stage partnerships, and the lower the use of IP protection and internal R&D as isolating mechanisms. Surprisingly, vertical rather than horizontal innovation partnerships emerged as the partner type associated with the highest imitation threat. The German manufacturing firms, we studied appear to anticipate such challenges with only 10.34% of them engaging in innovation partnerships with competitors. This might be indicative of careful partner selection, where preference is given to competitors with high levels of relational trust and a reputation for successful partnering. Contrary to our expectations, we also found that the strongest imitation threats emanated from continental rather than domestic or international partnerships. Although, this finding might be explained by the unique combination of sufficient partner-specific absorptive capacity, moderate institutional distance, and relative difficulties to establish informal safeguards that characterises continental partners, it appeared inconsistent with previous studies showing firm performance to be highest at low or moderate international diversity of the alliance portfolio (Cui and O'Connor, 2012; Goerzen and Beamish, 2005; Lavie and Miller, 2008). More generally, our study illustrates how the relative salience of appropriation concerns is contingent on the specific configuration of the partnership portfolio and the effectiveness of the isolating mechanisms employed. This insight will be of interest for the contingency literature on open innovation, which argues that the returns from innovation partnerships tend to be contingent on specific project, firm, and partnership characteristics (Laursen and Salter, 2006; Gesing *et al.*, 2015; Salge *et al.*, 2012), but has yet to examine the role of portfolio characteristics as contingency factors. As the locus of innovation gradually shifts to the level of the broader network of relationships, inter-organisational structures such as a focal firm's partnership portfolio move into the foreground. It is against this backdrop that our study demonstrates empirically how the structure of the innovation partnership portfolio affects the imitation threat with regards to (i) technical inventions, (ii) products and business models, (iii) brands and descriptions, and (iv) designs in a way that is statistically and practically significant. Importantly, we show that partnership portfolio structures are multifaceted processing not least a functional (i.e., partner type), geographical (i.e., partner location), and temporal (i.e., innovation phase) dimension. This inner structure of partnership portfolios is both complex and consequential in that it can reduce — or indeed amplify — the threat of collaboration-induced imitation. This highlights the need to carefully tailor the partnership portfolio grounded in an in-depth understanding of not only its benefits, but also its costs, risks, and tradeoffs.

As a third contribution, our study adds to resource-based theorising by unpacking how both portfolio structures and isolating mechanisms affect resource inimitability as a critical precondition for sustained competitive advantage (Ordanini *et al.*, 2008). As such, it illuminates a persisting puzzle in resource-based theorising. This puzzle pertains to the persistent ambiguity of the RBV regarding whether innovation partnerships facilitate resource imitation given the greater permeability of organisational boundaries or rather impede imitation attempts by means of greater causal ambiguity stemming from complex partnership portfolios that span multiple firms with different structures, capabilities, and cultures (Ketchen *et al.*, 2007). Our results support the former argument in that broad innovation partnership portfolios increase the risk of imitation. By providing partners with insights into its operations, the focal firm seems to violate a key principle of the RBV, according to which reducing the observability of knowledge is a necessary condition for protection against imitation (Barney, 1991; Liebeskind, 1996). Causal ambiguity, in contrast, is more likely to deter imitation attempts of outside actors (e.g., through reverse engineering), rather than of opportunistic innovation partners. In line with scholars emphasising the need to adopt a contingent RBV (Aragon-Correa and Sharma, 2003; Brush and Artz, 1999), this implies that the inimitability of a resource depends not only on its attributes and the environmental context in which it is used, but also on the characteristics of the imitating organisation. Inimitability is a critical attribute of resources needed to appropriate the economic rents accruing from novel products or services and sustain competitive advantages (Newbert, 2007). Succeeding with innovation partnerships therefore requires striking a subtle balance between knowledge revealing to enable fruitful collaboration and knowledge concealing to minimise the threat of imitation in a way that does justice to both the functional, geographical, and temporal dimensions of the partnership portfolio and the focal firm's strategic aspirations.

Limitations and future research directions

As with any study, our results should be considered in light of several limitations that suggest directions for future research. First, while the MIP offers a valuable complement to archival measures frequently used in alliance portfolio research, its reliance on single respondents' subjective accounts of a firm's partnership portfolio and imitation experience poses a limitation. Although, Harman's single-factor test revealed six factors with eigenvalues greater than 1, the most salient of which explained 25.5% of the total variance, indicating that substantial common method bias is unlikely (Podsakoff et al., 2003), future research might wish to integrate data from multiple data sources. Triangulating managers' subjective perception of being imitated by others with litigation data, for instance, would provide additional information about the extent to which a firm's IP rights are infringed upon and by whom. Moreover, by applying imitation measures that capture the intensity of imitation (actual number of imitation attempts), future studies could nicely complement our research that does not provide this information. Advancing the measurement of imitation appears particularly important given the lack of any established standard metric to quantify the imitation threat a focal firm is exposed to (Ordanini et al., 2008), with previous studies using, for example, dichotomous measures or assessing related aspects such as the time to imitation (Giachetti and Lanzolla, 2016).

Second, lack of data availability precluded us from examining certain aspects relevant to the research problem at hand. Among others, we were unable to explore the role of alternative, informal safeguards such as secrecy, partnership experience, and relational trust that manufacturing firms can rely upon to shield themselves against imitation. Moreover, a firm's partnership-induced imitation risks might be driven by partnership portfolio characteristics other than the value chain position, geographic origin, and temporal involvement of its innovation partners, as done in this study. In particular, future research might benefit from an appreciation of the intensity and history of these partnerships (Love et al., 2014). Unfortunately, lack of fine-grained dyadic data prevented us from examining this important aspect in sufficient detail. In order to understand the antecedents of imitation, exploring this qualitative and longitudinal dimension of a focal firm's innovation partnership portfolio is indeed a promising avenue for future research. Although, our *post hoc* analyses indicated that our main effect is robust to different model specifications including those that explicitly account for the potential endogeneity of partnership variety, longitudinal or experimental data promise to yield additional insights into the causal and temporal structure of this effect. As a case in point, it appears worthwhile to explore not only the extent to which the effect persists over time, but also how focal firms affected by imitation from partners adjust their partnership portfolio in response.

Third, there might be concerns regarding the generalisability of the findings to other contexts given our reliance on data from German manufacturing firms. Thus, future research in other countries is needed to corroborate our findings. Cross-national comparative studies would be most welcome, as they are well equipped to explore possible system-level contingencies such as differences in national appropriability regimes, which are likely to affect both the prevalence of partnership-induced imitation and the effectiveness of managerial remedies.

In conclusion, further research is required, to further strengthen the theory, evidence, and practical guidance available on how innovation partnership portfolios should be configured to reduce imitation threats and avoid getting caught on the wrong foot.

Appendix

Appendix A. Ordered probit regression analyses explaining financial performance.

Variable	Model A1	Model A2
Control variables		
Past financial performance	0.799 (0.061)***	0.800 (0.061)***
Past innovative performance	0.059 (0.032)*	0.068 (0.032)**
IP protection	−0.126 (0.099)	−0.086 (0.101)

(Continued)

Appendix A. (*Continued*)

Variable	Model A1	Model A2
R&D intensity	2.050 (1.625)	1.916 (1.614)
Firm size	−0.064(0.035)*	−0.059 (0.035)*
Export intensity	−0.208 (0.195)	−0.161 (0.196)
Human capital	−0.009 (0.027)	−0.011 (0.027)
Firm location	−0.077 (0.099)	−0.105 (0.100)
Industries	included	included
Main effects		
Imitation		−0.12 (0.047)**
Total observations	616	616
McFadden's R^2	0.265	0.267
Chi-squared	225.9 (0.000)***	230.4 (0.000)***

Notes: 616 total observations; ordered probit model; robust standard errors in parantheses; ***$p < 0.01$, **$p < 0.05$, *$p < 0.1$.

References

Ahuja, G and R Katila (2004). Where do resources come from? The role of idiosyncratic situations. *Strategic Management Journal*, 25(8–9), 887–907.

Alexy, O, G George and AJ Salter (2013). Cui bono? The selective revealing of knowledge and its implications for innovative activity. *Academy of Management Review*, 38(2), 270–291.

Almirall, E and R Casadesus-Masanell (2010). Open versus closed innovation: A model of discovery and divergence. *Academy of Management Review*, 35(1), 27–47.

Alvarez, SA and JB Barney (2004). Organizing rent generation and appropriation: Toward a theory of the entrepreneurial firm. *Journal of Business Venturing*, 19(5), 621–635.

Aragon-Correa, JA and S Sharma (2003). A contingent resource-based view of proactive corporate environmental strategy. *Academy of Management Review*, 28(1), 71–88.

Barney, JB (1991). Firm resources and sustained competitive advantage. *Journal of Management*, 17(1), 99–120.

Barney, JB (2014). How marketing scholars might help address issues in resource-based theory. *Journal of the Academy of Marketing Science*, 42(1), 24–26.

Barney, JB and W Hesterly (2012). *Strategic Management and Competitive Advantage*, 4th edition. NY: Pearson Education, Limited.

Ben Letaifa, S and Y Rabeau (2013). Too close to collaborate? How geographic proximity could impede entrepreneurship and innovation. *Journal of Business Research*, 66(10), 2071–2078.

Bercovitz, JE and MP Feldman (2007). Fishing upstream: Firm innovation strategy and university research alliances. *Research Policy*, 36(7), 930–948.

Boschma, R (2005). Proximity and Innovation: A critical assessment. *Regional Studies*, 39(1), 61–74.

Boudreau, K (2010). Open platform strategies and innovation: Granting access vs. devolving control. *Management Science*, 56(10), 1849–1872.

Brush, TH and KW Artz (1999). Toward a contingent resource-based theory: The impact of information asymmetry on the value of capabilities in veterinary medicine. *Strategic Management Journal*, 20(3), 223–250.

Carson, SJ, A Madhok and T Wu (2006). Uncertainty, opportunism, and governance: The effects of volatility and ambiguity on formal and relational contracting. *Academy of Management Journal*, 49(5), 1058–1077.

Cassiman, B and R Veugelers (2002). R&D cooperation and spillovers: Some empirical evidence from belgium. *The American Economic Review*, 92(4), 1169–1184.

Cassiman, B and R Veugelers (2006). In search of complementarity in innovation strategy: Internal R&D and external knowledge acquisition. *Management Science*, 52(1), 68–82.

Chesbrough, HW (2003). *Open Innovation: The New Imperative for Creating and Profiting from Technology*. Cambridge, MA: Harvard Business Press.

Chesbrough, HW, W Vanhaverbeke and J West (2006). *Open Innovation: Researching a new paradigm*. New York: Oxford University Press.

Cohen, WM and DA Levinthal (1990). Absorptive capacity: A new perspective on learning and innovation. *Administrative Science Quarterly*, 35(1), 128–152.

Cui, AS and G O'Connor (2012). Alliance portfolio resource diversity and firm innovation. *Journal of Marketing*, 76(4), 24–43.

Cyert, RM and JG March (1963). *A Behavioral Theory of the Firm*. NJ: Englewood Cliffs.

Dahlander, L and DM Gann (2010). How open is innovation? *Research Policy*, 39(6), 699–709.

De Leeuw, T, B Lokshin and G Duysters (2014). Returns to alliance portfolio diversity: The relative effects of partner diversity on firm's innovative performance and productivity. *Journal of Business Research*, 67(9), 1839–1849.

De Rond, M and H Bouchikhi (2004). On the dialectics of strategic alliances. *Organization Science*, 15(1), 56–69.

Dierickx, I and K Cool (1989). Asset stock accumulation and sustainability of competitive advantage. *Management Science*, 35(12), 1504–1511.

Diestre, L and N Rajagopalan (2012). Are all 'sharks' dangerous? new biotechnology ventures and partner selection in R&D alliances. *Strategic Management Journal*, 33(10), 1115–1134.

Duysters, G and B Lokshin (2011). Determinants of alliance portfolio complexity and its effect on innovative performance of companies*. *Journal of Product Innovation Management*, 28(4), 570–585.

Dyer, JH and K Nobeoka (2002). *Creating and Managing a High Performance Knowledge-Sharing Network: The Toyota Case*. Wiley online Library.

Dyer, JH and H Singh (1998). The relational view: Cooperative strategy and sources of interorganizational competitive advantage. *Academy of Management Review*, 23(4), 660–679.

Emden, Z, RJ Calantone and C Droge (2006). Collaborating for new product development: Selecting the partner with maximum potential to create value. *Journal of Product Innovation Management*, 23(4), 330–341.

Ethiraj, SK, D Levinthal and RR Roy (2008). The dual role of modularity: Innovation and imitation. *Management Science*, 54(5), 939–955.

Ethiraj, SK and DH Zhu (2008). Performance effects of imitative entry. *Strategic Management Journal*, 29(8), 797–817.

Garriga, H, G von Krogh and S Spaeth (2013). How constraints and knowledge impact open innovation. *Strategic Management Journal*, 34(9), 1134–1144.

Gesing, J, Antons, D, Piening, EP, Rese, M and Salge TO (2015). Joining forces or going it alone? On the interplay among external collaboration partner types, interfirm governance modes, and internal R&D. *Journal of Product Innovation Management*, 32(3), 424–440.

Giachetti, C and G Lanzolla (2016). Product technology imitation over the product diffusion cycle: Which companies and product innovations do competitors imitate more quickly? *Long Range Planning*, 49(2), 250–264.

Giarratana, MS and M Mariani (2014). The relationship between knowledge sourcing and fear of imitation. *Strategic Management Journal*, 35(8), 1144–1163.

Goerzen, A and PW Beamish (2003). Geographic scope and multinational enterprise performance. *Strategic Management Journal*, 24(13), 1289–1306.

Goerzen, A and PW Beamish (2005). The effect of alliance network diversity on multinational enterprise performance. *Strategic Management Journal*, 26(4), 333–354.

Greene, WH (2011). *Econometric Analysis: International Edition*, 7th edition, Harlow: Pearson Education Limited.

Grimpe, C and U Kaiser (2010). Balancing internal and external knowledge acquisition: The gains and pains from R&D outsourcing. *Journal of Management Studies*, 47(8), 1483–1509.

Harhoff, D, J Henkel and E von Hippel (2003). Profiting from voluntary information spillovers: How users benefit by freely revealing their innovations. *Research Policy*, 32(10), 1753–1769.

Helfat, CE (1994). Evolutionary trajectories in petroleum firm R&D. *Management Science*, 40(12), 1720–1747.

Hernandez, E, WG Sanders and A Tuschke (2015). Network defense: Pruning, grafting, and closing to prevent leakage of strategic knowledge to rivals. *Academy of Management Journal*, 58(4), 1233–1260.

Hoffmann, WH (2007). Strategies for managing a portfolio of alliances. *Strategic Management Journal*, 28(8), 827–856.

Hooley, GJ, GE Greenley, JW Cadogan and J Fahy (2005). The performance impact of marketing resources. *Journal of Business Research*, 58(1), 18–27.

Hsieh, KN and J Tidd (2012). Open versus closed new service development: The influences of project novelty. *Technovation*, 32(11), 600–608.

Huang, P, M Ceccagnoli, C Forman and DJ Wu (2013). Appropriability mechanisms and the platform partnership decision: Evidence from enterprise software. *Management Science*, 59(1), 102–121.

Hurmelinna, P, K Kyläheiko and T Jauhiainen (2007). The janus face of the appropriability regime in the protection of innovations: Theoretical re-appraisal and empirical analysis. *Technovation*, 27(3), 133–144.

Hussinger, K (2006). Is silence golden? Patents versus secrecy at the firm level. *Economics of Innovation and New Technology*, 15(8), 735–752.

James, SD, MJ Leiblein and S Lu (2013). How firms capture value from their innovations. *Journal of Management*, 39(5), 1123–1155.

Jensen, PH, R Thomson and J Yong (2011). Estimating the patent premium: Evidence from the australian inventor survey. *Strategic Management Journal*, 32(10), 1128–1138.

Jiang, RJ, QT Tao and MD Santoro (2010). Alliance portfolio diversity and firm performance. *Strategic Management Journal*, 31(10), 1136–1144.

Kale, P, JH Dyer and H Singh (2002). Alliance capability, stock market response, and long-term alliance success: The role of the alliance function. *Strategic Management Journal*, 23(8), 747–767.

Katila, R, JD Rosenberger and KM Eisenhardt (2008). Swimming with sharks: Technology ventures, defense mechanisms and corporate relationships. *Administrative Science Quarterly*, 53(2), 295–332.

Katsikeas, CS, D Skarmeas and DC Bello (2009). Developing successful trust-based international exchange relationships. *Journal of International Business Studies*, 40(1), 132–155.

Ketchen, DJ, RD Ireland and CC Snow (2007). Strategic entrepreneurship, collaborative innovation, and wealth creation. *Strategic Entrepreneurship Journal*, 1(3–4), 371–385.

Keupp, APMM, PCA Beckenbauer and O Gassmann (2010). Enforcing intellectual property rights in weak appropriability regimes. *Management International Review*, 50(1), 109–130.

King, AW (2007). Disentangling interfirm and intrafirm causal ambiguity: A conceptual model of causal ambiguity and sustainable competitive advantage. *Academy of Management Review*, 32(1), 156–178.

Klingebiel, R and C Rammer (2014). Resource allocation strategy for innovation portfolio management. *Strategic Management Journal*, 35(2), 246–268.

Kogut, B and U Zander (1992). Knowledge of the firm, combinative capabilities, and the replication of technology. *Organization Science*, 3(3), 383–397.

Kozlenkova, IV, SA Samaha and RW Palmatier (2013). Resource-based theory in marketing. *Journal of the Academy of Marketing Science*, 42(1), 1–21.

Laursen, K and A Salter (2006). Open for innovation: The role of openness in explaining innovation performance among UK manufacturing firms. *Strategic Management Journal*, 27(2), 131–150.

Lavie, D (2006). The competitive advantage of interconnected firms: An extension of the resource-based view. *Academy of Management Review*, 31(3), 638–658.

Lavie, D (2007). Alliance portfolios and firm performance: A study of value creation and appropriation in the U.S. software industry. *Strategic Management Journal*, 28(12), 1187–1212.

Lavie, D and SR Miller (2008). Alliance portfolio internationalization and firm performance. *Organization Science*, 19(4), 623–646.

Lawson, B, D Samson and S Roden (2012). Appropriating the value from innovation: Inimitability and the effectiveness of isolating mechanisms. *R&D Management*, 42(5), 420–434.

Leiponen, A and CE Helfat (2010). Innovation objectives, knowledge sources, and the benefits of breadth. *Strategic Management Journal*, 31(2), 224–236.

Lewin, AY, S Massini and C Peeters (2009). Why are companies offshoring innovation? The emerging global race for talent. *Journal of International Business Studies*, 40(6), 901–925.

Li, D, L Eden, MA Hitt and RD Ireland (2008). Friends, acquaintances, or strangers? Partner selection in R&D alliances. *Academy of Management Journal*, 51(2), 315–334.

Li, D, L Eden, MA Hitt, RD Ireland and RP Garrett (2012). Governance in multilateral R&D alliances. *Organization Science*, 23(4), 1191–1210.

Lieberman, MB and DB Montgomery (1988). First-mover advantages. *Strategic Management Journal*, 9(S1), 41–58.

Liebeskind, J (1996). Knowledge, strategy, and the theory of the firm. *Strategic Management Journal*, 17(S2), 93–107.

Love, JH, S Roper and P Vahter (2014). Learning from openness: The dynamics of breadth in external innovation linkages. *Strategic Management Journal*, 35(11), 1703–1716.

Mahoney, JT and JR Pandian (1992). The resource-based view within the conversation of strategic management. *Strategic Management Journal*, 13(5), 363–380.

Manzini, R and V Lazzarotti (2016). Intellectual property protection mechanisms in collaborative new product development. *R&D Management*, 42(S2), 579–595.

Martinez-Noya, A, E Garcia-Canal and MF Guillen (2013). R&D outsourcing and the effectiveness of intangible investments: Is proprietary core knowledge walking out of the door? *Journal of Management Studies*, 50(1), 67–91.

McEvily, SK and B Chakravarthy (2002). The persistence of knowledge-based advantage: An empirical test for product performance and technological knowledge. *Strategic Management Journal*, 23(4), 285–305.

McEvily, SK, KM Eisenhardt and JE Prescott (2004). The global acquisition, leverage, and protection of technological competencies. *Strategic Management Journal*, 25(8–9), 713–722.

Meuleman, M, A Lockett, S Manigart and M Wright (2010). Partner selection decisions in interfirm collaborations: The paradox of relational embeddedness. *Journal of Management Studies*, 47(6), 995–1019.

Miotti, L and F Sachwald (2003). Co-operative R&D: Why and with whom?: An integrated framework of analysis. *Research Policy*, 32(8), 1481–1499.

Newbert, SL (2007). Empirical research on the resource-based view of the firm: An assessment and suggestions for future research. *Strategic Management Journal*, 28(2), 121–146.

Nieto, MJ and L Santamaría (2007). The importance of diverse collaborative networks for the novelty of product innovation. *Technovation*, 27(6–7), 367–377.

OECD (2005). *The Measurement of Scientific and Technological Activities Oslo Manual: Guidelines for Collecting and Interpreting Innovation Fata*, 3rd edition. France: OECD Publishing.

Ordanini, A, G Rubera and R DeFillippi (2008). The many moods of inter-organizational imitation: A critical review. *International Journal of Management Reviews*, 10(4), 375–398.

Oxley, JE and RC Sampson (2004). The scope and governance of international R&D alliances. *Strategic Management Journal*, 25(8–9), 723–749.

Paasi, J, T Luoma, K Valkokari and N Lee, (2010). Knowledge and intellectual property management in customer–supplier relationships. *International Journal of Innovation Management*, 14(04), 629–654.

Parmigiani, A and M Rivera-Santos (2011). Clearing a path through the forest: A meta-review of interorganizational relationships. *Journal of Management*, 37(4), 1108–1136.

Peteraf, MA (1993). The cornerstones of competitive advantage: A resource-based view. *Strategic Management Journal*, 14(3), 179–191.

Phene, A and S Tallman (2014). Knowledge spillovers and alliance formation. *Journal of Management Studies*, published online before print.

Pisano, GP (1989). Using equity participation to support exchange: Evidence from the biotechnology industry. *Journal of Law, Economics, & Organization*, 5(1), 109–126.

Podsakoff, PM, SB MacKenzie, JY Lee and NP Podsakoff (2003). Common method biases in behavioural research: A critical review of the literature and recommended remedies. *Journal of Applied Psychology*, 88(5), 879.

Polidoro, F and PK Toh (2011). Letting rivals come close or warding them off? The effects of substitution threat on imitation deterrence. *Academy of Management Journal*, 54(2), 369–392.

Reed, R and RJ De Fillippi (1990). Causal ambiguity, barriers to imitation, and sustainable competitive advantage. *Academy of Management Review*, 15(1), 88–102.

Reitzig, M, J Henkel and C Heath (2007). On sharks, trolls, and their patent prey: Unrealistic damage awards and firms' strategies of 'being infringed'. *Research Policy*, 36(1), 134–154.

Ritala, P and P Hurmelinna-Laukkanen (2013). Incremental and radical innovation in coopetition: The role of absorptive capacity and appropriability. *Journal of Product Innovation Management*, 30(1), 154–169.

Robson, MJ, CS Katsikeas and DC Bello (2008). Drivers and performance outcomes of trust in international strategic alliances: The role of organizational complexity. *Organization Science*, 19(4), 647–665.

Roy, S and K Sivakumar (2011). Managing intellectual property in global outsourcing for innovation generation. *Journal of Product Innovation Management*, 28(1), 48–62.

Rumelt, RP (1984). Toward a strategic theory of the firm. In *Competitive Strategic Management*, R Lamb (Ed), Englewood Cliffs: NJ, Prentice Hall, pp. 556–570.

Salge, TO, TM Bohné, T Farchi and EP Piening (2012). Harnessing the value of open innovation: The moderating role of innovation management. *International Journal of Innovation Management*, 16(3), 1240005–12400026.

Salge, TO, T Farchi, MI Barrett and S Dopson (2013). When does search openness really matter? A contingency study of health-care innovation projects. *Journal of Product Innovation Management*, 30(4), 659–676.

Salomon, R and Z Wu (2012). Institutional distance and local isomorphism strategy. *Journal of International Business Studies*, 43(4), 343–367.

Schmiele, A (2013). Intellectual property infringements due to R&D abroad? A comparative analysis between firms with international and domestic innovation activities. *Research Policy*, 42(8), 1482–1495.

Sirmon, D, M Hitt and RD Ireland (2007). Managing firm resources in dynamic environments to create value: Looking inside the black box. *Academy of Management Review*, 32(1), 273–292.

Sivakumar, K, S Roy, JJ Zhu and S Hanvanich (2011). Global innovation generation and financial performance in business-to-business relationships: The case of cross-border alliances in the pharmaceutical industry. *Journal of the Academy of Marketing Science*, 39(5), 757–776.

Somaya, D (2012). Patent strategy and management: An integrative review and research agenda. *Journal of Management*, 38(4), 1084–1114.

Teece, DJ (1986). Profiting from technological innovation: Implications for integration, collaboration, licensing and public policy. *Research Policy*, 15(6), 285–305.

Thomä, J and K Bizer (2013). To protect or not to protect? Modes of appropriability in the small enterprise sector. *Research Policy*, 42(1), 35–49.

Vasudeva, G and J Anand (2011). Unpacking absorptive capacity: A study of knowledge utilization from alliance portfolios. *Academy of Management Journal*, 54(3), 611–623.

Vuong, QH (1989). Likelihood ratio tests for model selection and non-nested hypotheses. *Econometrica*, 57(2), 307–333.

Wang, HC, J He and JT Mahoney (2009). Firm-specific knowledge resources and competitive advantage: The roles of economic-and relationship-based employee governance mechanisms. *Strategic Management Journal*, 30(12), 1265–1285.

Wassmer, U (2010). Alliance portfolios: A review and research agenda. *Journal of Management*, 36(1), 141–171.

Wernerfelt, B (1984). A resource-based view of the firm. *Strategic Management Journal*, 5(2), 171–180.

Wind, J and V Mahajan (1997). Issues and opportunities in new product development: An introduction to the special issue. *Journal of Marketing Research*, 1–12.

Wuyts, S, S Dutta and S Stremersch (2004). Portfolios of interfirm agreements in technology-intensive markets: Consequences for innovation and profitability. *Journal of Marketing*, 68(2), 88–100.

Yang, H, C Phelps and HK Steensma (2010). Learning from what others have learned from you: The effects of knowledge spillovers on originating firms. *Academy of Management Journal*, 53(2), 371–389.

Zaheer, S (1995). Overcoming the liability of foreignness. *Academy of Management Journal*, 38(2), 341–363.

Zaheer, A and E Hernandez (2011). The geographic scope of the MNC and its alliance portfolio: Resolving the paradox of distance. *Global Strategy Journal*, 1(1–2), 109–126.

Part 2
Lean and Global Innovation

Part 2

Lean and Global Innovation

Chapter 7

Lean and Global Technology Start-ups: Linking the Two Research Streams*

Stoyan Tanev, Erik Stavnsager Rasmussen, Erik Zijdemans,
Roy Lemminger and Lars Limkilde Svendsen

In this paper, the authors introduce the concept of the lean global start-up (LGS) as a way of emphasising the problems for new technology start-ups when dealing separately with business development, innovation and early internationalisation. The paper has two components — an introductory conceptual part and an empirical part that should be considered as basis for the preliminary validation of the conceptual insights. The research sample includes six firms — three from Canada and three from Denmark. Two different early internationalisation paths have been identified: Lean-to-global (L2G start-ups) and lean-and-global (L&G start-ups). Both types of start-ups were found to have faced significant problems with the complexity, uncertainties and risks of being innovative on a global scale. They have however found ways of addressing these problems by a disciplined knowledge sharing and IP protection strategy and the efficient use of business and supporting and public funding mechanisms. The Danish firms have pivoted around the ways of delivering their value proposition and not around the specific value propositions themselves. The Canadian firms have actively pivoted their value proposition motivated by the degree of innovativeness of their products and the insights from business supporting organisations. The analysis of the results justifies the introduction of the LGS concept and opens the opportunity for future research focusing on the articulation of more practical LGS entrepreneurial frameworks.

Keywords: Lean global start-up; born global firm; international new venture; technology entrepreneurship; innovation management; business model; lean start-up.

Introduction

The focus of this research study is newly established technology-oriented firms, which have become both lean and global right from or near their inception. The

---

*Originally published in *International Journal of Innovation Management*, 19(3), pp. 1–41.

empirical study of the combination of lean and global characteristics allows for the integration of two different research streams which, unfortunately, have been separated so far. The first stream is well established and focuses on international new ventures (INVs) (Oviatt and McDougall, 1994) or born global (BG) firms (Rennie, 1993; Knight and Cavusgil, 1996). The second stream is in the process of emerging and deals with the specifics of lean start-ups (LSs) (Ries, 2011; Blank, 2013). It is our intention to show that the problems faced by LSs and BG firms during the early stages of their existence are to a large extent identical and could, from a theoretical point of view, be analysed in a unified way. The commonality of the problems is rooted in the challenge of dealing simultaneously with early internationalisation (starting or going global), business modelling, relationship management, resource allocation and innovation management under conditions of multiple uncertainties right from or near their founding. Integrating the two research streams should help the emergence of the LS research field as well as contribute to the articulation of design principles that would help the creation of more competitive lean global start-ups (LGSs).

In what follows we will offer a discussion of the two research streams and show how they could complement each other in studying the global start or the early internationalisation of newly created technology firms. We will also present empirical results derived from several case studies of technology start-ups in Denmark and Canada. The focus of the empirical research component will be on offering a first and preliminary substantiation of the LGS concept by examining how newly established firms combine the development and commercialisation of new technological products with their entry into global markets or access to global resources.

Combining the two perspectives above gives rise to a number of interesting questions that could be summarised as follows:

- How do new technology start-ups narrow down the scope of their business activities by effectuating the global dimensions of their businesses?
- What are the reasons for such firms to start looking for global resources, partnerships or markets right from their founding?
- What are the sources of innovation of such firms at the very early stages of their existence?
- How does the degree of innovativeness of their products or services affect their business model formulation and early internationalisation processes?
- How do the emerging relations and networks of such firms help in synergising their innovation and early internationalisation activities?
- How do they use international partnerships or networks to reduce the complexity and the uncertainties associated with global resource allocation?

Conceptual Insights Based on the Literature Review

The characteristics of born global firms and the entrepreneurial challenges of early internationalisation

For several decades, researchers have studied INVs and BG firms with a focus on their international markets, suppliers and networks from or near their founding. INVs have been characterised as "business organisations that, from inception, seek to derive significant competitive advantage from the use of resources and the sale of outputs in multiple countries" (Oviatt and McDougall, 1994: 49). Knight and Cavusgil (1996) labelled these global start-ups as BGs and many others have followed this path. Since then these research stream has gained significant traction by focusing on the various business aspects and the potential benefits and challenges associated with the early internationalisation of such firms. It should be pointed out however that there is a slight difference between the definitions of INVs and BG firms (Knight, 2014). Knight and Cavusgil (2004) defined BGs as "business organisations that, from or near their founding, seek superior international business performance from the application of knowledge-based resources to the sale of outputs in multiple countries." The BG concept emphasises young firms and SMEs as the unit of analysis and prioritises activities focusing on selling goods internationally. Thus, the emphasis is on the export activities of BG firms as their main foreign entry mode (Knight, 2014). The INV concept encompasses both young, internationalising firms, and new ventures launched in older, established multinationals. It includes a broader range of value chain activities such as internationally-based R&D or foreign manufacturing, as well as various entry strategies, including foreign direct investment (FDI).

According to a recent review of the relevant literature, the distinctive characteristics of BG firms are as follows (Tanev, 2012). *First*, BG firms are characterised by higher activity in international markets from or near the founding. The decision of a firm to engage into a systematic internationalisation process is usually determined by its nature — the type of technology that is being developed or the firms specialisation within the specific industry sector, value chain, or market (Jones et al., 2011). On the other hand, the extent to which a firm is "born global" rather than "born local" or "late global" depends on firms own early decisions (Moen and Servais, 2002). The founder's vision at the time of the incorporation is a key factor for a firms early internationalisation patterns (Gabrielsson and Pelkonen, 2008). *Second*, BG firms tend to be relatively small and have far fewer financial, human, and tangible resources as compared to large multinational enterprises that have been considered as dominant in global trade and investment. *Third*, many BG firms are technology firms although the BG phenomenon has been widely spread beyond the technology sector (Moen and Servais, 2002). *Fourth*, BG firms have managers possessing a strong international outlook and international entrepreneurial orientation. The skills of top management teams have been found very important for the enablement of a more intense internationalisation, particularly in the knowledge-based sectors (Johnson, 2004; Andersson and Evangelista, 2006; Loane et al., 2007). *Fifth*, BG firms tend to adopt differentiation strategies focusing on unique designs and highly distinctive products targeting niche markets, which may be too small for the tastes of larger firms (Cavusgil and Knight, 2009). *Sixth*, BG firms are often at the leading technological edge of their industry or of their product category. Typically, these firms do not operate in commodity markets (Cavusgil and Knight, 2009). *Seventh*, many BG firms leverage ICT to identify and segment customers into narrow global-market niches and skilfully serve highly specialised buyer needs. ICT allows them to process information efficiently and communicate with partners and customers worldwide at practically zero cost (Cavusgil and Knight, 2009; Maltby, 2012). *Eight*, many BG firms expand internationally through engaging into direct international sales or leveraging the resources of independent intermediaries located abroad (Cavusgil and Knight, 2009). Very often such firms cooperate with multi-national corporations (MNCs) by using their existing channels, networks, and Internet infrastructure to receive substantial revenues and cash flow rapidly (Vapola et al., 2008; Gabrielsson and Kirpalani, 2004). MNCs may act as systems integrators or distributors of BGs products and services providing opportunities for learning, technological infrastructure access, and evolutionary growth. *Last but not least*, recent studies have emphasised that the early internationalisation of BG firms and new ventures should be considered as an innovation process in itself and that innovation and internationalisation have a positive effect on each other. In this way, BG firms should be considered as possessing the unique capacity to both innovate and internationalise early by mastering advanced knowledge acquisition and networking capabilities as key innovation enablers (Zijdemans and Tanev, 2014).

Even though the BG phenomenon is not unique to the technology sector, many BG firms appear to be technology-driven companies. A recent research study identified a number of conditions for newly created technology firms that should be considered by early or rapid globalisation (Kudina et al., 2008):

- Operating in a knowledge-intensive or high-technology sector;
- A home country market that is not large enough to support the scale at which the firm needs to operate;
- Most of the potential customers are foreign, multinational firms;

- Competitive advantage is based on most technically advanced offering worldwide;
- The product or service category faces few trade barriers;
- The product or service has higher value relative to its transportation/logistics costs;
- Customer needs and tastes are fairly standard across the firms international markets;
- The product or service has a significant first-mover advantage or network effect;
- The major competitors have already internationalised or will internationalise soon;
- Executive managers who are experienced in international business (IB) development.

In addition to the above characteristics, researchers have identified another organisational capability that enables the internationalisation and enhances the international performance of BG firms — the ability to leverage different types of networks or ecosystems (Kudina *et al.*, 2008). Kudina *et al.* (2008) attribute the success of technology-based BG firms to the effective use of ecosystems comprising of: Universities and firms operating in the same industry as the focal firm which help in providing a flow of technological knowledge, experienced people, and contacts with local venture capitalists; foreign sales subsidiaries providing important sources of knowledge from experts that are spread out internationally and facilitating direct contacts between engineers and clients aiming at the specification of client needs in a way that provides a mechanism for winning additional business; foreign sales subsidiaries and local clients that are influential in the development of high-quality services on the basis of technological knowledge acquired through the clients or their business partners.

It should be pointed out that the increasingly growing studies focusing on BG firms have been naturally associated with research on the interface of IB and international entrepreneurship (IE) including the challenges of young SMEs interested in growing by leveraging resources, knowledge, networks and capabilities which are located abroad. "International entrepreneurship is the discovery, enactment, evaluation, and exploitation of opportunities — across national borders — to create future goods and services." (Oviatt and McDougall, 2005: 540) According to Sarasvathy *et al.* (2014) there are at least three characteristics of conducting cross-border business which call out for theories from entrepreneurship and effectuation that could be directly applicable to the context of BG firms. *The first one* is the need to explicitly deal with cross-border uncertainty. *The second one* is the need to leverage limited resources. Operating by leveraging limited resources within a context involving political, economic and sociocultural risks is particularly challenging for BG firms. *The third* characteristic is related to the challenges associated with taking into account network dynamics since "creating, maintaining, growing, and managing networks, whether at the individual, organisational, or inter-organisational level, becomes more challenging across borders because of geographic and cultural distance" (Sarasvathy *et al.*, 2014: 76).

The effectual approach to IE within the context of BG firms

Sarasvathy *et al.* (2014) emphasise the need for the adoption of an effectual approach to IE which is of particular interest within the context of BG firms. Results based on the effectuation theory approach to entrepreneurship (Sarasvathy, 2001) indicate that entrepreneurs tend to move away from prediction-based strategies by adopting a means-based approach by: Managing their particular level of affordable loss, forging partnerships, leveraging contingency and using a logic based on non-predictive control. More specifically, the uniqueness of the overall logic of the effectuation approach was articulated in the form of the following five principles (Sarasvathy, 2008): (i) bird-in-hand; (ii) affordable loss; (iii) crazy quilt; (iv) lemonade; (v) pilot-in-the-plane.

The bird-in-hand principle refers to the means-based approach which relies on the specific combination of entrepreneurial identity, previous knowledge, and access to networks as a way of generating new business opportunities. Such approach could be articulated in terms of firms resource-based capabilities (who I am), the number of years of industry experience (what I know), and the size and breadth of networks (whom I know) which are potentially available to a firm (Read *et al.*, 2009). Opportunity-specific managerial expertise reduces the effect of uncertainty which leads to better new market development abilities (Sarasvathy *et al.*, 2014; McKelvie *et al.*, 2011; Blume and Covin, 2011).

The affordable loss principle refers to entrepreneurs' attitude of focusing on what they can afford to lose rather than on the prediction of possible gains (Sarasvathy, 2008). It shifts the focus from the need to predict future returns and encourages the adoption of an experiential learning approach to opportunity creation as compared to undertaking a process of opportunity discovery and spending more time engaged in planning (Alvarez and Barney, 2007).

The crazy quilt principle extends the affordable loss logic by focusing on emerging partnerships as the dominant way of expanding resources. It refers to the tendency of entrepreneurs to create avenues for the self-engagement of stakeholders who can find an opportunity in co-creating the venture together with the original founders. Co-creation enables both opportunistic serendipity and genuine commitment of valuable partners through the emergence of relationships in which both parties share the risks of the venture but also benefit from its eventual success (Sarasvathy et al., 2014; Chandler et al., 2011).

The lemonade principle (based on the analogy of turning "lemons to lemonade") refers to the tendency of entrepreneurs to leverage uncertainty by treating surprises as opportunities (Sarasvathy et al., 2014). The multiple uncertainties force entrepreneurs to embrace new information as a basis to initiate, run or abandon experiments as well as to leverage emergent possibilities by bringing in serendipity and unintended discovery within the context of the opportunity development process (Chandler et al., 2011).

The pilot-in-the-plane principle "emphasizes the role of human beings rather than trends in determining the shape of things to come" (Sarasvathy et al., 2014: 74). It is an antithesis of seeing history as running on autopilot by considering entrepreneurs as co-pilots in the course of history. Under highly uncertain conditions, entrepreneurs are not just seeking to learn more in order to come up with better actions, but rather intervening in the event space itself to transform and reshape it in cooperation with other relevant players (Sarasvathy et al., 2013).

The integrative logic of the five effectuation principles coheres very well with the philosophy of actor-network theory (Latour, 2005), especially if it is taken in an entrepreneurial and innovation perspective. The link between effectuation and actor-network theories could be quite promising in opening new research opportunities in IE. What is more interesting here is the link between effectuation theory and the theory of BG firms and INVs. According to Sarasvathy *et al.* (2014), in the BG case, the effectual aspects of the internationalisation process are difficult to overlook. The value of effectuation theory for the study of the BG phenomenon has been also explicitly pointed out by Andersson (2011) according to whom effectuation theory provides new insights into entrepreneur's decisions regarding internationalisation and the early internationalisation pattern in BG firms. For example, it explicitly treats both the individual, firm and network level. In addition, it includes a pro-active entrepreneurial perspective that better describes the firms development than earlier theories focusing on the resource-based view (Knight and Cavusgil, 2004) and the role of networks (Coviello, 2006).

Andersson (2011) and Sarasvathy *et al.* (2014) clearly pointed out that effectuation theory is highly applicable to the uncertain environments which are typical of BG firms. Its relevance to the entrepreneurial marketing context of BG firms has been also pointed out recently by Mort *et al.* (2012) who referred to a specific "marketing under uncertainty" perspective substantiated by the effectuation logic (Stuart *et al.*, 2009). Mort *et al.* (2012) argue that the context of BG firms is highly appropriate for advancing the domain of entrepreneurial marketing and discuss the key marketing aspects that link strongly to BG performance in rapid internationalisation. For example, they found that BG firms achieve a superior market performance through specific strategies using an effectuation approach that substantially departs from the conventionally accepted marketing logic adopted by established firms. BG marketing actions are highly nonlinear but could be categorised in terms of four core marketing strategies including opportunity creation, customer intimacy-focused innovative products, resource enhancement and legitimacy.

Opportunity creation was found to be central to the growth of BG firms in both hi-tech and the low-tech sectors. The executive managers of BG SMEs engage into innovative product development on the basis of strong customer intimacy. This is what makes BGs niche marketers able to maintain a close relationship with their customers. It is the close linkage between customer intimacy and the focus on innovative product development that allows for enhanced performance outcomes. Firms' ability to enhance marketing communication resources was found to be particularly critical for increased effectiveness in market development as well as in gaining legitimacy, acceptance and trust. For example, BG technology-based firms strive for success in cooperating with lead users in order to enhance their legitimacy. This was found to be particularly relevant in business-to-business markets where potential buyers are looking for an assurance before committing to a specific technology. High-tech firms do also use international certification, prizes, industry awards and media recognition in order to authenticate their products (Mort et al., 2012). The focus of entrepreneurial marketing scholars on the effectual logic of BG firms is very promising. It should be pointed out however that the field of BG marketing focuses predominantly on SMEs and remains detached from the fields of technology entrepreneurship, global start-ups and new business creation (Tanev, 2012).

The LS approach

The term LS was coined by Erik Ries (2011). Its principles however have been intensively discussed within the last five years predominantly among practicing entrepreneurs and venture capitalists in the Silicon Valley. In this sense, the LS movement was initiated by practitioners who have pointed out a major problem in the way new start-ups have been evaluated by venture capitalists and business policy organisations (Stolze et al., 2014). Eisenmann et al. (2011) defined a LS as a firm that follows a hypothesis-driven approach to evaluate an entrepreneurial opportunity. The approach has now become widely known as the LS methodology. The essence of the methodology consists of the translation of a specific entrepreneurial vision into falsifiable hypotheses regarding the solution and business model that is going to be used to deliver it. The hypotheses are then tested using a series of well-thought product prototypes that are designed to rigorously validate specific product features or business model specifications. In this context, the entrepreneurial opportunity is based on the solution envisioned by the entrepreneur to solve a specific customer problem. The uniqueness of the methodology consists in its ability to explicitly take into account the numerous uncertainties regarding the suitability of a given solution towards a specific customer problem.

The LS process of validation was described initially by Blank (2003), through the introduction of a customer development model (CDM). It was later popularised by Ries through the articulation of several key paradigmatic principles as part of a build-measure-learn (BML-loop) framework which was described in his book "The LS" (Ries, 2011). These two methodological frameworks were complemented by the business model canvas (BMC) approach (Osterwalder and Pigneur, 2010) to form the basis for "the status quo" of the conceptual background of the LS approach (Stolze *et al.*, 2014).

The emergence of the LS approach is based on Blank's and Ries' observations of patterns of action and decision by successful entrepreneurs, including themselves, who tend to follow a customer learning and discovery process instead of a product-centric development model when founding a new company. According to Blank (2013), one of the key starting points is to emphasise that a start-up is not a smaller version of a large company: "One of the critical differences is that while existing companies execute a business model, start-ups look for one. This distinction is at the heart of the LS approach. It shapes the lean definition of a start-up: A temporary organisation designed to search for a repeatable and scalable business model." Eric Ries (2011) pointed out another important aspect of the LS by defining it as "a human institution designed to create new products and services under conditions of extreme uncertainty" (Ries, 2011). What is the meaning of "lean" in LSs? On a very intuitive level it appears to be associated with the traditional lean manufacturing concept of waste reduction since for a start-up "the biggest waste is creating a product or service that nobody needs" (Mueller and Thoring, 2012). The LS approach translates the well-known lean paradigm into the early stage business context by focusing on minimising the expenditure of resources for anything but the creation of value for the customer. Such approach to entrepreneurship favours experimentation over planning, customer feedback over intuition and iterative design over traditional business planning (Blank, 2013). The focus on experimentation as a source of customer knowledge is associated with the concept of minimum viable product (MVP) — a product consisting of a minimum set of features that is used, first, as a tactic to reduce wasted engineering hours; *second*, as a way of getting the product in the hands of early and visionary customers as soon as possible. The MVP is a way of selling the vision and delivering the minimum feature set to visionary customers and not to all customers. The MVP concept is the basis for another difference of LSs as compared to traditional ones — the need for the adoption of success metrics tolerating experimentation and productive failure.[1]

Eric Ries summarised the key principles of the LS approach as follows.[2] First, entrepreneurs are everywhere; one does not need to be in a garage in order to be in a start-up. Second, entrepreneurship is associated with a specific attitude to management; a start-up is an institution, not just a product, so it requires a new kind of management which is to be specifically geared to its specific context. Third, the LS approach is associated with a method of validated learning since start-ups do not exist just to make stuff, make money, or serve customers; they exist to learn how to build a sustainable business. This learning can be validated scientifically, by running experiments that allow entrepreneurs to test each element of their vision. Fourth, the specifics of the LS approach requires a new way of measuring progress, setting up milestones and prioritising work; this requires a new kind of accounting which is specific to start-ups. Last but not least, this is the adoption of the above mentioned BML-loop process which consists in turning ideas into products by measuring how customers respond, and then learn whether to pivot or persevere in the same way. Successful start-ups excel at the adoption of processes that are geared to accelerate that feedback loop.

The CDM and the BML-Loop frameworks propose that the start-ups' search for a business model should focus on insights from a special type of technology-enthusiasts and visionaries (Blank and Dorf, 2012). Technology-enthusiasts and visionaries, or innovators and early-adopters, are types of customers proposed by Everett Rogers in his diffusion of innovation theory (Rogers, 1983; Yadav *et al.*, 2006). The concept was revised and adapted to the context of technology start-ups (Moore, 1991; Yadav *et al.*, 2006). Moore (1991) validated the use of the term technology adoption life cycle (TALC) and argued that although technology adoption followed the pattern of generic innovation adoption proposed by Rogers (1983), there was a gap or chasm in the process of adoption. The chasm is defined by the moment when a technology start-up stops being a start-up focusing on technology enthusiasts and visionaries and transforms itself into a more structured organisation offering a "whole product" to a first niche market of early adopters in order to continue its progress towards the mainstream market.

---

[1]Disruptive entrepreneurs: An interview with Eric Ries, McKinsey & Company: http://www.mckinsey.com/Insights/High_Tech_Telecoms_Internet/Disruptive_entrepreneurs_An_interview_with_Eric_Ries.

[2]http://theleanstartup.com/principles.

The transition from a first niche market of early adopters to larger market segments and to the mass market has been referred to as "crossing the chasm" (Moore, 1991). According to Moore, there is a fundamental difference between the critical success factors of marketing to early and late adopters (mass markets) but most start-ups fail to make this transition. Therefore, the successful launch and diffusion of a technological innovation demand not only attention to traditional marketing issues such as the timing and positioning of the product, but also significant and dedicated efforts to manage early adoption in a way that could enable later adoption by the mass market.

The TALC framework has established itself as a systematic approach to technology marketing (Wiefels, 2002; Tidd, 2010) providing some of the conceptual apparatus of the LS approach. By focusing on start-ups, the LS approach focuses on the 16% of customers that compose the early-market of a new business (Blank and Dorf, 2012: xiv). In addition, the LS approach literature introduces another type of early market customer — the early vangelist (from early and evangelist). These customers have some special characteristics the identification of which helps in accelerating the viral growth of the business as these customers become eager disseminators of information and influencers within their closest peer groups (Blank and Dorf, 2012: 59). Thus, some of the key aspects of the TALC model have acquired a higher degree of sophistication as part of the LS approach. In other words, the LS approach has offered a specific operationalisation of the way start-ups deal with crossing the chasm by suggesting a systematic way of using experimentation and iterative learning to turn uncertainties into risks in the development of products at the very early stages of a new business.

Linking the two research streams

If one takes a closer look at the two research streams, a number of common research lines can be identified. The INV and BG firm research field has its focus on how SMEs can accelerate their entry into global markets while LS research has its focus on how new entrepreneurial firms can develop new products and services for a large number of customers in a shorter period of time. Both research streams stress the complexity of the process, the scarcity of resources, the innovation challenges and the risks and uncertainties the firms have to deal with. In many cases, the solution for the firms is to build relations and networks to minimise both complexity and uncertainty, as well as to optimise the use of external resources. The link between the two approaches can be found in their focus on entrepreneurship, since in both cases the entrepreneurs have to learn to operate in a complex and uncertain business ecosystem including suppliers, R&D partners, competitors, customers, etc. This is especially true in the case of high-tech firms which have to be active on a global scale right from the beginning.

In several cases technology entrepreneurship and innovation research studies have reached out to encompass themes that are typical of research focusing on BG firms. For example, Bailetti (2012) examines how new growth-oriented technology firms can (or must) operate in a global market right from their founding. The entrepreneurs behind these technology start-ups must plan the internationalisation of the firm in the right way from the very beginning. Moogk (2012) discusses the LS concept and how entrepreneurs can apply it to the process of new technology commercialisation. This is done in a context of extreme uncertainty and technology start-ups have to learn designing and using MVPs in order to be able to enter a market before potential competitors. MVPs offer the possibility for the technology to be tested in a way that could help the evaluation and the facilitation of a firms global growth opportunities. The MVP approach to new product development has been also discussed as part of a more generic cost-efficient perspective on entrepreneurship focusing on experimentation (Kerr *et al.*, 2014). According to Kerr *et al.* (2014) the MVP approach seeks to validate as many assumptions as possible about the viability of the final product before expending enormous effort and financial resources. According to them one of the key aspects of the lean start-up approach is the focus on how to make prototype-based experiments ever more cost-effective so that ventures do not need to raise as much money to pursue a range of alternative product ideas. It is not by accident that the LS philosophy "has coincided with the rapid rise of angel investors and crowd-funding platforms, particularly for consumer Internet start-ups" (Kerr *et al.*, 2014: 35). Other authors (Tanev, 2012; Tanev and Rasmussen, 2014; Lemminger *et al.*, 2014) have approached the technology start-up and the lean approach from the BG context. One of their key recommendations is that researchers should focus on defining design principles that incorporate the key attributes of born-global firms and use these design principles to launch and grow new technology firms (Tanev, 2012).

The emergent nature of strategic and business model design in INVs requires the integration of the strategic management, IE and internationalisation literature (Knight and Cavusgil, 2004). Trimi and Berbegal-Mirabent (2012) have discussed the emerging trends in business model design by focusing on open innovation, customer development, agile development and lean methodologies. According to them all these approaches converge in the use of quick iterations and the adoption of a trial-error philosophy for validating the hypotheses of the business model and the appropriateness of specific product or services. The logic of the LS model in particular could be substantiated by combining with business model frameworks that are able to integrate the entrepreneurial, innovation and internationalisation aspects of born global start-ups (BGSs) (Onetti et al., 2012). The framework of Onetti et al. (2012) defines a business model as the way a company structures its own activities in determining the focus, locus and modus of its business, where the "focus" of the business refers to the activities which provide the basis of the firms value proposition (the set of activities on which the company's efforts are concentrated); the "locus" refers the location or locations across which the firms resources and value adding activities are spread (i.e., local versus foreign based activities, inward-outward relationships, entry modes, etc.); and the "modus" refers to the specific business modes of operation with regards to the internal organisation and the network design (i.e., insourcing and outsourcing of activities along social and inter-organisational ties, inward–outward relationships with other players, strategic alliances, etc.). This is one of the few business model frameworks that allows for accommodating the global dimension of resources, partnerships and emerging technology markets. In addition, it emphasises that the process of business modelling should be detached from the efforts focusing on the formulation of the value proposition since the definition of a value proposition belongs to the strategic and not to the business process level. In a certain way, this emphasis corresponds to one of the key aspects of the LS approach which indicates that the emergence of a value proposition is the outcome of the business modelling process and not an initial pre-condition for it (Blank, 2013). As it appears, Onetti's framework could be easily integrated with the BMC approach (Osterwalder and Pigneur, 2010) through the examination of its nine building blocks with respect to the business focus, locus and modus. The framework could be additionally enhanced by exploring the internal fit of the business model elements as well the external configurational fit of the business model to the business models of all relevant stakeholders, including global partners, suppliers and customers (Nenonen and Storbacka, 2010). Last but not least, one could adopt existing models examining how the degree of firms internationalisation and innovation (for example, radical versus incremental) affect the allocation of resources, the assignment of activities and the nature of the relationships within the context of a specific global ecosystem (Chetty and Stangl, 2010).

Another link between the two fields of research is that they both offer an opportunity to further substantiate the potential of an effectuation-based view on entrepreneurship theory and practice by challenging the traditional way of seeing entrepreneurship as a rational, strategic process where opportunities are discovered through a well-planned search process (Sarasvathy et al., 2014; Sarasvathy, 2001). Effectuation means that entrepreneurs start with a generalised idea and then attempt to work towards that idea using the resources they have at their immediate disposal (i.e., who they really are, what and who they know best). The strategy of a new firm is not clearly envisioned at the beginning, and the entrepreneurs and firms using effectuation processes can to a large extent remain flexible, take advantage of new ideas and opportunities as they arise, and constantly learn.

It should be pointed out that there are also some important distinctions between the research trends related to BG firms and LSs (Table 1). Interestingly, these distinctions could offer additional opportunities for the exploration of potential synergies between the two research fields. For example, although in both cases there is a focus on niche markets as the main target of firms' products and services, the emphasis is completely different. While the LS approach focuses on the challenges associated with crossing the chasm between early enthusiasts and early adopters, i.e., on *developing* the first substantial market niche that would validate and economically fund the development of the whole product (Blank and Dorf, 2012; Wiefels, 2002), the research on BG firms seems to focus on market entry strategies (Burgel and Murray, 2000). On the other hand, the LS approach takes a definite new product development perspective and focuses on the challenges associated with moving across the stages of a specific TALC, while BG firm research rarely discusses the challenges associated with the development of new offerings and focuses on the global marketing impact of technology excellence and the degree of product innovation. In other words, BG firms seem to be considered in an entirely SME context which is different from the context of a start-up (Coviello, 2015). The reason for this difference is mainly historical since the BG concept emerged within the context of IB research focusing on retrospective studies of the process and antecedents of internationalisation. This fact is reflected in the ongoing critical discussion of the differences between BG firms and INVs

Table 1. Comparison between LS and BG firm research trends.

Dominant LS trends	Dominant BG firm trends
Technology startup context	SME context
Niche market development	Niche market entry
Technology life cycle marketing/ crossing the chasm	Global marketing
NPD/prototyping/experimentation/agile development	Innovative products/technology excellence
Business model emergence	Business model development/adoption
Hypothesis-driven entrepreneurship	Exploring the value of effectual entrepreneurship but seem to focus on causal, goal-driven internationalisation strategies
Ex ante perspective	*Ex post* perspective

(Coviello, 2015). According to Coviello, the influential article of Knight and Cavusgil (2004) has effectively popularised the original notion of a BG firm suggested by Rennie (1993) by referring to a BG as a young firm that is active through early export sales (see also Cavusgil and Knight, 2015). The focus on export sales is narrower than Oviatt and McDougall's (1994) definition of INVs as firms that coordinate multiple value chain activities across borders, including both upstream and downstream assets, processes and relationships. The upstream assets, processes and relationships are related to, for example, technology and product development, the production of working prototypes, etc. The downstream ones are related to final production, distribution, and service.

Coviello (2015) argues that using the terms INV and BG synonymously or interchangeably is inaccurate since these organisational forms are simply different. More importantly, "[i]f the study captures multiple and global value chain activities very close to birth, the term 'global start-up' from Oviatt and McDougall's (1994) typology of INVs is more appropriate." (Coviello, 2015: 21)

There is also a distinction in terms of business model perspectives. While the LS method focuses on the emergence of business models (Blank, 2013), the research on BG firms tends to speak about business model adoption or replication (Dunford *et al.*, 2010). Last but not least, the LS approach emphasises a hypothesis-driven entrepreneurial perspective (Eisenmann *et al.*, 2011), while BG scholars, although discussing the appropriateness of an effectual approach to BG entrepreneurship (Sarasvathy *et al.*, 2014; Andersson, 2011), seem to adopt a more causal, goal-driven perspective on early internationalisation (Chetty and Campbell-Hunt, 2004). A closer examination of Table 1 suggests that the LS approach employs an *ex ante* perspective on business development which is associated with firms anticipatory actions aiming at future sustainable growth, market development and ultimate business success, while BG research has been historically more retrospective by focusing on the antecedents of rapid internationalisation and the *ex post* characteristics of BG firms.

The LGS — A new type of firm?

The distinction between *ex ante* and *ex post* perspectives is an important aspect (Schmidt and Keil, 2013) which offers the opportunity to discuss two different paths in linking lean and BG strategies in new technology firms and thus helping the emergence of LGS as a new type of firms. The first path is associated with the opportunity for generic LS to go global by undertaking a rapid internationalisation strategy. Such L2GS establish themselves by using a generic LS approach on a local or national level and then engage into a more traditional BG journey by exploring internationalisation opportunities short after inception. The second path is associated with the opportunity for global start-ups (in the terminology of Oviatt and McDougall, 1994) to adopt the LS approach since their very inception by seamlessly synergising their global and lean activities. It might be appropriate for such new firms to be qualified as being both lean and global from the start (Tanev, 2012) or as L&GS. Since a G2L path does not seem to be probable in the case of start-ups, one could define the LGS by using the following symbolic equation: LGS = L2GS $\oplus$ L&GS.

On a more fundamental level, the LS approach appears to be inherently related to the relational and global business aspects of new technology firms. It includes firms ability to: (i) choose the operational focus, activities, internal resources, capabilities and assets that it is best at, and (ii) look for complementary external resources and partnerships (including global resources and partners) in order to complement their specific business and operational priorities. It is exactly their specific business and operational focus that predetermines the necessity and the relational nature of their external (including global) resource allocation process, as well as the wider lens operational, co-innovation and commercialisation perspectives (Adner, 2013). It is not by accident that the wide lens approach to innovation has introduced the concept of minimum viable ecosystem (MVE) — the simplest ecosystem possible that still creates new value (Adner, 2013: 198). The MVE aims at creating value as early as possible which allows for a staged expansion of the project and its corresponding ecosystem at later moment of time. It extends the concept of minimum viability to the business ecosystem level. In other words, the MVP approach naturally leads to the need of adopting a MVE approach which goes ones step further since by establishing a first base of customers it "reduces the demand uncertainty for partners and lowers the hurdles to bringing them on board" (Adner, 2013: 203). The distinction between the operational, co-innovation and adoption risks incorporated in the wide lens approach could be quite helpful in the context of LGS since it allows for the differentiation between the upstream and downstream aspects of the early internationalisation process. In addition, it allows for the integration of the LS focus on hypothesis-driven entrepreneurship (Eisenmann *et al.*, 2011) to the effectual entrepreneurial attitude of BG firms (Sarasvathy *et al.*, 2014; Andersson, 2011). Last but not least, it allows for the adoption of a common marketing perspective including insights from ongoing research in entrepreneurial (Mort *et al.*, 2012), effectuation-driven (Read *et al.*, 2009) and technology (Yadav *et al.*, 2006) marketing approaches.

The reflections suggested above suggest several different aspects that could be used as a way of conceptualising LGS as a special new type of firm: (i) the emergent nature of their business models where every specific business model framework is becoming just a template for the development of a viable business model; (ii) the inherently relational nature of the (global) resource allocation processes; (iii) the integration of the hypothesis-driven and effectual entrepreneurial perspectives; (iv) the integration of the entrepreneurial, effectuation-driven and technology marketing perspectives; (v) the need to deal with the high degree of uncertainties associated with the overall business, marketing, technology innovation and operational environment, including the uncertainty associated with cross-border business operations.

Uncertainty versus risk management in the context of LGSs

The focus on the high degree of uncertainties is particularly relevant within the context of new technology firms (Yadav et al., 2006). Moriarty and Kosnik (1989) identify two different types of uncertainty that could be directly related to the context of lean technology start-ups — market uncertainty and technology uncertainty. The market uncertainty is associated with issues such as: The kind of needs that are supposed to be met by the new technology and how these needs would change in the future; whether or not the market would adopt industry standards; how fast will the innovation spread and how large is the potential market (Yadav et al., 2006). The technology uncertainty is associated with issues such as whether or not: the product will function as promised; the delivery timetable will be met; the vendor will give high-quality service; there will be side-effects of the product; the new technology will make the existing technology obsolete (Yadav et al., 2006). A third type of uncertainty could be associated with the competitive volatility of high technology markets (Shanklin and Ryans, 1987). It refers to changes in the competitive landscape such as identifying emerging competitors, their product offerings or the tools they use to compete (Yadav et al., 2006). For example, quite often new technologies are commercialised by companies outside of a specific industry or sometimes a key partner at the early stages of a firm may later become a competitor (Cooper and Schendel, 1976). The new players are not viewed as disruptive and are frequently dismissed by both new firms and incumbents. However, they end up changing the ways of competing and may lead to reshaping the industry rules of competition for all other players (Hamel, 1997). Furr and Dyer (2014) have pointed out that many scholars do not realise the extent of the increase in uncertainty over the past thirty years. The greater degree of uncertainty has created the need to change the way most organisations and especially start-ups are managed. According to Furr and Dryer there are three types of uncertainty that influence firms ability to create customers: Demand uncertainty, technological uncertainty and environmental uncertainty which is associated with the overall macroeconomic environment and government policy. These three types could be considered as corresponding to the technology, market and competitive uncertainties that were discussed above (Yadav et al., 2006).

Adner (2013) suggests another way of conceptualising uncertainty and risk by emphasising that an overemphasis on the challenges associated with internal execution creates a blind spot by hiding key dependencies that are equally important in determining success and failure and makes companies fall victims to the innovator's blind spot by "failing to see how their success also depended on partners who themselves would need to innovate and agree to adapt in order for their efforts to succeed" (Adner, 2013: 4). By adopting an ecosystem perspective, Adner takes into account two other types of risk in addition to the one associated with the focus internal execution — the uncertainties rooted in co-innovation and adoption chain risks. Co-innovation risk is the extent to which the success of an innovation depends on the successful commercialisation of other innovations. Adoption chain risk is the extent to which partners will need to adopt firms innovation before end consumers have a chance to assess the full value proposition. Examining these additional types of uncertainty could help LGS to better address hidden dependencies. Interestingly, Girotra and Netessine (2014) have suggested recently a classification of risk as part of a business model development and innovation framework that could be related to the classification suggested by Adner (2013). According to them the key choices executive managers or entrepreneurs make in designing a business model either increase or reduce two characteristic types of risk — information risk, when one makes strategic operational decisions without enough information, and incentive-alignment risk, when entrepreneurs need to make assumptions about the expected incentives of all the relevant stakeholders involved in the company value creation network. Girotra and Netessine (2014) are fully aware that there are other types of risk such as financial and technological but believe that by mitigating information and incentive-alignment risk firms can improve their ability to tolerate all other risk categories. The reason to mention the recent work of Girotra and Netessine is two-fold. First, their framework relates the management of risk and business model development which is quite relevant for the context of LGS. Second, the explicit articulation of incentive-alignment risk provides additional support to the classification of risk as part of the wide lens approach to innovation suggested by Adner (2012) which specifies the need for the alignment of the relative benefits of all potential stakeholders involved in the business ecosystems of both LGS and existing firms. Table 2 summarises the aspects of risk/uncertainty that have been discussed so far in the context of LGS.

Table 2 could be used as a tool for the analysis of specific ways for LGS to deal with uncertainty and risk. The table includes some specific examples from the six firms described in greater detail in Appendices A–C. The main value of the suggested uncertainty/risk framework consists in decoupling the risks associated with the three typical types of business interactions (related to internal operations, external collaborative innovation and adoption) form the three types of generic uncertainties associated with technology development, market development and business competition. The two-dimensional presentation of risks versus uncertainties would definitely allow for a more systematic analysis of the relational aspects (including the global aspects) of LGS within the context of their specific technological, market and competitive issues in the way they appear as part of their cross-border business operations. The cross-border business aspect will only inject some additional degree of uncertainty that could be taken into account as part of

Table 2. Types of risks and uncertainties that could be used for the analysis of emerging LGs business models. Information about the different firms (cases) can be found in Appendices A (cases A & B), B (cases C & D) and C (cases E & F).

Risks	Uncertainty		
	Technology uncertainty	Market uncertainty	Competitive uncertainty
	Whether the product functions as promised; meeting the delivery timetable; the vendor provides quality service; product side-effects; technology makes the existing one obsolete.	What are the needs to be met by the technology; how these needs would change; adoption of industry standards; speed of innovation adoption; potential market size.	Newly emerging competitors; new competitive product offerings; new ways and tools used by others to compete.
Execution risk What it takes to deliver the right innovation on time, to specs and beat the competition?	**Case C:** Field tests with major global customers to mitigate both technological uncertainty and execution risk.	**Case B:** Accepted a longer time to market in order to better deal with execution risk. **Case D:** Multiple pivots to minimize market uncertainties and monetise VP.	**Case C:** Competition from Google and Apple forced the firm to focus their value proposition on platform independence and streaming speed.

(*Continued*)

Table 2. (*Continued*)

Risks	Uncertainty		
	Technology uncertainty	Market uncertainty	Competitive uncertainty
Co-innovation risk Extent to which the success of firms innovation depends on the successful commercialisation of other innovations.	**Case A:** Local co-innovation challenges and risks made the firm look for global partners. **Case C:** Global crowdfunding mechanism to reduce risks by enabling further development.	**Case B:** Using a global R&D office as a way of turning upstream partnership and resources into downstream global marketing opportunities.	**Case B:** Using intensive IP protection to reduce chances for partners to become competitors. **Case F:** Securing IP agreement with local university professor to gain first mover competitive advantage on a global scale.
Adoption chain risk Extent to which partners will need to adopt firms innovation before the end consumers have a chance to assess the full value proposition.	**Case A:** Using modular flexible design focusing on convenience and ease of use facilitated adoption by distributors and end users.	**Case A:** Using integrators for end user market development. **Case D:** Securing agreement with global e-commerce platform to drive adoption by target market niche. **Case F:** Using global contact to establish local partnership leading to funds allowing the cooperation with a major lead user.	**Case C:** Focus on platform independence and device operation speed to attract market adoption as well as position for a potential acquisition by big competitors.

the framework suggested Table 2. As Shrader *et al.* (2000) suggest, successful entrepreneurs should manage the multiplicity of risks involved in cross-border businesses by exploiting simultaneous trade-offs between these risks.

Finally, it should be pointed out that the specific focus on the systematic management of uncertainties has been identified as a major issue by all scholars and practitioners inspired by the LS paradigm. For example, Scott Antony (2014: 14) speaks about the need to adopt a scientific method to the management of strategic uncertainty by referring to the works of Mintzberg and Waters (1985) on the distinction between deliberate and emergent strategies, of McGrath and MacMillan (1995) on discovery-driven planning, of Blank (2013)[3] and Ries (2011) on the experimentation aspects of the LS approach, as valuable resources in the development of a more systematic way of addressing the multiple uncertainty issues which are typical of LGS. All analytical frameworks focusing on dealing with uncertainty should make a clear distinction between uncertainty and risk which appears to be highly relevant within the context of LGS (Schmidt and Keil, 2013). In a situation of risk, the future is known in terms of statistical probabilities expressed in the form of means and distributions. Uncertainty, on the other hand, characterises a situation where neither means nor distributions can be known. In an environment characterised by uncertainty, it is not possible to have more accurate information in a strict sense. What really matters is the interpretation of information. Firms and managers use pieces of information to classify future states subjectively and form beliefs about these states or estimates of them, essentially transforming the situation of uncertainty into one of risk (Schmidt and Keil, 2013). The way entrepreneurs do that is through their personal judgment, i.e., their ability to integrate many bits of information, to view objectively the multiplicity of the aspects of their particular situation, and to creatively conceptualise alternative feasible futures. The emphasis on the ability of managers to transform situations of uncertainty into situations of risk is particularly relevant within the context of LGS. It provides a very good reason to believe that the framework suggested in Table 2 could become a very useful tool for LGS managers.

Research Methodology

The first step in the research process was to examine the literature in order to identify some of the key crossing points between the two research streams focusing on BG firms and the LS approach. The outcome of this first step was not only the identification of specific research gaps and existing theoretical models but, before anything else, on conceptually justifying the need of introducing the LGS concept. In addition, the analysis of the literature resulted into a number of exploratory lenses that were used in the second (empirical) component of the study (Table 3), including: (i) business model emergence; (ii) characteristics of BG technology firms and the challenges of early internationalisation; (iii) degree of innovativeness and the classification of innovation; (iv) technology commercialisation strategies; and (v) characteristics of LS. The exploratory lenses were used as

---

[3] See also Blank and Dorf (2012).

Table 3. Exploratory lenses used in the construction of the interview guide.

Exploratory lenses	Categories	Key references
Business model emergence	Value proposition — market offer, target customer, job to be done Key resources Key partners/partnerships Key processes Key activities Distribution channels Revenue model	Johnson et al. (2008); Onetti et al. (2012); Osterwalder and Pigneur (2010); Trimi and Berbegal-Mirabent (2012)
Early internationalisation	Int. customer segment/target customer International partners International resources International activities Degree of internationalisation BG firm characteristics	Cavusgil and Knight (2015); Zijdemans and Tanev (2014); Tanev (2012); Chetty and Stangl (2010); Cavusgil and Knight (2009); Kudina et al. (2008); Gabrielsson and Kirpalani (2004); Knight and Cavusgil (1996); Rennie (1993)
Innovativeness	Incremental/radical innovation Innovation adoption activities R&D activities/spending Intellectual property rights (IPR)	Chetty and Stangl (2010); Garcia and Calantone (2002)
Technology commercialisation strategy	IPR protection strategy/degree of protection against competitors Appropriability regime Dependence on others to commercialise Commercialisation partnerships (upstream/downstream)	Cohendet and Pénin (2011); Teece (2006); Yadav et al. (2006); Gans and Stern (2003); Teece (1986)
LS characteristics	MVP Customer development process Pivot/iterations Prototyping and hypothesis testing Agile development	Kerr et al. (2014); Blank (2012); Blank and Dorf (2012); Moogk (2012); Moore (1991); Ries (2011); Stolze et al. (2014); Lemminger et al. (2014)

basis for the development of an interview guide. The focus of the empirical research component was on new technology firms that have been established within the last 5 years with a clearly expressed orientation to early internationalisation and global markets or on the first 5 years of older firms which have already internationalised within an upstream and downstream context. The sample included three firms from Denmark and three from Canada (Ottawa-Carleton region in Ontario).

The first Danish firm (Case A in Appendix A) develops customisable flexible robot arms helping firms to improve their productivity and profitability while providing a safer work environment for their employees. The second firm (Case B in Appendix A) focuses on developing advanced technology-based water purification solutions which are appropriate for the desalination of seawater through the use of industrial biotechnological techniques. The third firm (Case C in Appendix B) offers wireless streaming solutions enabling real time content sharing and easy collaboration with friends, colleagues or partners by streaming from nearly any device right up onto to a main screen at a specific customer location.

The focus of the Canadian firms is on cloud mobile solutions. The first firm (Case D in Appendix B) provides automated solutions enabling the instantaneous creation of e-magazines as a way of enhancing the experience of e-shoppers and increasing the revenue of e-commerce providers. The second firm (Case E in Appendix C) offers software solutions enabling the management and tracking of electronic signature workflows combined with an identity authentication method that allows online signing and authorisation anywhere in the world by computer, phone or any mobile device that is web-based or cloud-based. The third firm (Case F in Appendix C) delivers effective news media and government transcript monitoring in real-time together with a powerful analysis providing actionable insights for both individuals and organisations possessing a need to monitor web or government updates for the sake of crisis control or opportunity identification. The Canadian firms have been part of the lead-to-win business creation programme which is driven by the Technology Innovation Management Programme at Carleton University, Ottawa, Ontario. The six firms were intentionally selected from two different global locations and from a variety of technology sectors in order to enhance the validity of the research findings. The selection of the two global locations was additionally justified by the degree of access to information and the willingness of the executive managers to cooperate in the data collection.

The interview recordings were transcribed and the transcripts were examined in the light of the exploratory lenses in order to generate a list of key observations for each of the six cases. The key observations were used for the identification of insights that could help in substantiating some first answers to the initial research questions. It should be pointed out that at the present stage of the project, the case studies were used mostly as a way to validate the overall research methodology. In this sense, the main contribution of this paper is conceptual and the results summarised in the next section are of preliminary nature. A more systematic data analysis should be based on more detailed case studies that will be presented in future publications.

Summary of Research Findings

The findings discussed in this section are based on the observations about the six firms such as summarised in Appendices A–C. The discussion below will highlight several key themes by directly or indirectly referring to the information provided in these appendices.

A close relationship between lean and global

One of the main findings of our research consists in the realisation that in all the cases the technology start-ups had a clearly expressed international or global orientation in combination with different kinds of pivoting and an explicitly expressed strong desire to keep core competences, know-how and IP in-house. It appears to be natural for the technology start-ups included in the sample to adopt a lean approach to product and market development by focusing on the business processes and activities they are most good at, and it is exactly this focus which makes them look outside of their moving boundaries for both local and global resources and partners that could substitute, complement or enhance the ones that are available to them. This could be explained by the scarcity of resources and the uncertainty of the overall operational and business environment which is characteristic of both the start-up phase of the businesses and the early internationalisation attitude of INVs and BG firms (Chetty and Campbell-Hunt, 2004; Gabrielsson and Kirpalani, 2004; Knight and Cavusgil, 2004, 1996; Oviatt and McDougall, 1994). It could be also explained by the emerging LS culture leading to the adoption of an entrepreneurial logic focusing on experimentation (Blank, 2003; Kerr et al., 2014) as well as by the growing relevance of continuous business model innovation in both scholarly literature and entrepreneurial practices (Ries, 2011; Trimi and Berbegal-Mirabent, 2012). Last but not least, the openness, the relational and co-creative aspects of new technology firms seem to reflect the emergence of new innovation paradigms focusing on collaboration, user involvement and co-creation (Knudsen et al., 2013). The adoption of the LS approach appears to be inherently associated with the adoption of a relational or network approach to innovation and commercialisation through the access of international or global resources, partners and networks. The preliminary results of our analysis suggest that the current excitement with the LS approach naturally leads to a focus on global. This is a point that was also emphasised recently in the blog of Steve Blank who introduced the slogan "Born global or die local".[4] In other words, all the above points seem to justify our motivation to link the two research streams by considering the LGS as a new type of firm.

The role of IP

The relationship between lean and global opens up the discussion of an additional issue which is related to the role of IP in LGS. In all of the cases, there was a clear tendency to keep IPR within the control of the firm. This finding might be found to be counter intuitive to the above mentioned relational aspects of the start-ups since the traditional view of patents is usually emphasising their importance in excluding imitators and preserving the incentives to invent on the individual firm level. Such understanding however is quite restrictive and runs counter to some of the research conducted in the past few decades indicating that one should acknowledge a dual role for patents — they can increase incentives to innovate but they can also mitigate the specific coordination difficulties linked to open innovation (Cohendet and Pénin, 2011). It is true that LGS exchange knowledge and technologies collaborating and co-innovating formally and informally with both local and global partners. They could therefore very likely encounter challenges in negotiating the IPR with their partners. It is also true however that a properly developed patent system could help in dealing with partnership development problems. The main role of the patent system in this situation could be found in providing the proper ground for including all the relevant stakeholders in the innovation process (Cohendet and Pénin, 2011). This ground emerges out of a fundamentally important feature of a patent — it both protects and discloses an invention. The coupling of these two properties mediates or facilitates technology transfer through the exchange of licenses on markets for ideas and technology (Gans and Stern, 2003; Cohendet and Pénin, 2011) but also helps in framing collaborations, partnerships and alliances among heterogeneous organisations.

What is more interesting however is how the specific IPR attitude relates to a particular early internationalisation strategy. Two examples could be relevant to mention. In Case B (Appendix A) the firm adopted a very aggressive IP strategy

---

[4] http://steveblank.com/2014/10/31/born-global-or-die-local-building-a-regional-startup-playbook/.

resulting in more than 40 patents worldwide positioning the firm as world leader in the field of water filtration. The patents were the result of an intense upstream global cooperation soon after its inception. The success of the IP strategy led to the emergence of a complementary business model based on revenue from offering knowledge-based services and expertise even before the product launch. The initial idea to sell 100% rights of using the technology but now there are multiple other options resulting in additional joint ventures. Today, 10 years after inception, the firm is finally moving globally downstream working on field trials with major global customers to prepare the commercial launch of their product. This example illustrates a close interrelation between IP strategy, lean and global aspects. The second example is related to Case E where the Canadian firm focusing on e-signatures used a local IP agreement with a university professor which automatically made it a competitive first mover on a global scale by becoming the only firm meeting the hard regulation requirements in the EU. This is another example of how an IP strategy could become a global enabler.

The distinction between upstream and downstream global resource allocation

The analysis of the six cases suggests that in discussing LGS, one should make a clear distinction between upstream and downstream resources and globalisation efforts. This distinction is directly associated with the definition of INV as firms "seeking to derive significant competitive advantage from the use of resources and the sale of outputs in multiple countries" (Oviatt and McDougall, 1994: 49). It is just another illustration of the fact that INV terminology appears to be better suited to address the characteristics of truly BG firms (Coviello, 2015). The distinction appears to be quite appropriate in addressing the context of LGS. For example, none of the firms included in our sample have started globally by the simultaneous actualisation of both upstream and downstream resources. This of course is not surprising for start-ups but it demonstrates the value of the distinction. Only two of the firms (Cases C and D, see Appendix B) could be qualified as both L&GS (see the definition of L&GS suggested above) and it is in terms of their upstream resources. The other four firms could be qualified as having started in a lean way and later engaging into a lean-to-global journey (see the definition of L2GS suggested above). They have however also followed an upstream to downstream path. An interesting example represents Case B (Appendix A) where the firm used its global upstream subsidiary in order to launch a downstream (sales) office. In other words, the empirical findings suggest that without such distinction it would be impossible to describe the challenges of LGS.

Global by necessity rather than through preliminary articulated deliberate strategy

We have found that in some of the cases the adoption of a LGS approach is not the result of a deliberate strategy from the inception of the firms but could be rather considered as a "desperate" move motivated by the inability of the firms to find reliable local partners. It appears that firms need to "struggle" locally before they realise the need and the benefits of early internationalisation. For example, the Danish robotic firm (Case A) engaged into partnerships with international partners in Germany and the USA only after it failed at reaching its specific production requirements with local partners. The difficulty in identifying reliable local partners was due to the production challenges associated with ensuring tight tolerance precision machining in combination with a high mechanical flexibility in the operation of the robot arms. In this sense the ultimate reason was based on the innovative characteristics of the robot arm system which combined both flexibility and precision in order to get automation down to the people on the production floor, so that they could repair, program and adjust the system on their own instead of relying on experts. This is an example of how the degree of innovation could become a driver for internationalisation but also how a more deliberate early internationalisation could have been a better strategy. The example shows that a more radical innovation might be associated with faster upstream but slower downstream globalisation. In all cases globally oriented firms should deliberately enhance their global network relationships in order to achieve more radical internationalisation and innovation (Chetty and Stangl, 2010).

Our findings indicate the importance of resource complementarity for the proper selection of both upstream and downstream global partners (Schmidt and Keil, 2013; Wilson and Doz, 2011). The ability of LGS to conceptualise resource complementarity as part of their global partnership development process should become the subject of future studies focusing on the development of practical LGS entrepreneurial frameworks. As a final point in this section, one should also point out that a more careful approach to early internationalisation could be considered as quite natural for a lean start-up, i.e., as part of a lean globalisation approach which is trying to avoid waste by effectuating the minimum possible resources very soon after firms inception.

The value of early business supporting programs

Another interesting finding identified the tendency of LGS to use the multiplicity of regional, national and IB supporting programs and funding mechanisms to help their financial and R&D needs. The Danish firms have greatly benefited from their involvement in EU projects supporting the R&D and the commercialisation of new products and services. The Canadian firms have enjoyed the benefits of the national R&D tax return programme, the resources of local business accelerators and the expertise of public organisations offering entrepreneurial support. For example, the Canadian firm focussing on the development of cloud-based mobile tools facilitating online shopping through the creation of e-magazines (Case D) became global by starting a firm in Canada on the basis of existing subsidiary in Cambodia. Since its inception in 2011, the newly created Canadian firm has had several pivots and almost all of them were conceptualised through the active participation in the lean-to-win program which is driven by the Technology Innovation Management Master's program at Carleton University in Ottawa, Ontario. Another Canadian firm (Case F) has been successful in using the Build in Canada Innovation Program which helps companies in bridging the pre-commercialisation gap by offering them the opportunity to test their new products or services by ensuring a first entry to the Canadian Federal Government marketplace. Interestingly, the link to the Canadian Government organisation which supported the firm in getting this important funding was made possible through a contact of the CEO in Denmark. This is an interesting example of how a global resource could become the trigger of local funding. It could be pointed out that business supporting programmes should be conceptualised as an important part of the overall effectual entrepreneurial environment of LGS (Sarasvathy et al., 2014; Sarasvathy, 2001).

Differences between Canadian and Danish start-ups

We have found an interesting difference between the Danish and the Canadian firms. The pivoting of Danish firms did not affect their overall value proposition which was not the case in the Canadian start-ups. All Danish firms were founded around a well-thought initial value proposition which did not change very much over time, i.e., their business model evolved only in terms of the specific ways used to deliver the initial value proposition (Onetti et al., 2012). This finding could be explained by the difference in the nature of the businesses in the two samples. The Canadian firms were all within the ICT domain with a major focus on cloud and mobile technologies. Such technologies allow for a significant degree of flexibility and quick changes of the value propositions and the characteristics of the target customers. On the other hand, all Danish firms have engaged into the development of more tangible physical products which could be associated with less flexibility in pivoting. One should also point out the role of the Canadian lean-to-win program which has an explicit focus on supporting the early internationalisation of its start-ups and have definitely influenced the pivot frequency of the Canadian firms. The higher pivoting frequency suggests the Canadian firms may provide a more appropriate empirical base for the integration of the hypothesis-driven and effectual entrepreneurship perspectives (Eisenmann et al., 2011; Sarasvathy et al., 2014).

Conclusion

The present study suggests that for a newly established technology firm being global and innovative at the same time must be seen as part of one single process. This is why we suggest the introduction of the concept of LGS as a way of emphasising the impossibility of dealing separately with innovation, business development and early internationalisation. As it appears, for most technology start-ups being lean necessarily leads to being global. It is true that going global has its own characteristic time periods and challenges but they cannot be considered aside of the establishment of the firm as such. We suggested two types of early internationalisation paths: Lean-to-global (L2G start-ups) and lean-and-global (L&G start-ups). Two of the firms in the research sample were found to be of L&G type. The other four firms were found to be of the L2G type. The analysis showed that it is important to distinguish between the upstream and downstream global resources — a distinction without which it is impossible to conceptualise the early internationalisation of LGS. The firms included in the research sample have had problems dealing with the complexity, uncertainties and risks of being innovative on a global scale. Some of the specific ways of addressing these problems include a disciplined knowledge and IP protection strategy, the efficient use of business support and public funding mechanisms, and pivoting around the ways of delivering a value proposition and not around the value proposition itself. In addition, all the firms have managed the different types of uncertainties by moving one step at a time in a way that they could maximise the value of newly emerging relationships. The results provide an empirical basis for further integration of the hypothesis-driven logic of the LS approach with effectuation-based theoretical perspectives of BG firms.

The findings will be relevant for company founders, executive managers, investors and venture capitalists or anyone dealing with innovative and global firms which are in the early stages after their founding. Some of the key managerial implications would be that executive managers of LGS have to learn to change roles and management styles when the firm is moving from an innovation (upstream) to marketing and sales (downstream) mode of operation. Second, they need to learn doing that under the conditions of multiple uncertainties which should be managed in a systematic way. The uncertainty versus risk framework suggested in this study provides opportunities for future research focusing on more practical global entrepreneurial frameworks that could be used by both LGS and business development agencies supporting their globalisation efforts. The results will be also of interests for scholars interested in innovation management, global technology entrepreneurship and IB management.

Acknowledgments

The authors express their gratitude to Professor Tony Bailetti, Director of the Technology Innovation Management Program at Carleton University, Ottawa, Canada, for the fruitful discussions and for enabling the data collection from the Canadian firms discussed in this study. We are also grateful for the financial support from the Danish Entrepreneurship Foundation which allowed one of us to participate in and present a first version of this paper at the ISPIM Americas Innovation Forum 2014, Montreal, Canada, 5–8 October, 2014.

Appendix A. Summary of Information About Cases A and B

Case	Nature of business	Year of inception	Year of international/global engagement			Innovativeness	IP protection	LS aspects
			Upstream	Downstream	Internationalisation			
A	Small, flexible and affordable easy-to-use handling robots based on a modular design. Focus on automation for manufacturing and assembly industries. Target global customers, who develop, integrate or customise solutions for end users. Software is free. Money is made on robots only.	2005	2007: International and global partnerships with component suppliers. Inability to keep component suppliers in close proximity due to problems meeting high quality production requirements. Deliberately avoiding consultant firms. Substantial use of EU R&D funding.	International sales start in 2009. Global sales start in 2011.	Rapid internationalisation since 2009. As of 2014, a global network of distributors with 200 sales partners in 50 countries. Today 95% of sales go to export. International experience of managers critical for globalisation.	Medium degree of innovation. Programming is large part of the innovation. Key focus of innovation on quick change time, ease-of-use, customisation, programmability & convenience. OS software innovation practices.	Special care of keeping know how in house. R&D and production kept locally because "knowledge could move out with production." Relying on national and EU strategy to keep advanced manufacturing in Europe.	Early prototyping practices including multiple iterations. No major pivot since inception. Continuous improvements of technology and software. Starting local and then moved to global. Open source software practices.

(Continued)

Appendix A. (*Continued*)

Case	Nature of business	Year of inception	Year of international/global engagement			Innovativeness	IP protection	LS aspects
			Upstream	Downstream	Internationalisation			
B	Advanced biotechnology for water purification and desalination. First mover in this technology area. Focus on large customers as end users and thus on systems vendors who could integrate the technology into solutions for the larger firms.	2005	EU funding in 2006 involving 10 other universities in intense R&D activities. Major funding and R&D collaboration with Singapore in 2009. Using strategic global partnership agreements to consolidate the position of the firm as market leader in its technology area. Choosing top global suppliers with highest degree of business model complementarity and potential for synergy.	Not yet but R&D office in Singapore prepared to launch Asian sales soon. A successful field tests in 2010 with major future global customers allowing the adjustment of specs to customer needs. Using word-of-mouth by global experts & consultants to attract potential new customers.	In a product development phase with a high degree of upstream internalisation. Using upstream resources to prepare global market development. Activating subsidiary in Asia to prepare access to major markets in India, China and Japan. Activities are managed through EU head office.	High degree of innovation requiring intense and longer R&D involving multiple international and global partners. 2012: Ambition to move from an R&D focused firm to a global commercial organisation rooted in R&D and ground-breaking innovation.	Disciplined international IP protection strategy following "the pharma" way and positioning the firm as world leader in the field of water filtration. Interested to follow an "Intel inside" strategy for the technology. Application for trademark to enhance international market value of the firm. Currently more than 40 patents worldwide.	2010: Major pivot with respect to the core technology but not with overall value proposition. Emerging complementary business model capitalising on offering knowledge-based services and expertise even before product launch. Initial idea was to sell 100% rights to use the technology. Now there are other options resulting in multiple joint ventures.

Appendix B. Summary of Information About Cases C and D

Case	Nature of business	Year of inception	Year of international/global engagement			Innovativeness	IP protection	LS aspects
			Upstream	Downstream	Internationalisation			
C	Platform independent wireless HDMI streaming solutions enabling real time sound, image, video and screen sharing between mobile devices and HDTV screen monitors. Competing with big players like Google and Apple by offering platform independence, ease-of-use and better functionality. Targeting both consumers and enterprises by developing a pro-product version. Customer homogeneity based on similarity of their mobile technology.	Initial idea in 2011 and incorporation by the end of 2013.	2014: Establishing through an extremely successful crowdfunding on a global scale. 8th most successfully funded project on a well-known crowdfunding platform. Crowdfunding success ensures support from major platform players. Both founders and partners have had international experience.	Plans for future production in China. A focus on large scale global adoption starring with the USA and then moving to local market. The target is to have 98% international and global sales.	Currently, the product is under development. Global efforts focusing on upstream resources. Public release expected in Q2 of 2015.	Medium degree of innovation with a focus on user-friendly interface. Most innovation is in the software. Best Startup of CES 2014.	Medium protection efforts with a focus on keeping software in house as core asset. Hardware production is outsourced restricting suppliers from production. Focus on the innovative environment for employees.	Using a MVP approach, multiple prototypes and open source software. Most product features were crowdsourced before implementation. "Listening to your customer real time from the very start."

(*Continued*)

Appendix B. (Continued)

Case	Nature of business	Year of inception	Year of international/global engagement					
			Upstream	Downstream	Internationalisation	Innovativeness	IP protection	LS aspects
D	Automated solutions enabling the instantaneous creation of e-magazines and one-click online brochures for e-commerce businesses. Focus on platform independence for both mobile and web-based solutions. Exploring market value from multiple dimensions— content, marketing and events. Design is not done by the end user but by the engine itself.	2012: Inception after participating in a business startup programme in 2011.	2012: Global product development from inception based on ownership of software development firm in Cambodia. The business startup program encouraged a born global start but it was very hard to make it work. Finally, leveraged the resources of a global e-commerce platform.	First deals in 2014. Created an App with a plugin for a global e-commerce platform enabling the one click creation of electronic brochures.	A global start in terms of upstream resources and growing downstream globalisation through e-commerce platform. Could get global market access by being lean alone. The lean approach needed to be enhanced by global aspects.	First idea was too radical in terms of technology, amount of funding needed and user behaviour change. The next idea was to rethink the web by separating the data domain from the presentation domain and be able to change the look on the fly. "That's radical innovation." Using existing open source assets with appropriate licenses and innovating on top of them. Disruption based on convenience and speed.	Keeping IP and innovation in the head office focusing on user experience and overall architecture design. Only software coding was kept outside in a global location.	Multiple pivots due to the radicalness of the offer, the lack of funding or business mentorship aiming at better monetisation of the value proposition. The nature of the technology allows for pivoting. Realised that content marketing is too big and focused on servicing e-commerce. Teaming up with other entrepreneurs in the business startup program to start a new venture based on the existing global resources. The revenue model evolved to subscription.

Lean and Global Technology Start-ups 233

Appendix C. Summary of Information About Cases E and F

Case	Nature of business	Year of inception	Year of international/global engagement					
			Upstream	Downstream	Internationalisation	Innovativeness	IP protection	LS aspects
E	Electronic signature workflows combined with an identity authentication method. Targeting heavy duty signing people in government, banks or law firms. "In 5–10 years this should become commodity." The business model is based on monthly subscription fee so there is real need to retain customer satisfaction. The focus is on cloud, mobile and ease-of-use independently of how difficult it is in the back end. A reasonable focus on customisation. Main focus on law firms in English speaking countries — USA, UK and Australia.	Late 2010.	Needed a partner who does both cloud and security and found it locally as co-founder. There are multiple needs but there is no money for that. Small 2011: US-based development partner (friendship based) with relevant expertise — cloud, mobile and security. Using local business startup program for mentorship and building a board of directors helping with partnership development.	The cloud-based solution is a global offer. It is accessible by anyone. Global internet sales start in 2013. All international/global partners are marketing and sales oriented. Using a local company with offices in the USA and UK. Addressing a repetitive pattern in law firm customers on a global scale.	Currently developing in one continent and selling worldwide but only in English. Sales in French to be opened soon.	Radical innovation that was not available before. Focus on technological differentiation expressed in a inscription patent offering legal advantage in EU. Cloud businesses are very different in terms of pricing and marketing. This is not just software, it is a service application.	Two key patents one of which is break through made possible through collaboration with a local university. It offers security protection and a unique competitive advantage by meeting both US and EU regulations. Nobody else has that.	Trying first collaborative cloud-based mobile solutions. Then moved to signatures where the key is in regulation. Since then the goal was set straight forward. Using resources suggested by board of directors to examine potential opportunities. Focus on English speaking countries first in order to minimise development resources.

(*Continued*)

Appendix C. (Continued)

Case	Nature of business	Year of inception	Year of international/global engagement			Innovativeness	IP protection	LS aspects
			Upstream	Downstream	Internationalisation			
F	Effective news media and government transcript monitoring and analysis in real-time. Focus on information sources that customers really care about and would pay for. Helping analysts to automate the process of cracking the media. It also helps customers to track and monitor the PR services by journalists they have paid for news releases. Emerging focus on younger companies who really need and search for competitive information. Differentiation from Google by focusing on Government customers. License based pricing.	2011	Local development resources only. Future global sales expansion will require inclusion of global media sources. One man start expanding to three partners by the end of 2011 one which was an angel investor who gave funding soon after inception. Additional funding in 2012 allowing the development of better user interface. Using Canadian market to step in the US one. Good chances for acquisition as potential exit strategy.	National sales focus. First local sale to Government in 2013 to bring some cash and helping to frame differentiation. More than five local customers in 2014 together with another 80 leads. First global customers in 2014. Support by business accelerators and R&D tax return programs was critical. Need to take into account the specifics of B2B marketing and help champions to sell internally.	Currently developing locally in English only. Moving to French soon to address both Canadian bilingualism and global opportunities. One global client so far. In 2014 used personal acquaintance in EU university to establish link to Canadian Gov. Department resulting in funds from innovation program focusing on Gov. users.	Medium to high innovation focusing on providing automation, lower costs, good information processing workflows, customised source setup and individual search index.	Search engine is based on patents filed in Canada and the USA allowing for high quality search, quick automation and customisation.	First idea was to help bloggers identify other bloggers. Then the focus was on enterprise solutions in order differentiate and be able to monetise. Started with a monitoring app for all the Canadian news. First prototype right after inception. Beta version from Fall 2012 to Summer 2013. Mistake to start with customers who were already paying for something similar where there is already a lot of competition.

References

Adner, R (2013). *The Wide Lens. What Successfull Innovators See that Others Miss*. Revised edition. New York: Portfolio/Penguin.

Alvarez, S and J Barney (2007). Discovery and creation: Alternative theories of entrepreneurial action. *Strategic Entrepreneurship Journal*, 1, 11–26.

Andersson, S (2011). International entrepreneurship, born globals and the theory of effectuation. *Journal of Small Business and Enterprise Development*, 18(3), 627–643.

Andersson, S and F Evangelista (2006). The entrepreneur in the Born Global firm in Australia and Sweden. *Journal of Small Business and Enterprise Development*, 13(4), 642–659.

Antony, S (2012). *The First Mile. A Launch Manual for Getting Great Ideas into the Market*. Boston, MA: Harvard Business Review Press.

Bailetti, T (2012). What technology startups must get right to globalize early and rapidly. *Technology Innovation Management Review*, October Issue on Born Global Firms, 5–16.

Blank, S (2013). Why the lean start-up changes everything. *Harvard Business Review*, May, 1–9.

Blank, S and B Dorf (2012). *The Startup Owner's Manual. The Step-by-Step Guide for Building a Great Company*. Pescadero, California: K&S Ranch Publishing Division.

Blank, S (2003). *The Four Steps to the Epiphany: Successful Strategies for Products that Win*. Pescadero: K&S Ranch.

Blume, B and J Covin (2011). Attributions to intuition in the venture founding process: Do entrepreneurs actually use intuition or just say that they do? *Journal of Business Venturing*, 26(1), 137–151.

Burgel, O and G Murray (2000). The international market entry choices of start-up companies in high-technology industries. *Journal of International Marketing*, 8(2), 33–62.

Cavusgil, S and G Knight (2009). *Born Global Firms: A New International Enterprise*. New York: Business Expert Press.

Cavusgil, S and G Knight (2015). The born global firm: An entrepreneurial and capabilities perspective on early and rapid internationalization. *Journal of International Business Studies*, 46, 3–16.

Chandler, G, D DeTienne, A McKelvie and TV Mumford (2011). Causation and effectuation processes: A validation study. *Journal of Business Venturing*, 26(3), 375–390.

Chetty, S and L Stangl (2010). Internationalization and innovation in a network relationship context. *European Journal of Marketing*, 44, 1725–1743.

Chetty, S and C Campbell-Hunt (2004). A strategic approach to internationalization: A traditional versus a "born-global" approach. *Journal of International Marketing*, 12(1), 57–81.

Cohendet, P and J Pénin (2011). Patents to exclude vs. include: Rethinking the management of intellectual property rights in a knowledge-based economy. *Technology Innovation Management Review*, December, 12–17.

Cooper, A and D Schendel (1976). Strategic responses to technological threats. *Business Horizons*, 19(1), 61–69.

Coviello, N (2015). Re-thinking research on born globals. *Journal of International Business Studies*, 46, 17–26.

Coviello, N (2006). The network dynamics of international new ventures. *Journal of International Business Studies*, 37(5), 713–731.

Dunford, R, I Palmer and J Benveniste (2010). Business model replication for early and rapid internationalisation: The ING Direct experience. *Long Range Planning*, 43(5–6), 655–674.

Eisenmann, T, E Ries and S Dillard (2011). Hypothesis-driven entrepreneurship: The Lean Startup. Harvard Business School Entrepreneurial Management Case No. 812–095. Available at SSRN:http://ssrn.com/abstract=2037237 (last accessed on 5 May 2015).

Furr, N and J Dyer (2014). *The Innovator's Method. Brining the Lean Start-up into Your Organization*. Boston: Harvard Business Review Press.

Gabrielsson, M and T Pelkonen (2008). Born internationals: Market expansion and business operation mode strategies in the digital media field. *Journal of International Entrepreneurship*, 6(2), 49–71.

Gabrielsson, M and V Kirpalani (2004). Born globals: How to reach new business space rapidly. *International Business Review*, 13, 555–571.

Gans, J and S Stern (2003). The product market and the market for "ideas": Commercialization strategies for technology entrepreneurs. *Research Policy*, 32, 333–350.

Garcia, R and R Calantone (2002). A critical look at technological innovation typology and innovativeness terminology: A literature review. *Journal of Product Innovation Management*, 19, 110–132.

Hamel, G (1997). Killer strategies that make shareholders rich. *Fortune* 135(12), 70–84.

Johnson, J (2004). Factors influencing the early internationalization of high technology start-ups: US and UK evidence. *Journal of International Entrepreneurship*, 2(1–2), 139–154.

Johnson, M, C Christensen and H Kagerman (2008). Reinventing your business model. *Harvard Business Review*, December issue.

Jones, M, N Coviello and Y Tang (2011). International entrepreneurship research (1989–2009): A domain ontology and thematic analysis. *Journal of Business Venturing*, 26(6), 632–659.

Kerr, W, R Nanda and M Rhodes-Kropf (2014). Entrepreneurship as experimentation. *Journal of Economic Perspectives*, 28(3), 25–48.

Knight, G (2014). Born Global Firms — Retrospective and a Look Forward. Paper presented at the *International Mini-Conference "New Research Themes in International Entrepreneurship"*, Department of Marketing and Management, University of Southern Denmark, Odense, 22–23 May 2014.

Knight, G and S Cavusgil (2004). Innovation, organizational capabilities, and the born-global firm. *Journal of International Business Studies*, 35, 124–141.

Knight, G and S Cavusgil (1996). The born global firm: A challenge to traditional internationalization theory. *Advances in International Marketing*, 8, 11–26.

Knudsen, M, S Tanev, T Bisgaard and M Thomsen (2013). How do new innovation paradigms challenge current innovation policy perspectives? *Contemporary Perspectives on Technological Innovation, Management and Policy*, Bing Ran, Ed., Charlotte, NC, USA: Information Age Publishing, 61–100.

Kudina, A, G Yip and H Barkema (2008). Born Global. *Business Strategy Review*, 19(4), 38–44.

Latour, B (2005). *Reassembling the Social. An Introduction to Actor-Network-Theory*. Oxford, UK: Oxford University Press.

Lemminger, R, L Svendsen, E Zijdemans, E Rasmussen and S Tanev (2014). Lean and global technology start-ups: Linking the two research streams. *Proceedings of the ISPIM Americas Innovation Forum* (October, 2014, Montreal, Canada), Huizingh, K, Conn, S, Torkkeli, M and Bitran, I, eds., p. 13.

Loane, S, J Bell and R McNaughton (2007). A cross-national study on the impact of management teams on the rapid internationalization of small firms. *Journal of World Business*, 42(4), 489–504.

McGrath, R and I MacMillan (2009). *Discovery-Driven Growth: A Breakthrough Process to Reduce Risk and Seize Opportunity*. Boston, MA: Harvard Business Press.

Mintzberg, H and J Waters (1985). Of strategies, deliberate and emergent. *Strategic Management Journal*, 6, 257–272.

Moen, Ø and P Servais (2002). Born Global or Gradual Global? Examining the export behavior of small and medium-sized enterprises. *Journal of International Marketing*, 10(3), 49–72.

Moogk, D (2012). Minimum viable product and the importance of experimentation in technology startups. *Technology Innovation Management Review*, March Issue on Technology Entrepreneurship, 23–26.

Moore, G (1991). *Crossing the Chasm: Marketing and Selling High-Tech Products to Mainstream Customers or Simply Crossing the Chasm*. Harper Business.

Moriarty, R and T Kosnik (1989). High-tech marketing: Concepts, continuity, and change. *Sloan Management Review*, 30(4), 7–17.

Mort, G, J Weerawardena and P Liesch (2012). Advancing entrepreneurial marketing. *European Journal of Marketing*, 46(3/4), 542–561.

Mueller, R and K Thoring (2012). Design Thinking vs. Lean Startup: A comparison of two user-driven innovation strategies. In: Erik Bohemia, Jeanne Liedtka and Alison Rieple (Eds.). *Leading Innovation Through Design: Proceedings of the DMI 2012 International Research Conference*, 8–9 August, 2012, Boston, MA, USA, 151–161, p. 151.

Nenonen, S and K Storbacka (2010). Business model design: Conceptualizing networked value co-creation. *International Journals of Quality of Services Sciences*, 2(1), 43–59.

Onetti, A, A Zucchella, M Jones and P McDougall-Covin (2012). Internationalization, innovation and entrepreneurship: Business models for new technology-based firms. *Journal of Management & Governance*, 16(3), 337–368.

Osterwalder, A and Y Pigneur (2010). *Business Model Generation: A Handbook for Visionaries, Gme Changers, and Challengers*. John Wiley & Sons.

Oviatt, B and P McDougall (1994). Toward a theory of international new ventures. *Journal of International Business Studies*, 25(1), 45–64.

Read, S, M Song and W Smit (2009). A meta-analytic review of effectuation and venture performance. *Journal of Business Venturing*, 24(6), 573–587.

Read, S, N Dew, S Sarasvathy, M Song and R Wiltbank (2009). Marketing under uncertainty: The logic of an effectual approach. *Journal of Marketing*, 73, 1–18.

Rennie, M (1993). Global competitiveness: Born global. *The McKinsey Quarterly*, 4(1), 45–52.

Ries, E (2011). *The Lean Startup: How Today's Entrepreneurs Use Continuous Innovation to Create Radically Successful Businesses*. New York: Crown Business.

Rogers, E (1983). *Dufusion of Innovations*. Third edition. The Free Press (First edition in 1962).

Sarasvathy, S, K Kumar, J York and S Bhagavatula (2014). An effectual approach to international entrepreneurship: Overlaps, challenges and proactive possibilities. *Entrepreneurship Theory and Practice*, 38(1), 71–93.

Sarasvathy, S, A Menon and G Kuechle (2013). Failing firms and successful entrepreneurs: Serial entrepreneurship as a temporal portfolio. *Small Business Economics*, 40(2), 417–434.

Sarasvathy, S (2001). Causation and effectuation: Toward a theoretical shift from economic inevitability to entrepreneurial contingency. *The Academy of Management Review*, 26(2), 243–263.

Sarasvathy, S (2008). *Effectuation: Elements of Entrepreneurial Expertise*. Cheltenham, U.K.: Edward Elgar.

Schmidt, J and T Keil (2013). What makes a resource valuable? Identifying the drivers of firm-idiosyncratic resource value. *Academy of Management Review*, 38(2), 206–228.

Shanklin, W and J Ryans (1987). *Essentials of Marketing High Technology*. Lexington, MA: Lexington Books.

Shrader, R, B Oviatt and P McDougall (2000). How new ventures exploit trade-offs among international risk factors: Lessons for the accelerated internationalization of the 21st century. *Academy of Management Journal*, 43(6), 1227–1247.

Stolze, A, T Arnsfeld, L Kelly and Ch Lüdtke (2014). The Lean Startup status quo: Deconstructing the Lean Startup movement to assess its validity as a strategic planning tool for entrepreneurs. Working Paper No. 3. Faculty of Business Management and Social Sciences, University of Applied Sciences Oshabruck. Accessible at: http://www.econbiz.de/Record/the-lean-startup-status-quo-deconstructing-the-lean-startup-movement-assess-its-validity-strategic-planning-tool-for-entrepreneurs-stolze-audrey/10010458686.

Tanev, S (2012). Global from the start: The characteristics of born-global firms in the technology sector. *Technology Innovation Management Review*, March Special Issue on Technology Entrepreneurship, 5–8. Accessible at: http://timreview.ca/article/532.

Tanev, S and E Rasmussen (2014). The lean global start-up — a new type of firm? Paper presented at the *International Mini-Conference "New Research Themes in International Entrepreneurship"*, Department of Marketing and Management, University of Southern Denmark, Odense, 22–23 May, 2014.

Teece, D (1986). Profiting from technological innovation: Implications for integration, collaboration, licensing and public policy. *Research Policy*, 15, 285–305.

Teece, D (2006). Reflections on Profiting from innovation. *Research Policy*, 35, 1131–1146.

Tidd, J, Editor (2010). *Gaining Momentum. Managing the Diffusion of Innovations*, Series on Technology Management, Vol. 15, London: Imperial College Press.

Trimi, S and J Berbegal-Mirabent (2012). Business model innovation in entrepreneurship. *International Entrepreneurship and Management Journal*, 8, 449–465.

Vapola, T, P Tossavainen and M Gabrielsson (2008). The battleship strategy: The complementing role of born globals in MNC's new opportunity creation. *Journal of International Entrepreneurship*, 6(1), 1–21.

Wiefels, P (2002). *The Chasm Companion — A Fieldbook to Crossing the Chasm and Inside the Tornado*, HarperBusiness, 2002.

Wilson, K and Y Doz (2011). Agile innovation: A footprint balancing distance and immersion. *California Management Review*, 53(2), 6–26.

Yadav, N, S Swami and P Pal (2006). High technology marketing: Conceptualization and case study. *Vikalpa*, 31(2), 57–74.

Zijdemans, E and S Tanev (2014). Conceptualizing innovation in born-global firms. *Technology Innovation Management Review*, 4(9), 5–10. Accessible at: http://timreview.ca/article/826 (last accessed on October 16, 2014).

Chapter 8

Global Innovation: An Answer to Mitigate Barriers to Innovation in Small and Medium-Sized Enterprises?[*]

Stephan Buse, Rajnish Tiwari and Cornelius Herstatt

Innovations have acquired a key role in the growth and competition strategies of firms today. They are regarded as an essential tool for stimulating growth and enabling firms to master the competition brought about by the forces of globalization. However, in particular, many small and medium-sized enterprises (SMEs) in Western countries are facing insurmountable barriers to innovation. To what extent might the concept of global innovation and, more specifically, the internationalization of research and development (R&D) be an answer to deal with these problems, is discussed in this paper. Based on empirical studies conducted by the authors in Germany, the paper presents results from research-in-progress and proposes a reference model for chances and challenges of global innovation activities.

Keywords: SMEs; global innovation; internationalization of R&D.

Introduction

In the era of globalization, the economic wealth of advanced nations like Germany is largely based on the innovativeness of their firms and individual entrepreneurs. Referring to this, small and medium-sized enterprises (SMEs) are of vital importance since they constitute the largest business block and provide the bulk of employment in most countries. However, numerous studies have revealed that the innovativeness of SMEs is affected by several barriers to innovation. To overcome or at least mitigate these barriers is a major challenge for this group of companies. To what extent might the concept of global innovation and, more specifically, the internationalization of research and development (R&D) provide a solution to this challenge, is the topic of this paper. Based on empirical studies conducted by the

---

[*]Originally published in *International Journal of Innovation and Technology Management*, 7(3), pp. 215–227.

authors in Germany, the paper gives results from research-in-progress and presents a reference model for chances and challenges of global innovation activities.

This paper is structured on the following lines. After this brief introduction, the terms "innovation" and "SMEs" are defined and their relation to each other established in the Innovation and SMEs section. The Barriers to Innovations in German SMEs section presents barriers to innovation in German SMEs. The Global Innovation as a Chance section introduces the concept of global innovation and how it may be used as an instrument to mitigate the effects of innovation barriers in SMEs. Challenges associated with the chances of global innovation are discussed in the Difficulties in Global Innovation section. The Implications and Research Outlook section presents general implications and the research outlook. Finally, the Conclusion section contains a brief conclusion.

Innovation and SMEs

This section defines the terminological base for this paper and establishes the need for innovation in SMEs, while elaborating the crucial role that SMEs play in the economy.

Innovation

Innovation, according to Rogers [2003], is "an idea, practice, or object that is perceived as new by an individual or other unit of adoption." This "newness" need not involve "new" knowledge, thereby effectively implying that the "newness" may also concern advancement or modification of existing knowledge. For the purpose of this paper, we may regard innovation as invention and commercialization of new (or betterment of existing) products, processes and/or services [Tiwari (2007)].

Innovations usually do not take place in a given, static environment. They are, rather, a result of a dynamic process in an organization that involves the interplay of several internal and external factors. R&D constitutes a major — though not exclusive — part of the "innovation process." It encompasses several systematic steps, such as requirement analysis, idea generation, idea evaluation, project planning, product development, product testing and product marketing [Verworn, Herstatt and Nagahara (2006)].

Small and medium-sized enterprises

The European Commission defines SMEs as enterprises which employ fewer than 250 persons and which have an annual turnover not exceeding 50 million euros

[EC (2003)]. This definition inherently implies that SMEs have less human and financial resources at their disposal than large firms. Contrasted against most large corporations that generate billions of euros in annual sales and employ hundreds of thousands of workers, SMEs are by definition equipped with much less resources. This implication is a supported in the academic literature relating to SMEs; see e.g., [Herstatt et al. (2001)]. This resource constraint exists even though SMEs play an important role in the national economy, as discussed in the following.

According to a report by the European Commission, there existed in 2003 some 23 million SMEs, which represented 99% of all enterprises in the enlarged European Union of 25 countries, providing around 75 million jobs [EC (2003)]. In Germany, according to the Bonn-based Institute fuer Mittelstandsforschung (IfM),[1] SMEs accounted for 99.7% of all enterprises in 2005 and provided employment to 70.9% of all employed persons (20.42 million) in 2006. These data exemplarily demonstrate the key role which SMEs play in Germany's economy.

Connecting SMEs to innovation

Notwithstanding their large share in all enterprises and the overall employment generated, SMEs in Germany continue to remain weak on the revenue front when compared with their large counterparts. For instance, only 39.1% of the total turnover generated by all enterprises in Germany in 2005 went into SMEs' account [IfM (2008)].

At the same time, the increasing globalization is bringing more competition to the home market, the traditional stronghold of many SMEs. In Germany as well as in many other EU member countries, SMEs usually operate under high overhead costs, such as labor costs, and find themselves faced with tough price-oriented competition from low-cost producers from emerging economies in Asia and eastern Europe.

Besides, the globalization not only brings in challenges but also presents an opportunity to internationalize sales in new, rapidly growing markets and thereby to generate additional revenues. However, new markets (may) also require products and services which are adapted to the local needs and tastes of those markets.

Providing innovative products with enhanced utility may help firms strengthen their competitive position in home as well as international markets. This necessitates innovation efforts to bring new and/or better products into the market while developing organizational and manufacturing processes that enable more

---

[1] IfM Bonn works with a definition of SMEs that differs from the definition used in this paper. It defines SMEs as firms that employ less than 500 workers and whose annual turnover does not exceed 50 million euros.

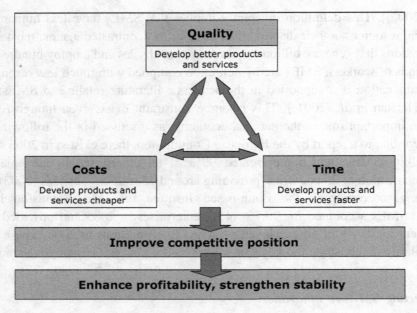

Fig. 1. The BCF model for innovation in SMEs.

efficient and cost-effective production, distribution and after-sales services; see e.g., [Dangayach *et al.* (2005)]. Figure 1 shows a "goal model" for innovation activities in SMEs. This model may be referred to as a "BCF model for innovation" (BCF = better, cheaper and faster).

SMEs frequently operate in niches and have direct contact with customers, thereby potentially gaining valuable impulses in the form of customer feedback. Acting often in a more informal manner and confronted with fewer intrafirm hierarchy levels than large firms, SMEs seem to be, in many respects, better placed for innovations than their large counterparts.

In practice, however, the resource constraints coupled with market uncertainties (and a few other factors) limit the ability of SMEs to indulge in dedicated R&D and to experiment with the purpose of new product development, as demonstrated by many studies, some of which are discussed in the next section.

Barriers to Innovations in German SMEs

Barriers to innovation in SMEs have been the object of investigation in a large body of national and international studies. A few are mentioned here. Acs and Audretsch [1990] worked on this topic in the US, Ylinenpää [2008] in Sweden, while Mohnen and Rosa [1999] as well as Baldwin and Gellatly [2004] researched on it in Canada. In Germany the Centre for European Economic Research (ZEW)

Table 1. Previous studies on barriers to innovation in SMEs.

Barriers to innovation in SMEs	Studies (amongst others)
Financial bottlenecks – hindered access to external finance, – high innovation costs (and therefore) – high economic risks	[Acs and Audretsch (1990); Baldwin and Gellatly (2004); Rammer et al. (2006)]
Shortage of and hindered access to qualified personnel	[Ylinenpää (1998); Rammer et al. (2005); Rammer et al. (2006); FES (2004)]
Limited internal know-how to manage the innovation process effectively and efficiently	[Mohnen and Rosa (1999); ZEW and DIW (2004); Rammer et al. (2005)]
Missing market know-how – to meet customer's needs – to enter foreign markets	[Ylinenpää (1998); FES (2004); HWWA (2004)]
Bureaucratic hurdles – long administrative procedures – restrictive laws and regulations	[Acs and Audretsch (1990); Rammer et al. (2005); Rammer et al. (2006); HWWA (2004)]
Lack of intellectual property rights	[Baldwin and Gellatly (2004); ZEW and DIW (2004)]

has conducted several studies in recent years; see e.g., [ZEW and DIW (2004); Rammer et al. (2005); Rammer et al. (2006)]. Further studies dealing with the German situation have been conducted by the Friedrich Ebert Stiftung [2004] and Hamburg Institute of International Economics [2004].

Comparing the findings of the aforementioned surveys, it would not be an unreasonable assumption that SMEs in the respective countries or regions often face similar barriers to innovation. The most dominant problems are listed in Table 1.

An empirical study,[2] conducted by the authors, revealed that financial bottlenecks and problems in finding suitable and qualified personnel are also the most severe barriers to innovation by SMEs in the Metropolitan Region of Hamburg, Germany. Bureaucratic hurdles and problems in finding "right" cooperation partners were ranked next; see Fig. 2.

Despite a small response rate of only 70 SMEs from different industries the findings seem to be significant because they correspond strongly to the results of the aforementioned studies. In addition, they were confirmed in various interviews with experts from the selected, industries, like firm representatives, representatives of industry associations and cluster managers.

---

[2]The methodology and the complete results of the survey are contained in [Herstatt et al. (2008)].

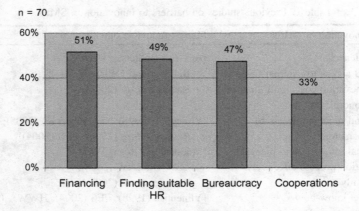

Fig. 2. Barriers to innovation in Hamburg's SMEs.

To overcome or at least mitigate these barriers to innovation is a major challenge for SMEs. However, to gain or sustain innovativeness and therefore competitiveness, they have to solve these problems. To what extent might the internationalization of R&D be a way to deal with this challenge, is discussed in the next section.

Global Innovation as a Chance

Based on the results of several studies and reports, e.g., [Boutellier *et al.* (2000); DIHK (2005); UNCTAD (2005); Ernst (2006); OECD (2006); LTT Research (2007); Tiwari *et al.* (2007)], it seems reasonable that global innovation activities, including internationalization of R&D, may at least help mitigate the effects of the barriers faced even if not completely overcome them.

It is therefore not surprising that many SMEs have started recognizing the opportunities that the globalization offers not only in production but also in R&D. In many instances, SMEs have set up R&D centers abroad, as a survey by DIHK revealed. The survey, with a sample base of over 1600 firms, 77% of them SMEs, showed that as of February 2005 one-third (33%) of all German firms were engaged in offshore (international) R&D. Interestingly enough, over 25% of the surveyed SMEs also engaged in offshore R&D. Some had their own R&D facilities abroad, while others forged cooperation with firms and R&D institutions abroad [DIHK (2005)].

Global innovation activities, particularly when conducted in emerging, fast-growing markets such as China and India, may offer tremendous opportunities, such as vast pools of qualified human resources in science and technology, cheaper labor costs and access to new, fast-growing consumer markets with substantial purchasing power and/or infrastructural needs.

In the following, we discuss the "motivators" of global innovation, which can be divided into three main categories:

- Access to knowledge;
- Cost advantages;
- Market opportunities.

Leveraging access to knowledge

Quantitative availability of skilled labor

The demographically disadvantageous factor of an aging population in many Western countries, including Germany, is coupled with another challenge, namely the decline in the number of science and technology (S&T) students. In contrast, countries such as China and India are producing a large number of S&T graduates. In China, 61% of undergraduates are studying for a science or engineering degree. Also, as far as the quality of higher education is concerned, many "emerging" countries around the globe, especially in Asia and eastern Europe, are able to produce world class graduates [EIU (2004)]. India and China alone are reported to produce 350,000 and 600,000 engineers a year, respectively, in contrast to 70,000 in the USA and nearly 33,000 in Germany [DBR (2005); Farrel and Grant (2006); Farrel et al. (2005); BMBF (2007)].

Setting up offshore R&D centers provides firms with an opportunity to tap into a larger talent pool which is otherwise not accessible, for instance owing to restrictive immigration policies, which are often caused by social, political and/or security-related concerns and are not rarely supported by strong pressure groups in society. On the need for immigration in Germany and the refusal of the political establishment to accede to it, see [BITKOM (2007)]. A report by the American Electronics Association [Kazmierczak and James (2007)] documents this angle from an American perspective.

Reducing bottlenecks in the product pipeline

Global innovation activities may ensure that work can be carried out simultaneously from multiple locations and on multiple projects, if needed. Several independent modules of a single project may be worked on at the same time to shorten the time-to-market. Following the same logic, even a single step of a project may be worked on round the clock in changing shifts the world over, whereby the data are transmitted electronically from one center to the next. Such a step could be of crucial importance for time-critical projects, such as for pharmaceutical firms while conducting clinical trials. The shortened time-to-market may be crucial for ensuring large-scale competitive advantage [BCG (2006)].

Proximity to production centers

The globalization has moved production centers of many industries to emerging countries, where new industry clusters have grown up. Some industry-specific innovation activities, e.g., in the automotive sector, may require close interaction with the production department. It may be useful to locate R&D facilities in the proximity of the production center, unless other factors (e.g., availability of knowledge resources, of affordable costs) threaten to hamper the process. In particular, German firms seem to locate their R&D activities in the close vicinity of their production centers. Whereas many international studies found "access to knowledge resources" as the leading reason for many international offshore R&D activities, a DIHK survey in Germany revealed that "proximity to production centers" proved German firms most often to offshore R&D [DIHK (2005)]. Another study, by LTT Research [2007], confirmed this finding.

Learning from lead markets

Unsaturated, emerging economies in Asia are rapidly taking over the role of "lead markets" by their openness to consumption and their willingness to spend money on technological innovation. The Asian consumers already play a key role in the electronics industry, today.

Leveraging cost advantages

Innovation activities too generate costs which need to be minimized in order to compete with other "innovators," especially since the outcome and the ensuing commercial success of innovation efforts remain to a large extent uncertain. Global innovation, in addition to access to skilled labor, may also contribute to reducing the costs of innovation.

Lower costs for (skilled) labor

Global activities, particularly in emerging countries, may lead to significant reduction in the costs [EIU (2004)]. According to Müller [2004], the starting salary of a software developer working for the German software firm SAP in India was reported as 8000 euros per annum in 2004, while the salary for a similarly qualified person at the headquarters in Germany was reported to be five times higher, at 40,000 euros. The DIHK survey in Germany found that 41% of all offshore R&D activities of German firms were motivated by the incentives of the lower costs abroad [DIHK (2005)]. The labor costs are, however, going up, see the Difficulties in Global Innovation section.

State-induced incentives

There may be lucrative state-induced incentives to indulge in R&D activities abroad if the host country offers significant financial support, such as in the form of "tax holidays," subsidies and/or other tax incentives. In India, for example, expenditure incurred on R&D may be deducted from corporate taxes with a weighted average of 150% [DSIR (2006)]. Moreover, there might be high barriers, or altogether restrictions, on carrying out R&D in certain fields, e.g., genetics. Such restrictions either increase the R&D cost in order to fulfill the legal requirements (in case of high barriers) or may cause high opportunity costs in the form of lost business opportunities (in case of prohibition). If R&D in that particular field is allowed in another country or is possible with significantly lower restrictions, then it may make sense to locate R&D efforts in that country.

Leveraging market opportunities

In addition to knowledge and cost factors, there might be significant market opportunities abroad in the form of demand for localized products in fast-growing markets. More and more people in emerging economies are having the financial resources to buy high-end products [EIU (2004)], and the number of middle class consumers is growing rapidly in emerging countries, particularly China and India.

A McKinsey study predicts that China will become the third-largest consumer market worldwide, surpassing Germany and behind Japan and the US, by 2025. The urban incomes are set to rise significantly, providing ample opportunities for the manufacturers to sell not only basic necessities but also items of a "discretionary" nature [MGI (2006)].

The situation in India looks similar. The Indian middle class, comprising an estimated 200–250 million people, is believed to be one of the largest worldwide. According to estimations by McKinsey, India is expected to become the fifth largest consumer market by 2025, moving up from its 12th position [MGI (2007)]. The study forecast that India will have a 583-million-strong middle class by 2025. It also observed a shifting focus in the consumer behavior which is connected with increasing income levels: as a smaller share of the income is spent on basic necessities, more "discretionary items" are being purchased [MGI (2006)].

Difficulties in Global Innovation

The section above has given us an overview of how global innovation may be used as a chance to mitigate the effects of innovation barriers prevalent in Germany and other advanced economies. Such chances — though realistic — are fraught with certain challenges that need to be addressed and mastered in order to realize the full potential of global innovation. In the following we describe some primary challenges.

Finding "qualified" personnel

The access to knowledge may be fraught with difficulties, as the "global war for talents" gets more intense murkier. Even China and India are reportedly experiencing shortage of skilled labor with international knowledge standards; see e.g., [Farrel and Grant (2006)]. Many firms, including as reputed names such as Google and Infosys, are complaining of a shortage of suitable candidates. This shortage leads to a high attrition rate in firms, which sometimes reaches 30–40%. The shortage of qualified personnel is also felt in the booming economy of China, where German firms are finding it increasingly difficult to recruit local technicians, as a McKinsey study found [McKinsey (2006)].

Cost explosion in booming economies

The cost advantage of many "emerging" nations with booming economies is disappearing in many respects; for instance, wages of highly skilled labor in India reportedly grow by 10–15% and above per annum on average [Hein (2004)]. The wage growth in senior positions, like project manager, has been even higher, at about 25% per annum [Farrel and Grant (2006)]. Wage costs for semiskilled or unskilled labor, however, remain significantly lower than in Western, industrialized nations.

Protection of intellectual property rights

The protection of intellectual property rights (IPRs) remains a concern, even if to varying degrees, in most emerging countries, particularly when seen in conjunction with often-delayed judicial processes and/or often-prevalent corruption. Fabian and Schmidli [2005] reported problems related to IPR protection and the fulfillment of contractual obligations in China.

Market uncertainties

Local adaptation of products may cause financial constraints if the size of the target market does not provide scale effects. R&D efforts for local adaptation can be justified only in the presence of a large market. At the moment there are not many such markets if one excludes China and India, and probably some eastern

European countries. This effectively means that the global activities are actually "Asian" or "eastern European" activities. This problem may, however, be overcome by concentrating on regional markets, such as East Asia or eastern Europe.

Startup and operational costs

It is possible that some firms, particularly SMEs, may not have sufficient financial resources to set up and operate an innovation center abroad. Hence, the financial effects of global innovation activities may not be equal for all firms. As a 2004 study by McKinsey suggests, German companies save 0.52 euros for every euro of corporate spending on IT jobs offshored to India, whereas their US counterparts save 0.58 cents for every dollar they spent on jobs in India. The higher costs for German firms' operations in India are caused by "differences in language and culture," which "raise the cost of coordinating offshoring projects" [Farrel and Grant (2006)].

Cross-cultural issues and communication

Global innovation invariably involves multidisciplinary teams with international backgrounds. The resultant disparity requires a high degree of social competences, and a sound understanding of cross-cultural interactions. For instance, McKinsey [2006] reported several incidents of intercultural nuisance in Indo-German software development work. Fabian and Schmidli [2005] reported similar problems in Sino-Swiss projects.

Acceptance issues

The parent unit (headquarters) tends to interfere in the innovation work being carried out at the foreign location, which often limits the flexibility of the subsidiaries "to bring their innovation initiatives fully in line with host country best practices," as a large-scale empirical study by Sofka [2006] revealed.

Additionally, there might be reservations/biases in certain quarters at the headquarters regarding R&D capabilities of the colleagues abroad. An example has been cited by Hein [2004], wherein Jürgen Schubert, Chief Executive Officer (CEO) of Siemens India, a German national himself, is quoted with the complaint that products designed by his engineers in India were often rejected by the central R&D unit on flimsy grounds. Schubert narrated that those very same products, however, passed the test without any hassles whatsoever, once they were labeled "Made in Germany." "The quality of the products was identical," recalled Schubert, "only India's image was not befitting."

Furthermore, some employees in the R&D units at the headquarters tend to see the new location as a potential threat to their job security, leading to resentments, antagonism and even noncooperation [Hein (2004)].

Implications and Research Outlook

As the discussion above has demonstrated, global innovation activities, especially the internationalization of R&D, may lead to mitigation of the effects of innovation barriers faced by SMEs in their home country. However, going abroad with a sensitive function like internal R&D and/or other functions from the innovation process requires a deep and thorough understanding of internal business processes and of business environment conditions in the country concerned.

Based on these considerations, firms, need to decide which form of R&D internationalization ("captive offshoring," "offshore cooperation" or "offshore outsourcing") is more suitable for their needs. Additionally, firms should bear the following factors in mind:

To get access to local knowledge abroad, firms, especially those which are facing financial or managerial constraints, should initially focus on those forms of internationalization which do not require a high level of capital investments. Potential strategies, for instance, could be:

- Cooperative agreements with local research institutions and/or firms.
- Outsourcing of parts of the innovation process.
- To limit the financial burden of setting up and maintaining their own international R&D facilities, firms might consider sharing resources (facilities, etc.) with partners. These partners might be other domestic firms within interest in global innovation, firms from other countries with an interest in the target country, or local firms and research institutions in the target country.

In the case of any kind of partnering, the involved parties must find ways to protect their individual core competences and to share the intellectual property generated by such a joint venture, in a justified manner.

If companies enter foreign markets that require local adaptation of products (and therefore local R&D), they need to be sure that the potential of the target market is sufficient to achieve a favorable cost structure. If companies have reasons to expect problems in achieving needed experience curves (economies of scale and learning curve effects), they should reconsider the market entry.

Firms need to pay attention to cultural aspects and should provide their employees involved in international activities with cross-cultural training.

This sensitization to mutual cultural issues may play a key role in the success of an international venture.

The motivation (potential benefits) as well as the necessity for global innovation activities (such as tapping new markets and reducing the time-to-market) must be explained and discussed with existing R&D units so as to secure their benevolent cooperation with overseas operations.

The above discussed measures may play a crucial role in meeting the challenges of global innovation management. However, exact modalities of global innovation activities, particularly for SMEs, need further ascertainment and are set to be examined by our current research, which is basically focused on following issues:

- What is the role of outsourcing and/or offshoring in the R&D strategy of firms today?
- What are the advantages and possible risks of internationalization of innovation activities, particularly R&D? Which are the factors that are critical to success?
- Which are the most attractive R&D locations for particular branches? What are the reasons for their attractiveness?
- Could firms be at a disadvantage if they choose not to internationalize their innovation activities, especially R&D?
- How to implement and coordinate international R&D activities at the organizational level while securing the cooperation of all the parties involved?
- What are the lessons that SMEs can learn from the success/failure of international innovation/R&D activities of multinational firms?

Conclusion

The discussion in the sections above has established that global innovation opens up new arenas for firms, especially SMEs, to strengthen their innovation capabilities and thereby to increase their competitiveness in a globalized world. In this respect the internationalization of R&D seems to be a useful instrument for mitigating the effects of barriers to innovation often faced by SMEs in Germany, the EU or anywhere else in industrialized economies.

At the same time, these "global" opportunities are invariably associated with challenges that need to be met in order to fully exploit the chances of global innovation. A thorough understanding of internal business processes, organizational backing not only by senior management but also by other employees, especially in R&D departments, as well as a profound analysis of business environment conditions of the target offshore country, are prerequisites for a successful global operation.

References

Acs, Z. and Audretsch, D. (1990). *Innovation and Small Firms*, Cambridge University Press.

Baldwin, J. R. and Gellatly, G. (2004). "Innovation strategies and performance in small firms," Ottawa.

Boutellier, R., Gassmann, O. and von Zedtwitz, M. (2000). *Managing Global Innovation: Uncovering the Secrets of Future Competitiveness*. Heidelberg: Springer-Verlag.

BCG (2006). "Harnessing the power of India: rising to the productivity challenge in biopharma R&D," The Boston Consulting Group.

BITKOM (2007). "Standpunkte zur Zuwanderung Hochqualifizierter," Berlin: Bundesverband Informationswirtschaft, Telekommunikation und neue Medien e.V. (BITKOM), 24.Aug.2007.

BMBF (2007). "Bericht zur technologischen Leistungsfähigkeit Deutschlands 2007," German Federal Ministry of Education and Research.

Dangayach, G. S., Pathak, S. C. and Sharma, A. D. (2005). "Managing innovation," in *Asia Pacific Tech Monitor*, 22, 3: 30–33.

DBR (2005). '*India Rising: A Medium-Term Perspective*, Frankfurt: Deutsche Bank Research.

DIHK (2005). "FuE-Verlagerung: Innovationsstandort Deutschland auf dem Prüfstand," Berlin: Deutscher Industrie- und Handelskammertag e.V.

DSIR (2006). "Research and development in industry: An overview, Department of Scientific and Industrial Research," Ministry of Science and Technology, Government of India.

EC, Commission Recommendation of 6 May 2003 concerning the definition of micro, small and medium-sized enterprises, Annex Title I — Definition of Micro, Small and Medium-sized Enterprises Adopted by the Commission, European Commission (2003/361/EC).

EIU (2004). "Scattering the seeds of invention: The globalization of research and development," Economist Intelligence Unit.

Ernst, D. "Innovation offshoring: Asia's emerging role in global innovation network." East–West Center Special Reports, No. 10/2006.

Fabian, C and Schmidli, C. (2005). "Problems of R&D internationalization of small and medium companies," in *Proc. European Academy of Management Annual Conference 2005* (Munich, 2005).

Farrel, D. and Grant, A. (2006). "Addressing China's looming talent shortage." San Francisco: McKinsey Global Institute.

Farrel, D., Kaka, N. and Stürze, S. (2005). "Ensuring India's offshoring future," in *The McKinsey Quarterly, 2005 Special Edition: Fulfilling India's Promise*, pp. 74-83.

FES (2004). "KMU und Innovation — Stärkung kleiner und mittlerer Unternehmen durch Innovationsnetzwerke," Friedrich-Ebert-Stiftung (FES), Paper 10/2004, Bonn.

Hein, C. "Auch Indiens Ingenieure sind nicht mehr billig," in *Frankfurter Allgemeine Zeitung* (Frankfurt am Main, 23.Dec.2004), p. 20.

Herstatt, C., Buse, S. Tiwari, R. and Stockstrom, C. (2008). "Innovationshemmnisse in KMU der Metropolregion Hamburg: Ergebnisse einer empirischen Untersuchung in ausgewählten Branchen." Hamburg University of Technology, http://www.tuhh.de/tim/ris-hamburg/befragung.html (last accessed 27.03.2008).

Herstatt, C., Lüthje, C. and Verworn, B. (2001). "Die Gestaltung von Innovationsprozessen in kleinen und mittleren Unternehmen." In: J.A. Meyer (ed.), *Innovationsmanagement in kleinen und mittleren Unternehmen — Jahrbuch der KMU-Forschung* pp. 149–169, München, Vahlen.

HWWA (2004). "Die Position Norddeutschlands im internationalen Innovationswettbewerb." Hamburg: Hamburg Institute of International Economics.

IfM (2008). "Schlüsselzahlen des Mittelstands in Deutschland." Institut für Mittelstandsforschung, http://www.ifm-bonn.org/index.htm?/dienste/schluesselzahlen_des_mittelstands.htm (last accessed 28.03.2008).

Kazmierczak, M.F. and James, J. (2007). *"We are Still Losing the Competitive Advantage: Now Is the Time to Act."* American Electronics Association.

LTT Research (2007). "The implications of R&D off-shoring on the innovation capacity of EU firms." Study on behalf of the PRO INNO Europe initiative of the European Union.

McKinsey (2006). "Deutschen Unternehmen in China gehen einheimische Fachkräfte aus." McKinsey & Company, press release (Düsseldorf, 17.Dec.2006).

MGI (2006). "From 'Made in China' to 'Sold in China': The rise of the Chinese urban consumer." San Francisco: McKinsey Global Institute.

MGI (2007). "The 'Bird of Gold' — The rise of India's consumer market." San Francisco: McKinsey Global Institute.

Mohnen, P. and Rosa, J. (1999). "Barriers to innovation in service industries in Canada," Science and Technology Redesign Project, Research Paper No. 7, Ottawa.

Müller, O. (2004). "Walldorf spiegelt sich Indien," in *Handelsblatt*, Stuttgart (17 Apr. 2004).

OECD, (2006). "OECD science, technology and industry outlook 2006," Organization for Economic Cooperation and Development (Paris, 2006).

Rammer, C., Löhlein, H., Peters, B. and Aschhoff, B. (2005). "Innovationsverhalten der Unternehmen im Land Bremen." Mannheim: Zentrum für Europäische Wirtschaftsforschung (ZEW).

Rammer, C., Zimmermann, V., Müller, E., Heger, D., Aschhoff, B. and Reize, F. (2006). "Innovationspotenziale von kleinen und mittleren Unternehmen." Mannheim: Centre for European Economic Research (ZEW).

Rogers, E. M., (2003). *Diffusion of Innovations*, 5th ed. New York: Free Press.

Sofka, W. (2006). "Innovation activities abroad and the effects of liability of foreignness: Where it hurts," Centre for European Economic Research (ZEW), Discussion Paper No. 06-029.

Tiwari, R., (2007). "The early phases of innovation: Opportunities and challenges in public-private partnership," *Asia Pacific Tech Monitor*, **24**, 1: 32–37.

Tiwari, R., Buse, S. and Herstatt, C. (2007). "Innovation via global route," in *Proc. Second Int. Conf. Management of Globally Distributed Work*, (Indian Institute of Management, Bangalore), pp. 451–465.

UNCTAD, (2005). "Globalization of R&D and developing countries," in *Proc. Expert Meeting United Nations Conference on Trade and Development* (Geneva, 2005).

UNCTAD, (2005). "UNCTAD survey on the internationalization of R&D: Current patterns and prospects on the internationalization of R&D." Occasional Note, United Nations Conference on Trade and Development (Geneva, 2005).

UNCTAD, (2005). "World Investment Report 2005: Transnational corporations and the internationalization of R&D." United Nations Conference on Trade and Development (Geneva, 2005).

Verworn, B., Herstatt, C. and Nagahara, A. (2006). "The impact of the fuzzy front end on new product development success in Japanese NPD projects" (conference proceedings), in *R&D Management Conference*, (Manchester, 2006).

Ylinenpää, H. (2008). "Measures to overcome barriers to innovation in Sweden," Paper EFMD European Small Business Seminar in Vienna (16.09.1998), http://www.iesluth.se/ org/Rapporter/AR9826.pdf (last accessed 25.03.2008).

ZEW and DIW (2004). "Innovationsbarrieren und internationale Standortmobilität," a joint study by Centre for European Economic Research (ZEW), Mannheim, and German Institute for Economic Research, Berlin.

Chapter 9

International Corporate Entrepreneurship with Born Global Spin-Along Ventures — A Cross-Case Analysis of Telekom Innovation Laboratories' Venture Portfolio*

Sarah Mahdjour and Sebastian Fischer

This study investigates a special kind of corporate ventures, so called spin-along ventures, and their motivations to internationalise early. Insights are built from a multiple case study approach, investigating the spin-along program of Telekom Innovation Laboratories (T-Labs). Our results show that early internationalisation can avoid or reduce challenges that spin-alongs face when entering the domestic market. Four major motivations for early internationalisation could be identified: (1) avoid termination based on the parent's perceived threat of cannibalisation of existing products, (2) enable a venture's collaboration with competitors, (3) overcome restrictions of parental assets in the domestic market, and (4) address markets that offer greater chances for success than the domestic market does. Based on our findings we derive concrete implications for practitioners and academics in the field of innovation management.

Keywords: Born globals; corporate venturing; spin-along; case study; spin-out; international entrepreneurship; Telekom Innovation Laboratories.

Introduction

In order to secure their long-term competitiveness, established companies are challenged to continuously stay innovative. Literature on the incumbent's curse has revealed that incumbency can lead firms to assume a lower level of market pressure which decreases innovation incentives. With this in mind, large companies apply corporate venturing to commercialise inventions by establishing new internal or external units (Burgelman, 1983; Keil, 2004; Sharma and Chrisman, 1999). If the transfer of R&D results into the parental organisation is accompanied by high internal barriers (Riege, 2007), then the creation of independent

---

*Originally published in *International Journal of Innovation Management*, 18(3), pp. 1–18.

organisational units can become an attractive alternative path. Companies can decide whether to create own ventures from the inside out or to acquire external start-ups from the outside in. In recent years, practitioners as well as scholars have identified organisations that consolidate both internal and external venturing activities and called this phenomenon the "spin-along approach" (Rohrbeck et al., 2007). Companies that follow a spin-along approach build on ideas or R&D results to spin out new ventures as "innovation speedboats" to reduce barriers within the internal innovation process. At the core of the spin-along approach is the maintenance of a relevant stake in the new ventures (e.g., via ownership of shares or key assets) to flexibly decide whether to spin them in or off in the future.

Ventures that internationalise early after inception are called "born globals" (Knight and Cavusgil, 1996) or "international new ventures" (Oviatt and McDougall, 1994). Callaway (2008) linked born global research with the concept of internal corporate venturing (ICV) and presented "global corporate ventures" as measures to quickly internationalise internal corporate ventures.

In this paper, we will highlight early internationalisation as a strategy of spin-alongs to avoid internal and external barriers of domestic commercialisation. We build our insights on an explorative multiple case study of five ventures from the spin-along program of Telekom Innovation Laboratories (T-Labs) that aimed for a launch in non-domestic markets early after their inception and present potential to become born globals. We will thereby link the spin-along approach with the concept of international entrepreneurship. Our results suggest that if corporates do not accept that their new ventures cannibalise own business models, engage in business with competitors or may require key parental assets, early internationalisation can be an attractive new direction when spinning-out new ventures successfully.

Related Literature

The spin-along approach

In 2007, Rohrbeck et al. first introduced the *spin-along approach* as a novel concept of corporate venturing. Previously, literature on corporate venturing emphasised two major foci. While *ICV* deals with the creation of new business within an established company, *external corporate venturing* focusses on sourcing ideas and concepts that are already in the market (Burgelman, 1983; Keil, 2004; Sharma and Chrisman, 1999). Corporate venturing allows companies to create new business, achieve growth, and diversify their product portfolio. Indirect benefits can include the development of new competencies and technologies, the promotion of an innovative corporate culture and the opportunity to learn through exploration (Backholm, 1999).

Companies applying the spin-along approach source promising ideas or generate new technology via internal R&D and commercialise them as spin-outs in separate new business organisations. A novel aspect of the concept is a close linkage and monitoring of these spin-alongs after their market launch, followed by a regular evaluation and decision to either reintegrate them, keep them at a distance or to reduce ownership of shares completely (Rohrbeck *et al.*, 2009). In this context it is important to differentiate the terms spin-off, spin-in and spin-out: *Spin-offs* are completely separated from the corporate parent, although the parent might be a partner or customer of the new company. *Spin-ins* are external ventures that are acquired and integrated into an organisational structure. Spin-along ventures are *spun out* (not "*off*") by their corporate parent in order to keep them at a distance over a longer period of time (Rohrbeck *et al.*, 2009).

While several case study-based publications have presented the relevance of the spin-along approach in management practice (Mahdjour and Fischer, forthcoming; Klarner *et al.*, 2013; Michl *et al.*, 2012; Rohrbeck *et al.*, 2007, 2009), the number of publications exploring and sharpening the concept to date is scarce (Michl *et al.*, 2012). Especially the detailed characteristics that distinguish the approach from other existing theoretical concepts in the area of corporate venturing need to be further investigated.

Born globals

Born globals or *international new ventures* are young companies that internationalise early after their inception (Knight and Cavusgil, 1996; Oviatt and McDougall, 1994). They challenge the widely established notion that internationalisation measures are only relevant for large established firms that seek to enhance their existing business with opportunities in international markets (Chang, 1995; Johanson and Vahlne, 1977; Vernon, 1966). Different definitions of the term *born global* have been proposed in management literature (e.g., Knight and Cavusgil, 1996; Knight, 1997; McDougall and Oviatt, 2000; Wright and Ricks, 1994; Wurster, 2011). By taking previous definitions into account, Wurster (2011) proposed that *"Born globals are companies that operate in international markets from the earliest days of their establishment and derive a substantial proportion of their revenues from sales in these markets."* (Wurster, 2011:41). To identify born globals, Holtbrügge and Enßlinger (2005) and Holtbrügge and Wessely (2007) proposed the following criteria: First, a born global makes a first attempt towards internationalisation within the first three years of its lifecycle and a second attempt within the following three years, and second, the internationalisation activities take place in five or more countries, two different cultural clusters or two different geographic regions. In order to enter new markets, born globals exploit various opportunities to access networks, partners and build strategic alliances, e.g., via licensing, joint ventures or franchising. Key success factors of born globals are an entrepreneurial orientation, the ability to enter international markets early, to protect intangible assets and to penetrate market niches (Wurster, 2011).

Global corporate ventures

The interest in linking international business research with entrepreneurship is growing (McDougall and Oviatt, 2000; Young *et al.*, 2003). Moreover internationalisation plays an increasingly important role in corporate entrepreneurship (Hoskisson *et al.*, 2011). Callaway (2008) linked the concept of born globals with ICV. By presenting examples of ING Direct and HSBC Direct he showed how two large financial service companies were able to create international ventures via internal corporate venturing. He introduced the concept of *global corporate ventures* as vehicles to quickly bring new products to the international market place. Innovations introduced by global corporate ventures are often niche products or platform technologies which require a global reach to be successful. Internal ventures may internationalise early, regardless of the degree of internationalisation of the corporate parent and can therefore, become a vehicle to internationalise the parent itself (Callaway, 2008). Furthermore, global corporate ventures face a trade-off between independence and access to parental resources. The more independently a venture acts, the lower is the likelihood of receiving proper access to corporate resources and vice versa. However, more access to resources is linked with stronger parental monitoring which hinders entrepreneurial freedom of the venture (Callaway, 2008). This notion is in line with Burgelman's (1984) assessment that ICV top management "*should learn to assess better the strategic importance of ICV projects to corporate development and their degree of relatedness to core corporate capabilities*" (p. 44). Additionally high-technology ventures in the ICT industry can specifically benefit from network effects. Global corporate ventures that are able to scoop network effects and achieve technological lock-ins have the chance to increase internationalisation speed (Callaway, 2008) and thereby set de-facto standards (Wurster, 2011).

Research questions

We follow Callaway's (2008) suggestion to enhance research on global corporate ventures and to extend it to sectors outside of the financial services industry.

Moreover, we agree with the need to identify motivations for corporations to internationalise their ventures. Additionally, Callaway underlines the need for richer case studies to test his propositions and develop new ones.

As opposed to Callaway's research in which he focusses on chances of early internationalisation for internal corporate ventures, we aim to reveal challenges of spin-alongs that trigger early internationalisation. By analysing motivations and benefits of five early internationalising spin-alongs in the ICT industry, the following research questions will be addressed:

(1) What were the motives of the T-Labs' spin-along ventures to aim for early internationalisation?
(2) How did early internationalisation help the spin-along ventures of T-Labs to overcome internal barriers?

Methodology

Case studies are a suitable method to answer complex "how" and "why" questions and can serve to explore *"holistic and meaningful characteristics of real-life events"* (Yin, 2009:4). Furthermore, multiple case studies analyse and compare a set of cases and can therefore set a suitable level of abstraction with greater generalisability (Eisenhardt and Graebner, 2007). To answer our research questions we conducted a multiple case study with five of T-Labs' spin-along ventures.

Data collection

The sample for our analysis consists of five ventures from T-Labs' spin-along program, which were under development in 2013, and aimed for early internationalisation.

Semi-structured qualitative expert interviews were conducted in 2013, with eight members of the management staff of these spin-alongs (240 transcribed pages) and with four T-Labs managers (115 transcribed pages). Being members of the parental organisation ourselves, we had the chance to benefit from practitioner research (Jupp, 2006), and could include our observations of the spin-alongs' activities over a 12-month period into our research project. To ensure objectivity of this study, internal documents (e.g., guidance documents of the spin-along program, target pictures, success metrics, and plan versus reality comparisons), as well as public documents (e.g., press releases and websites of the ventures) were also collected and analysed.

Sample

The following paragraphs will provide a short overview of the five spin-along ventures. For the purpose of this publication the ventures' names have been changed to maintain confidentiality and to provide a descriptive title.

IPcall offers an internet protocol-based communication solution. As IPcall provides a technology that has the potential to cannibalise the parental core business it had to deal with low levels of trust internally.

SECphone delivers a security solution for mobile phones. A highly complex architecture of the product requires close collaboration of several highly-specialised individuals at the parent company. Since, thematically, SECphone was close to the parental core business, a business unit of the parent became its first customer.

TrafficNET offers a technology that enhances transparency of network utilisation. The underlying idea arose from the parent's own needs which it could address with the technology. TrafficNET commercialises this technology and builds on highly sensitive parental assets making a close collaboration with the parent necessary.

JinglePhone offers a mobile entertainment application which enables users to enrich phone calls with specific sounds. The idea resulted from a prior ideation project. Although it appeared trivial at first, it soon became apparent that the technological realisation is highly complex and requires access to sensitive parental assets.

InfoSense offers an analytics service based on complex data sources. Data is the core parental asset which InfoSense requires to run its business. InfoSense is challenged to fulfil strict legal requirements in the domestic market which limits its business potential.

Data analysis

Literature recommends the construction of category systems to analyse and interpret qualitative data. These categories can be built either deductively, based on existing theory, or inductively, constructed from the collected data (Saunders *et al.*, 2007). To analyse our data, the following steps based on Eisenhardt (1989) were taken:

(1) Following an inductive approach the transcribed interview data was reduced by coding all instances that reveal contexts and motivations for early internationalisation. In order to manage the large data volume Atlas.ti was used as coding software. In an iterative manner, the resulting categories were

regrouped to receive a comprehensive set of codes to categorise the identified motivations.
(2) From this first analysis step we shaped hypotheses about our research object. These were reviewed and validated in interviews with representatives of T-Labs' spin-along program.
(3) The resulting set of categories was reviewed with reference to similar and diverging findings in existing literature, until theoretical saturation was achieved.

Early Internationalisation Motives of T-Labs' Spin-Along Ventures

Spin-along executives are regularly confronted with challenges related to the fact that they need to manage their venture's positioning in the external market (moving from being an internal corporate venture to becoming a spin-along venture) and at the same time facing pressure to satisfy the goals and expectations of the parent firm. The continuous endorsement of the parent can be crucial to ensure the availability of sufficient support and financing. If a venture fails to manage this balancing act well, conflicting goals and activities can cause an untimely termination of ventures, regardless of their chances for success. Additionally, parent organisations are not always able to provide all of their ventures with the resources they need to run their business successfully, and detrimental domestic market conditions can hinder the profitable commercialisation of a venture's offer.

Many of T-Labs' spin-alongs faced these challenges when developing their business strategy. Our study focusses on the subset of spin-alongs that chose early internationalisation as a proper balancing mode (Table 1).

Avoidance of cannibalisation

IPcall developed a technical solution that had the potential to cannibalise the parental core business of traditional mobile telephony. When applying for internal funding the spin-along faced much internal resistance. From the parent's point of view the offering was too close to the current business and therefore, too risky. With this in mind, great chances were seen in commercialising IPcall's technology in markets in which the parent had only a weak footprint. This allowed building on already established resources, e.g., sales channels, as well as partner and customer networks in the target market, without threatening existing core revenue sources. At the point in time when our study was conducted, IPcall had not brought its

Table 1. Motives for early internationalisation.

Motives	Effects	Reference cases
Avoidance of cannibalisation	• Reduction or avoidance of competition with parental offerings in currently served market	• IPcall
Business with competitors	• Provision of offer to competitors in other markets	• TrafficNET • SECphone
Access to assets	• Faster utilisation of assets from subsidiaries in other markets	• JinglePhone
Market selection	• Selection of markets that provide beneficial conditions for successful launch	• InfoSense • TrafficNET

offering to the market yet. The consideration of an early internationalisation approach, however, offered the team a promising chance to avoid complete termination based on fears of cannibalisation, and allowed them to develop their product further.

The case of IPcall shows that spin-alongs, which commercialise technology from their parent's own R&D portfolio, may develop products that are rather strongly related to current parental products. This strong relatedness can raise corporate fears of disrupting currently stable revenue sources. Early internationalisation can offer chances for launching spin-alongs successfully, without cannibalising current parental business and can thereby, enable the parent to invest in promising initiatives while securing current business.

Business with competitors

TrafficNET initially started as an internal development project. It was based on the parent's own need of receiving high-quality information on its network utilisation. The team realised that the technology would also meet a strong demand in the external market. However, other companies that had a demand for this solution would be current competitors of the parent in the domestic market. In order to avoid critical conflicts at the management level, which risk termination of the venture, the team approached international customers who operated in markets where the parent had no footprint with the affected product. Thereby, TrafficNET could offer its service to foreign competitors of the parent while avoiding internal conflicts. If internationalisation would not have been identified as a suitable strategy, the venture might have never been started.

Furthermore, an early internationalisation can help to avoid parental lock-in effects that are not favoured from a venture's perspective. **SECphone** was one of the first ventures to be created as part of T-Labs' spin-along program. It benefitted from the chance to work closely with a parental business unit that provided it with its valuable distribution channels. The utilisation of these channels allowed attracting large business customers, however, the close collaboration with the internal partner also created barriers for SECphone. Since, the business unit assumed an important role in the venture's business activity and organisational structure, it was able to block aspirations of approaching other external customers in the domestic or international markets in which the business unit had an established footprint. Due to the exclusive focus on the parent as an internal customer, SECphone was not able to realise its whole business potential at the time when our study was conducted. While the early affiliation with the parental business unit helped to leverage SECphone's business, an early internationalisation could have taken place in parallel to avoid the creation of barriers and retain SECphone's independence. We believe that an expansion to other markets in which the business unit currently has no footprint may still be a feasible approach at this point.

Spin-along ventures that originate from internal development projects may solve problems with which the corporate parent is itself concerned. In this case, a pilot implementation at the parent can help ventures to deliver a proof-of-concept which may be a prerequisite for attracting prospective external customers. The early integration with the parent can also create organisational linkages which restrict the venture's approach of other target customers. This can especially hold true in cases where other potential customers to the solution could be current competitors of the parent. An early internationalisation can circumvent perceived threats of collaboration with competitors by serving these customers in markets where the parent currently has no footprint.

To reach a relevant market size, it is not advisable for spin-alongs to exclusively focus on internal customers but to also address external customers which in some cases may be competitors of the parent. Early internationalisation in markets where the parent organisation is not active is a valid approach to overcome resulting conflicts. However, in today's globalised world, distances between separated markets decrease constantly, which should be monitored and managed to avoid future conflicts. The European telecommunications industry is a rather specific example as it is shaped by a rather constant number of network operators in each country.

Access to assets

JinglePhone commercialised a technology that required utilisation of one of the parental core assets: the mobile telephony network. It soon became apparent to the team that the venture was too small to be granted the needed resources from the parent. The team looked for other countries in which the parent was active, to explore possibilities for partnerships with national network providers of the parent. By intensifying contacts to an international subsidiary of the parent, JinglePhone could gain access to the needed resources more easily. An adjustment of the business model to the specifications of the target market (e.g., pricing) was needed to create an offer that accounted for the local conditions.

In comparison to independent start-ups, corporate ventures can build on the assets of a large company. Needed assets may be tangible and intangible properties, such as IT infrastructure, sales channels, partner and customer networks and data, intellectual property rights, and also brand and market power. In a broader sense parental assets are resources that are needed to fulfil current business activity. Many of the T-Labs spin-alongs were built on parental assets, but in order to ensure continuous operation, the parent needed to carefully weigh its decisions of which ventures to grant its resources to and which not to. Despite their strategic importance, spin-alongs were not always considered high-priority candidates for receiving access to corporate assets. They needed to compete for resources with internal units and projects that already had established revenue streams and were therefore, often prioritised. In our research we found that international offices of the corporate parent were in some cases more willing to share their assets with new spin-along ventures. Due to the smaller size and stronger market orientation, these subsidiaries have a stronger resemblance to the structures of new ventures, making collaboration easier. Furthermore, spin-alongs can build on experiences of international offices and may in many cases be prioritised higher there.

To conclude, an early internationalisation can enable ventures to profit from access to resources of parental subsidiaries in other markets. Even if the corporate parent has no internationalisation experiences, building on other key partners' assets can serve as a competitive strategy of born global spin-alongs.

Market selection

Technology-push oriented ventures may trade highly specialised solutions. In many cases such specialised solutions are offered to narrow niche markets which can have a very small volume on a national scale, but may be large internationally. TrafficNET delivered a solution to a problem which only participants of a very narrow niche market had. In order to reach a relevant market size, the venture considered an international market launch and engaged in an early exchange with other market participants in international markets. An international market launch could, therefore, extend the size of the addressed market and create a promising outlook for future financial returns. Early internationalisation was a rational decision for TrafficNET, because it was the only way to commercialise the very specific invention in a rather narrow but international niche market.

InfoSense offered a service that was difficult to deliver in the domestic market due to strictly regulated legal conditions. The spin-along therefore, decided to launch a limited version of its offer in the domestic market with the intention to soon expand to an international market with less restrictive legal conditions. Internationalisation, thereby, provided a plan B for launching the product even though it was not in line with domestic legal restrictions.

In the two above-mentioned cases, an early internationalisation strategy was pursued because of the beneficial conditions or higher level of opportunities in foreign markets. Most of T-Labs' spin-alongs originated from a technology-push orientation, therefore, only few were based on a strong identified market demand. In fact, the fit of developed technologies with domestic market conditions was explored at a rather late stage. The capability to identify and select the right market can be vital for spin-along ventures. Foreign markets can serve as trial environments, where a spin-along can learn and sharpen its product offering. Markets generally differ from one another and products always need to be adapted to specific market conditions, however, early internationalisation can improve chances for success and enable a staged roll-out into different attractive markets. After all, the effort it takes to alter their own business model to local needs may be much lower than the effort it takes to overcome barriers which are in place in the domestic market.

Discussion

Positioning of T-Labs' born global spin-along ventures

In this section we will analyse how the envisaged internationalisation strategies of the five aspiring born global spin-alongs differ from one another. Recalling the definition of Wurster (2011), we identify born globals as firms that operate internationally from their earliest days on and generate a large share of their revenues from these international markets.

In Fig. 1 we adapted the 4-field matrix of types of international new ventures by Oviatt and McDougall (1994). It was modified to reflect the context of corporate venturing and we positioned the five ventures along the dimensions "number of activities across countries" and "number of countries involved." Since the analysed T-Labs ventures were only started in 2012, it is still open how they will evolve and if they are really going to become born globals by executing the internationalisation strategies that have been defined during their development. Their positioning within the matrix is therefore, based on the strategies that their leaders envisioned at the time when our research was conducted.

TrafficNET's sophisticated product can be used by all kinds of telecommunication companies over the world. If the venture launches internationally, not only

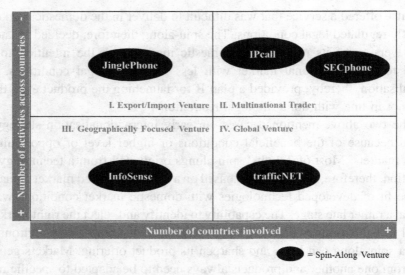

Fig. 1. Types of born global spin-along ventures (adapted from Oviatt and McDougall, 1994).

sales activities take place on an international level but also customer specific software implementation at the local telecommunication service provider's infrastructure. In our matrix, this places TrafficNET in the category **Global Venture**. According to Oviatt and McDougall (1994), ventures in this category have the potential to derive *"significant competitive advantage from extensive coordination among multiple organizational activities, the locations of which are geographically unlimited"* (p. 59).

JinglePhone's and InfoSense's business models are also designed to conduct most of their business activities across different countries, nevertheless, both ventures' geographical reach is focused on countries in which their corporate parent has an already established footprint. This is based on the fact that both ventures build strongly on parental assets which need to be in place continuously in each country that they want to operate in. In the long run they may have the opportunity to decouple their business from parental assets and open up for other enabling assets. However, JinglePhone and InfoSense differ from each other in the amount of cross country operating business activities.

JinglePhone is based on a technology that is not expected to need to adapt extensively to specific local conditions in foreign markets (e.g., domestic legal regulations or enabling technical infrastructures) and may therefore be classified as an **Export/Import Venture** with only few activities dispersed across countries. A long-term competitive advantage of international new ventures in this category depends on *"(1) unusual abilities to spot and act on (...) emerging opportunities*

(...), *(2) knowledge of markets and suppliers, and (3) the ability to attract and maintain a loyal network of business associates*" (Oviatt and McDougall, 1994, :58).

Due to the different national regulations for procession and commerce with data (the core business of InfoSense), we expect that InfoSense will need to adapt more strongly to local conditions in foreign markets than JinglePhone does and may therefore, need to conduct more activities in its target markets, classifying it as a **Geographically Focussed Venture**. This type of international new ventures may secure its advantages by "*a close and exclusive network of alliances in the geographical area served*" (Oviatt and McDougall, 1994:59).

IPcall and SECphone can operate in many different countries as they are not restricted to countries in which their parent already has a footprint. As they are also not expected to be required to adapt their activities strongly to local contexts in the target markets, their activities in foreign markets may be limited (e.g., to sales activities). IPcall and SECphone can be categorised as **Multinational Traders**, who operate mostly from their domestic market and reach into many different markets.

Reflection on theory

Callaway (2008) proposes that if a corporate parent has strong international experiences and a widely established footprint in global markets, global corporate ventures may exploit this and internationalise faster than more independent and decoupled ventures. Our analysis underlined that this does not always hold true. Especially for spin-along ventures, a more nuanced viewpoint should be taken. Spin-along ventures that commercialise parental assets and have the ability to cannibalise the parental business should, under specific circumstances, avoid commercialisation in parental markets. Although, a willingness to cannibalise has been named in management literature as a key success factor for remaining innovative (Nijssen *et al.*, 2005), its realisation in management practice is difficult when the fulfilment of ongoing contracts and target agreements is at risk. This also holds true when a venture's business with parental competitors is necessary to generate sufficient revenue. Spin-along ventures need to find the right (niche) market to develop their products and to scale the business up from there. Compared to Callaway's findings, our research does not primarily focus on faster internationalisation, but gives antecedents for how a venture can be spun-out despite internal barriers.

Furthermore, we agree with Callaway that the degree of a venture's independence is directly related to the support given by the parent, and that a high level of support (e.g., granting access to resources) is coupled with more monitoring and guidance. When it comes to business with competitors it seems that only less integrated ventures are likely to attract customers that are competitors of the corporate parent. This is not only the case because customers may want more independent spin-alongs, but also because the parental unit may prohibit business with competitors if the level of integration, and by that, influencing power is high. Nevertheless, we suggest that the importance of parental assets may decrease over time. At the beginning, parental assets build a competitive advantage compared to independent start-ups, however a born global spin-along should aim for building own distinct assets and capabilities to increase independence from the corporate parent in the long run. In this regard, the selection of the right market plays a key role. Our findings are in line with Jolly *et al.* (1992) by saying that for a few new ventures internationalisation is the only way to start a business based on high initial investments in R&D. Furthermore, we agree with scholars that new ventures, which satisfy a need of a niche market need to internationalise early after inception, because the domestic market is too small to gain significant revenues (Hordes *et al.*, 1995). From the perspective of the spin-along approach, internationalisation measures do not primarily aim for returning R&D investments, but rather for expanding into global niche markets to make commercialisation generally possible.

Teece (1986) claims that "*in almost all cases, the successful commercialization of an innovation requires that the know-how in question be utilized in conjunction with other capabilities or assets*" (p. 288). This means that the single invention is not enough to successfully turn it into an innovation. For T-Labs' spin-along ventures complementary assets like the parental brand reputation, sales channels, funding capacities, network embeddedness, and also knowledgeable and experienced staff and shared services from the corporate parent may leverage their success. But in general, how can a corporate judge the suitability of an invention to be commercialised by a corporate venture? One explanation can be the above-mentioned degree of dependence on complementary assets. One can assume (and our study showed evidence for that), that if an innovation needs complementary assets which (only) the corporate parent can offer, then a new venture has a significant competitive advantage in the market, in case it is granted exclusive access to these assets.

Conclusion

With our research questions we sought to identify what motivations T-Labs' spin-alongs had to aim for early internationalisation and how this approach helped to overcome internal barriers. By analysing the experiences of five new ventures of T-Labs ventures, we could narrow down the core motivations for internationalisation and could identify early internationalisation as a strategy that can circumvent specific challenges.

Four major motivations for early internationalisation could be identified: (1) to avoid termination based on the parent's perceived threat of cannibalisation of existing products, (2) to enable a venture's collaboration with competitors, (3) to overcome restrictions of parental assets in the domestic market, and (4) to address markets that offer greater chances for success than the domestic market does. We thereby identify that internationalisation can serve as

(1) a measure taken to avoid or reduce conflicts with the corporate parent,
(2) an approach to increase likelihood of access to parental assets by sourcing from international subsidiaries, and
(3) a measure to select and address those markets in which the offer had the greatest chances to succeed.

Our research aims to enhance literature on the spin-along approach and provides the research community with cases that serve to further analyse the concept and its challenges. Moreover, it highlights the legacy of the spin-along approach as a distinctive approach of corporate venturing. By linking the spin-along approach with born global theory, potentials could be identified that current literature has not yet focussed on.

Limitations and future research

Our study focussed on the benefits that early internationalisation offers to corporate spin-alongs of T-Labs, reflecting cases in the ICT industry. The ICT industry, and especially the telecommunication service provider market, is exceptional. Additionally, telecommunication providers, have a rather constant footprint in several markets. To build or buy a telecommunication network infrastructure or to acquire a whole operator, requires high investments and takes place only occasionally. Other industries might face much more dynamic international activities and case studies in those fields may show other characteristics.

Moreover, our study is limited in a sense that it incorporates the findings of only a small number of cases, however we aimed to show deep insights from multiple cases and did not aim for generalisable findings. Last, it should be noted that the T-Labs spin-along program was still at an early stage of its implementation at the time when this study was conducted. Therefore, it is still too early to draw conclusions from the effects of the early internationalisation aspirations and the realisation of respective strategies in the long run. Future research should therefore, include long-term effects of early internationalisation on spin-alongs.

Our study aimed to broaden the scope of the spin-along approach by linking it with another theoretical concept — born globals. Future studies may aim at deepening the spin-along approach by explaining its foundation with the help of theories like resource-based view or transaction cost economics. Mahdjour and Fischer (forthcoming) took a first step in that direction.

Acknowledgement

We would like to thank all of our interviewees who took the time and effort to provide us with the information that the findings in this paper are based on. Also we would like to thank our reviewers for the very helpful feedback that was incorporated to improve this publication.

References

Backholm, A (1999). *Corporate Venturing: An Overview*. Espoo: Helsinki University of Technology.

Burgelman, RA (1983). A process model of internal corporate venturing in the diversified major firm. *Administrative Science Quarterly*, 28(2), 223–244.

Burgelman, RA (1984). Managing the internal corporate venturing process. *Sloan Management Review*, 25(2), 33–48.

Callaway, SK (2008). Global corporate ventures: A new trend of international corporate entrepreneurship. *Multinational Business Review*, 16(3), 1–22.

Chang, J (1995). International expansion strategy of Japanese firms: Capability building through sequential entry. *Academy of Management Journal*, 38(2), 383–407.

Eisenhardt, KM (1989). Building theories from case study research. *The Academy of Management Review*, 14(4), 532–550.

Eisenhardt, KM and ME Graebner (2007). Theory building from cases: Opportunities and challenges. *Academy of Management Journal*, 50(1), 25–32.

Holtbrügge, D and B Enßlinger (2005). Initialkräfte und Erfolgsfaktoren von Born Global Firms (Working Paper No. 2/2005). University of Erlangen-Nürnberg.

Holtbrügge, D and B Wessely (2007). Initialkräfte und Erfolgsfaktoren von Born Global Firms. In Oesterle M.-J. (ed.), *Internationales Management im Umbruch*, 169–205. Germany: Deutscher Universitätsverlag.

Hordes, MW, JA Clancy and J Baddaley (1995). A primer for global start-ups. *Academy of Management Perspectives*, 9(2), 7–11.

Hoskisson, RE, J Covin, HW Volberda and RA Johnson (2011). Revitalising entrepreneurship: The search for new research opportunities. *Journal of Management Studies*, 48(6), 1141–1168.

Johanson, J and J-E Vahlne (1977). The internationalization process of the firm — A model of knowledge development and increasing foreign market commitments. *Journal of International Business Studies*, 8(1), 23–32.

Jolly, VK, M Alahuhta and J-P Jeannet (1992). Challenging the incumbents: How high technology start-ups compete globally. *Strategic Change*, 1(2), 71–82.

Jupp, V (2006). *The SAGE Dictionary of Social Research Methods*. Thousand Oaks, CA: Sage Publications.

Keil, T (2004). Building external corporate venturing capability. *Journal of Management Studies*, 41(5), 799–825.

Klarner, P, T Treffers and A Picot (2013). How companies motivate entrepreneurial employees: The case of organizational spin-alongs. *Journal of Business Economics*, 83(4), 319–355.

Knight, GA (1997). Emerging paradigm for international marketing: The born global firm. University Microfilms International Dissertation Service, Michigan State University.

Knight, GA and ST Cavusgil (1996). The born global firm: A challenge to traditional internationalization theory. In Madsen TK (ed.), *Advances in International Marketing Vol. 8*, 11–26. Bingley, UK: Emerald Group Publishing Limited.

Mahdjour, S and S Fischer (forthcoming). Implementing the spin-along approach — A capability analysis of Telekom Innovation Laboratories' corporate venturing program. *International Journal of Technology Marketing*.

McDougall, PP and BM Oviatt (2000). International entrepreneurship: The intersection of two research paths. *Academy of Management Journal*, 43(5), 902–906.

Michl, T, B Gold and A Picot (2012). The spin-along approach: Ambidextrous corporate venturing management. *International Journal of Entrepreneurship and Small Business*, 15(1), 39–56.

Nijssen, EJ, B Hillebrand and PAM Vermeulen (2005). Unravelling willingness to cannibalize: A closer look at the barrier to radical innovation. *Technovation*, 25(12), 1400–1409.

Oviatt, BM and PP McDougall (1994). Toward a theory of international new ventures. *Journal of International Business Studies*, 25(1), 45–64.

Riege, A (2007). Actions to overcome knowledge transfer barriers in MNCs. *Journal of Knowledge Management*, 11(1), 48–67.

Rohrbeck, R, M Döhler and H Arnold (2009). Creating growth with externalization of R&D results — The spin-along approach. *Global Business and Organizational Excellence*, 28(4), 44–51.

Rohrbeck, R, M Döhler and HM Arnold (2007). Combining spin-out and spin-in activities — The spin-along approach. Paper presented at *ISPIM 2007 Conference: "Innovation for Growth: The Challenges for East & West"*, pp. 1–12, June 17, Warsaw, Poland.

Saunders, M, P Lewis and A Thornhill (2007). *Research Methods for Business Students*. Pearson Education Limited.

Sharma, P and JJ Chrisman (1999). Toward a reconciliation of the definitional issues in the field of corporate entrepreneurship. *Entrepreneurship: Theory and Practice*, 23(3), 1–16.

Teece, DJ (1986). Profiting from technological innovation: Implications for integration, collaboration, licensing and public policy. *Research Policy*, 15(6), 285–305.

Vernon, R (1966). International investment and international trade in the product cycle. *The International Executive*, 8(4), 190–207.

Wright, RW and DA Ricks (1994). Trends in international business research: Twenty-five years later. *Journal of International Business Studies*, 25(4), 687–701.

Wurster, S (2011). Born Global Standard Establishers: Einfluss- und Erfolgsfaktoren für die internationale Standardsetzung und -erhaltung. Germany: Gabler Verlag.

Yin, RK (2009). *Case Study Research: Design and Methods*. Bickman, L and Je. Rog, D (eds.), Thousand Oaks, CA: Sage Publications.

Young, S, P Dimitratos and L-P Dana (2003). International entrepreneurship research: What scope for international business theories? *Journal of International Entrepreneurship*, 1(1), 31–42.

Chapter 10

Innovative Born Globals: Investigating the Influence of Their Business Models on International Performance[*]

Sascha Kraus, Alexander Brem, Miriam Schuessler,
Felix Schuessler and Thomas Niemand

Internationalization is a hot topic in innovation management, whereby the phenomenon of "Born Globals" is still limited to research in the domains of Entrepreneurship and International Management. As business model design plays a key role for Born Globals, we link these two concepts. For this, we propose hypotheses about the influence of efficiency-centered and novelty-centered business model design on international firm performance. To test these hypotheses, we performed a quantitative survey with 252 founders of international companies in Germany, Switzerland and Liechtenstein. Additionally, we gained further insights through a case study analysis of 11 Born Globals. The results show that business model design matters to international firm performance and the business model design of Born Globals tends to be more efficiency-centered. Based on a multiple case study, we analyzed business models in a more sophisticated way and derived propositions that yielded in an archetype of a Born Global's business model.

Keywords: Born global; international new venture; business model; internationalization; international performance.

Introduction

During the past decades the volume of global business has increased dramatically and has forced more and more small companies to start their international activities right after their inception (Knight and Cavusgil, 2004; Oviatt and McDougall, 1994; Rennie, 1993; Welch and Luostarinen, 1988). The expansion of *global networks and alliances* was fostered by the rapid development of advanced communication technologies which accelerate the transfer of information around the globe (Knight and Cavusgil, 2004; Moen, 2002). In this context, the term *Born*

---

[*]Originally published in *International Journal of Innovation Management*, 21(1), pp. 1–54.

Global (BG) is gaining momentum in business practice as well as research and due to the increasing number of firms that can be classified as BGs and their potential to become a leading species in the ecosystem of international trade, the phenomenon BG has become an important driver of international entrepreneurship research (Andersson, 2011; Knight and Cavusgil, 2004; Moen, 2002). But what is the phenomenon of BGs, why it is important and what do we know about them?

We know that the distinguishing characteristics of BGs are that they start internationalizing rapidly, right after their foundation or soon afterwards (Chetty and Campbell-Hunt, 2004; McDougall *et al.*, 1994; Moen and Servais, 2002) and "sell a high share of their output abroad" (Hennart, 2014, p. 117). The debate about theories and models, which explain BGs best, is still ongoing. Blurred boundaries between different countries, new technologies and cheaper exchange of goods were seen as the main drivers that provide opportunities for young fast growing firms (Moen, 2002; Oviatt and McDougall, 1997; Solberg, 1997). On the other hand, especially new ventures are facing specific problems "when dealing separately with business development, innovation and early internationalization" and have to find an appropriate way to handle "complexity, uncertainties and risks of being innovative on a global scale" (Tanev *et al.*, 2015, p. 1). But the economic conditions have changed for all companies around the globe (for the old and the new ones) and so the question remains unanswered why BGs obviously exploit new international chances and technological advantages better than the majority of the existing companies (Trkman *et al.*, 2015).

What do we not know? In this context, the term *business model* gets rising attention which provides a framework that commercializes innovations, but more important it can be the decisive accelerator for a company's internationalization and thus represents a new theoretical approach within the BG research (Mahdjour and Fischer, 2014).

Despite the growing interest in the potential influence of the business model on international performance, empirical findings are rare and not clear-cut and the research field so far lacks a comprehensive framework that supports entrepreneurs in their endeavor to design their firms' business models. At the spearhead of research on BGs, business models play an increasingly important role, but it is still undetermined which specific features lead to outstanding their performance in international markets (George and Bock, 2011; Hagen *et al.*, 2014; Onetti *et al.*, 2012; Rask, 2014). In this context, research is needed because business models are seen as a source of innovation and continuous improvement that can lead to more informed decisions and thus help the entrepreneur to increase the chances of success due to the usefulness and predictable power of business models (Trimi and Berbegal-Mirabent, 2012).

In attempt to advance the current understanding of BGs' business models and its impact on international performance this study has the overarching aim to address the research questions of how the design of the business model does influence the international performance of BGs? Further, what is the dominant design theme that connects the elements of a BG's business model? And if BGs distinguish themselves through a specific configuration of business model components?

The main contribution of this study is to provide an empirical linkage between the business models of BGs and their influence on international performance in terms of both theory and empirical findings. By combining the methodological approaches of a systematic literature review with an empirical mixed methods approach, we are able to overview, analyze and critically evaluate the scientific field of study and to add empirical research findings of both quantitative and qualitative nature which yields in a comprehensive conceptual framework that helps as well researchers and practitioners in their endeavor to investigate the phenomenon of BGs by illustrating the important design themes and components of a BG business model archetype.

In the following section, we provide a broad and multifaceted view of relevant literature on BGs and business models in the context of international performance and develop a set of hypotheses intended to assess the particular design theme of a BG business model. We next explain our underlying research design followed by the detailed description of the single steps conducted in the mixed method approach, the sample and data collection. In Chapter 4 we present the empirical results of the study of both the quantitative and qualitative approach by outlining the statistical analysis of the large-scale survey and developing a conceptual framework by providing an archetypical business model profile of the BGs within the overall sample. Finally, we report on empirical findings, providing discussion and conclusions along with some important directions for further BG business model research yielding to implications for theory and practice.

Theoretical Foundation and Hypotheses

Innovative born globals

Theoretical explanations about why and how BGs seek to internationalize vary. Researchers already discussed the phenomenon in conjunction with the role of innovative culture, knowledge and capabilities (Knight and Cavusgil, 2004) and the acting individual; i.e., the manager or entrepreneur, seems to be the key element for a rapid and successful internationalization of BGs (Park and Rhee, 2012). Based on our research, conceptual explanations of accelerated internationalization of BGs can be divided into three levels that lead us to the following conclusions regarding BGs:

- The *individual-level* puts the *entrepreneur*, his *experience* and *knowledge* gained through specific learning strategies in the center to explain the rapid internationalization of BGs while entrepreneurial orientation, and especially the innovativeness component, is positively related to international performance (Carayannis *et al.*, 2015; Freeman and Cavusgil, 2007; Kim and Min, 2015; Park and Rhee, 2012; Smith, 2015; Taran *et al.*, 2015).
- The *firm-level* puts the *networks and alliances* in the center which can accelerate their entrance to foreign markets by using unique firm-specific *resources* (of technological, organizational, relational and human nature) and *hybrid structures* (e.g., extensive outsourcing to partners) to explain how BGs overcome their liability of smallness and reach potential markets simultaneously through their shared knowledge of niche products (Autio *et al.*, 2000; Barney, 1991; Bouncken *et al.*, 2015b; Chroneer *et al.*, 2015; Hennart, 2014; Trkman *et al.*, 2015; Young, 1987).
- The *national-level* puts small *countries* with advanced economies (in which also reductions in communication and transportation costs have played an important role) in the center because BGs appear frequently in these surroundings (Almor, 2013; Fan and Phan, 2007; Schuessler *et al.*, 2014).

The research outcomes were not sufficient to explain the success story of BGs and new theoretical approaches need to be developed to explore what we do not know yet about the phenomenon of BGs. In this context, Fan and Phan (2007), Hennart (2014) and Bouncken *et al.* (2015b) argue that the uniqueness of BGs is related to their *business models*. Onetti *et al.* (2012) predicate that the survival and success of a BG depends on effective business model design. The term business model has emerged as one of the key drivers to commercialize innovations (Chesbrough and Rosenbloom, 2002; Chesbrough, 2010; Teece, 2010) and provides "the framework for a firm to create and capture value out of an innovative idea or technological development" (Schneider and Spieth, 2013, p. 1).

International performance of BGs

Ranging in the firm-level, especially BGs internationalization models which include the processes, actions and outcomes and so the configuration of their internationally dispersed resources reflect the business model and is considered to be the key to the firm's international performance (Coviello, 2015; Zander *et al.*, 2015). Due to the increasing internationalization that leads to globalized competition, international performance is of particular importance in economics and business literature, and research has developed various measures of unidimensional and multidimensional nature to assess the firm's *internationalization* and *performance* (Glaum and Oesterle, 2007; Ietto-Gillies, 1998; Ramaswamy et al., 1996; Sullivan, 1994a,b, 1996). Additionally to the discussion regarding BGs international performance, Hennart (2014) describes BGs even as accidental internationalists because in "acquiring foreign customers is [...] no different than acquiring domestic ones" (p. 118). So far, empirical definitions for BGs vary significantly. While Evers (2010) suggests 25% of total sales in foreign countries in the first year of trading, Chetty and Campbell-Hunt (2004) set a benchmark of 75% export intensity within two years after inception. Already, McDougall and Oviatt (1996) stress that measuring the international performance of BGs is a difficult and complex endeavor and suggest to use the return on investment and relative market share also in a follow-up study considering a longitudinal view. However, Glaum and Oesterle (2007) point out that unidimensional measures which mainly focus on the share of exports as a percentage of total sales revenue or the number of subsidiaries in foreign countries are not sufficient to assess internationalization with its multi-dimensionality.

In this context, recent BG research conducted by Cesinger et al. (2012) concludes that "rapid internationalization manifests itself in the degree of internationalization" (p. 1830) which represents a crucial measurement that has been discussed for long time. The degree of internationalization (DOI) can be derived through the well-accepted transnationality index (TNi) that is based on the World Investment Report by the UNCTAD (1995) which focuses on the relationship between the activities in the domestic and foreign country for any particular company and represents the first and one of the most sophisticated multi-dimensional DOI indicators (Dörrenbächer, 2000; Ietto-Gillies, 1998). The calculation of the TNi is conducted through the average of the three ratios of foreign to total elements based on the sales, assets and employment which is outlined in the following formula by Ietto-Gillies (1998, p. 19).

$$TNi = \frac{Ai + Si + Ei}{3}$$

Ai = Foreign Asset index = foreign assets/total assets
Si = Foreign Sales index = foreign sales/total sales
Ei = Foreign Employment index = foreign employment/total employment

Besides the general accepted measure of the DOI, research proposes for measuring international performance to further include international orientation, skills, innovativeness and market orientation representing all together the international business competence that influences positively the international market share, sales growth, profitability and export intensity all leading finally to international performance (Knight and Kim, 2009). As many BGs are involved in advanced niche markets or knowledge-intensive industries which implies that they have to cooperate with internationally dispersed and specialized suppliers and customers (representing their international market orientation) by using their international skills and innovativeness, they have from their nature a high international business competence (Cavusgil and Knight, 2015; Hennart, 2014; Knight and Kim, 2009; Zander *et al.*, 2015). In this context, recent research agrees that BGs start international activities early after founding and have a high speed, scale and scope of internationalization as core characteristics (Cesinger *et al.*, 2012; Freeman and Cavusgil, 2007; Kuivalainen *et al.*, 2007; Madsen *et al.*, 2012). Based on a comprehensive review of the current literature a definitional corridor explaining the BGs international nature can be derived as follows (Table 1).

Moreover, researchers have observed that many BGs are forced into globalization because of their high specialized products (Trkman *et al.*, 2015) and that their export performance depends on *innovations*: the more a firm innovates the more innovations will be exported (Pels and Kidd, 2015). As reflection of the business model their product differentiation strategy influences positively the internationalization process (Gudiksen, 2015) and technological intensity as well as R&D intensity, representing the BGs international innovativeness, are

Table 1. Key dimensions to define BGs.

Dimension	Measurement	References
Speed	The internationalization process has to happen within the first five years after foundation.	McDougall (1989); Rennie (1993); McAuley (1999); Zahra *et al.* (2000); Shrader (2001); Bell *et al.* (2001); Andersson and Wictor (2003); Knight and Cavusgil (2004)
Scale	≥25% of a company's total revenue has to be gained through international sales.	Andersson and Wictor (2003); Bell *et al.* (2003); Knight and Cavusgil (2004); Belso-Martínez (2006); Evers (2010); Mandl and Celikel-Esser (2012)
Scope	At least four foreign markets have to be conquered — these markets have to dispose of a high psychic, cultural and geographic distance.	Oviatt and McDougall (1994); McAuley (1999); Zahra *et al.* (2000); Moen (2002); Andersson and Wictor (2003); Bell *et al.* (2003); Knight and Cavusgil (2004); Laanti *et al.* (2007); Kuivalainen *et al.* (2012)

significantly associated with export performance (Amshoff *et al.*, 2015; Dhanaraj and Beamish, 2003).

Whereas traditional firms focus on the domestic market for many years and gradually internationalize their operations by building up experience and commitment step by step, BGs have an international orientation with a global view of their markets right from the beginning (Bell *et al.*, 2003; Knight and Cavusgil, 2004). In contrast to the Uppsala counterpart with its typical phases, BGs jump straight from the domestic to the global phase, a phenomenon also known as leapfrogging, and psychic distance plays only a minor role in the internationalization process (Fink *et al.*, 2008; Gabrielsson *et al.*, 2012). In this context, by targeting international and emerging markets Cavusgil and Knight (2015) highlight the importance of developing "new business models" (p. 13). Also, recent research from Zander *et al.* (2015) state that studies of the business model could promise important and helpful information about the phenomenon of the BGs and their unique system.

In research there has been a growing academic interest in how firms design their business models, especially in the fields of *innovation management* and *entrepreneurship* (George and Bock, 2011; Schneider and Spieth, 2013; Trimi and Berbegal-Mirabent, 2012). In this case, business models are seen as a *source of innovation and continuous improvement* that can lead to more informed decisions and thus help the entrepreneur to increase the chances of global success due to the usefulness and predictable power of business models (Trimi and Berbegal-Mirabent, 2012). Moreover, Onetti *et al.* (2012) argue that the survival and success of a firm depends on effective *business model design*, where decisions about core activities and about where to focus investments are interconnected to decisions concerning location of activities, and on inward and outward relationships with other players.

In a nutshell, the design of the business model has received rising attention by practitioners as a means for firms to achieve *superior performance* (Frankenberger *et al.*, 2013; George and Bock, 2011; Nair *et al.*, 2013; Trimi and Berbegal-Mirabent, 2012). Furthermore, several researchers argue that the uniqueness of BGs is related to their design of the business model (Bouncken *et al.*, 2015b; Fan and Phan, 2007; Hennart, 2014; Onetti *et al.*, 2012). Hence, Bouncken *et al.* (2015b) conclude that the business model can be seen as main accelerator of the international performance of BGs. Therefore, considering a positive effect of the business model design on international performance, we expect differences between BGs and traditional internationalizing firms in their business model designs.

H1: *The business model design of BGs differs from the business model design of traditional internationalizing firms.*

Business models

Developing the term

Our review of the literature on business model research reveals that very few authors focus on internationalization and the term did not see widespread use for decades (Gray and Farminer, 2014; Onetti *et al.*, 2012; Osterwalder and Pigneur, 2005). However, with the emergence of Internet companies and the development of information and communication technologies, the term *business model* quickly gained prominence among both business scholars and practitioners (DaSilva and Trkman, 2014). The business model literature came out of empirical settings in e-business and entrepreneurship and finally has been argued to be a relatively new and potentially powerful generic concept in strategic and innovation management literature (Alt and Zimmermann, 2001; Mason and Spring, 2011; Onetti *et al.*, 2012; Richardson, 2008; Teece, 2010; Timmers, 1998; Zott and Amit, 2008, 2010, 2013).

To clarify the term business model in more detail it is advisable to make a clear separation between the business model and the strategy concept. Following Casadesus-Masanell and Ricart (2010), "business models are reflections of the realized strategy" (p. 204). Additionally, according to Ritala *et al.* (2014), the business model has been defined as a generic platform between strategy and practice. However, DaSilva and Trkman (2014) argue that strategy shapes the development of capabilities which can alter current business models in the future and is about building dynamic capabilities. They argue that strategy (a long-term perspective) sets up dynamic capabilities (a medium-term perspective), which then constrain possible business models (present or short-term perspective to face both upcoming and existing contingencies). Furthermore, they emphasize that "strategy reflects what a company aims to become, while business models describe what a company really is at a given time" (p. 5).

Proposal for a common definition

In a competitive environment dominated by turbulence and complexity the business model gains further importance and is becoming more and more popular but only few studies are concerned with early internationalization and growth factors (e.g., Dunford *et al.*, 2010; Sleuwaegen and Onkelinx, 2014). However, there is an ambiguity of definitions because these are fairly heterogeneous and none appears to be generally accepted. Therefore, Bouncken *et al.* (2015b) compared previous definitions, synthesized the characteristics of the term business model and implemented an international view of the business model by arguing that through its adaption to international markets it can represent an important growth component and suggest the following definition:

> "A business model is a strategic and dynamic value-creation process among a value network that is characterized by the way the type of product or service is linked to a particular group of customers using a specific communication and delivery method and accelerates, by adaptation, the early internationalization process"(p. 250).

Operationalizing a business model

As so far, the academic literature on business models is fragmented and lacks of coherence. Therefore, some scholars have concentrated on the way to conceptualize the business model (George and Bock, 2011; Morris et al., 2005; Tracey and Jarvis, 2007; Trimi and Berbegal-Mirabent, 2012; Zott and Amit, 2007, 2008).

Operationalizing the business model on a high level of granularity

According to George and Bock (2011) the work conducted by Zott and Amit (2007, 2008) represents an exceptional work in business model research because unlike most of the other works it builds on prior research within a coherent framework. Thereby, it examines the relationship between business model design and firm performance on a *transactive* level that is considered by Trimi and Berbegal-Mirabent (2012) "the approach that has generated the greatest interest among academics" (p. 453). Zott and Amit (2007, 2008) derived the design themes *novelty* and *efficiency* from previous work conducted by Miller (1996) and developed them to simplify and describe the conceptualization and measurement of the business model on a high level of granularity, as they are not mutually exclusive and represent the corresponding themes to the product market strategy level of product differentiation and cost leadership. Anchored in the Schumpeterian innovation regarding novelty, or in transaction cost economics regarding efficiency, they argue that the business model is dominant to value creation and hence the overall firm performance is influenced by the specific business model characteristics that lead to competitive advantages (Zott and Amit, 2007).

In this context, within the *novelty-centered business model design* the focal firm represents the innovator with a focus on innovation in its business model whereby either a new market or new opportunities and transactions in existing markets can be created, developed and exploited through an innovative business model (Amit and Zott, 2008; Amit and Zott, 2012; Zott and Amit, 2007). Recent studies have emphasized the role of country- and culture-specific influences on a business model's design when it comes to internationalization of new product development (Brem and Freitag, 2015). Thereby, innovative opportunities can be co-created through the entrepreneur who is acting as designer by bridging "factor and product markets in new ways" (Zott and Amit, 2007, p. 184) based on new information and communication technologies. Research on entrepreneurial teams with a multicultural background has shown that internationality can be a wellspring of creativity and innovativeness (Bouncken et al., 2015a).

Among other things, it can be constituted that BGs tend to conduct *novel* ways of transactions to overcome constraints of smallness (e.g., by using non-equity modes of entry) (Acedo and Jones, 2007; Oviatt and McDougall, 1994). Furthermore, as part of the firm-level variables that lead to rapid internationalization, the RBV can again be addressed because BGs use unique and thus novel organizational, relational and firm-specific resources to compete in international markets (Autio, 2005; Bouncken et al., 2015b; Jones et al., 2011; Knight and Cavusgil, 2004; Rialp et al., 2005). Hence, we expect in accordance with the mentioned insights:

H2: *Having a novelty-centered business model design is positively associated with the BG's international performance.*

Besides the *value creation* also the *value appropriation* among several participants needs to be considered (Amit and Zott, 2012; Zott and Amit, 2007). By connecting before unconnected parties, by linking transaction participants in new ways or by designing new transaction mechanisms, the novelty-centered business model focus relays on creating novel types of economic exchanges among the transactions partners representing the focal firms *network* (Zott and Amit, 2007, 2008). In this context, the empirical analysis conducted by Zott and Amit (2007) concludes that the novelty-centered business model design has a positive effect on the entrepreneurial firm's performance even when the necessary resources are limited that facilitate the implementation of steps regarding business model innovation.

Furthermore, BGs overcome hurdles by using *networks and alliances* frequently which enhances their competitive power due to gained knowledge and resources that accelerates their *transactions* and thus the market entry (Autio, 2005; Bouncken et al., 2015b; Freeman and Cavusgil, 2007; Freeman et al., 2006; Moen, 2002). As the novelty-centered business model design aims to create novel transactions among the network partners, it can be referred to the theory concerning the mediating role of network relationships on speed of internationalization, which is stressed by several authors such as Madsen and Servais (1997), Oviatt and McDougall (2005) and Bouncken et al. (2015b) and brought into line with the previously described impact of the business model design on international performance (Zott and Amit, 2007, 2008). Therefore, we expect that:

H3: *The network mediates the relationship between a novelty-centered business model design and the BG's international performance.*

The *efficiency-centered business model design* focuses on conducting transactions among the participants in a more efficient way by reducing for example transaction costs or complexity to enhance the information flow (Amit and Zott, 2012; Zott and Amit, 2007, 2008). Thereby, all participants shall benefit from the reduction of information asymmetries and transaction costs (Zott and Amit, 2007, 2008). Furthermore, Zott and Amit (2007) argue that due to a reduction of transaction costs related to search, transportation and coordination activities the firm gets more competitive because potential customers, partners and suppliers are more replaceable and the focal firm can act in a more dynamic and efficient way.

In accordance with the theoretical insights, BGs tend to sell niche products which implies a cooperation with the transaction partners in a more *efficient* way, based on their domain-specific familiarity, hence potential markets may be reached simultaneously and thus also more efficient (Fan and Phan, 2007; Hennart, 2014). Thereby, referring to the concept of *dynamic capabilities* it can be constituted that BGs conduct efficiency enhancements by adapting and integrating external knowledge (Eisenhardt and Martin, 2000; Jantunen *et al.*, 2008; Teece, 2007). Indeed, when resources are used to improve the efficiency among transactions the firm performance tends to be higher also in times of limited resources (Zott and Amit, 2007). According to the mentioned insights and the fact that BGs represent global acting entrepreneurial firms that are limited in their resources, we expect that:

H4: *Having an efficiency-centered business model design is positively associated with the BG's international performance.*

Based on the empirical results from Zott and Amit (2007) they determine that in times of limited resources firms have a higher performance when efficiency enhancements are included in the business model design. Hence, the value proposition as well as the value appropriation among customers, partners and suppliers are enhanced through a simplified and faster procedure of transactions that leads to a reduction of the related costs (Zott and Amit, 2007, 2008).

Furthermore, Hennart (2014) argues that there is no difference for BGs between conducting foreign and domestic transactions because they make full use of low-cost rapid communication and distribution methods that *facilitate transactions among participants* who are spread all over the world. Hence, the interaction with the network partners is enhanced due to the reduction of information asymmetries which enables faster and more cost-efficient transactions (Zott and Amit, 2007, 2008). Therefore, considering a positive effect of the network as mediating element, we suggest that:

H5: *The network mediates the relationship between an efficiency-centered business model design and the BG's international performance.*

So, the survival and success of a firm, or in other words, the firm performance depends on an effective business model design (as an alternative or complement to product or process innovation) that furthermore gets increasing attention due to the rising global competitive pressures (Amit and Zott, 2012; Onetti *et al.*, 2012; Turcan and Juho, 2014; Zott and Amit, 2007).

Operationalizing the business model on a fine level of granularity

According to the analyzed literature, Morris *et al.* (2005) propose an integrative framework with six components (Fig. 1) with the purpose to design, assess and characterize the business model that is comprehensive in its kind and simplified in a way that researchers as well as entrepreneurs can adapt it for any type of company as a general guideline for analysis. In this context, the framework underpins the view of Afuah and Tucci (2001) that the business model is a "system that is made up of components, linkages between components and dynamics" (p. 4). Furthermore, the framework is adapted to the entrepreneurial mode and is according to the literature of representative nature regarding its point of view because it represents a new unit of

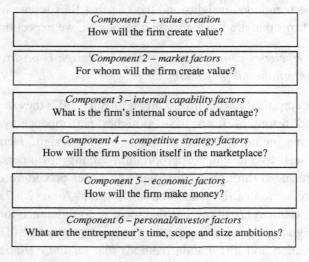

Fig. 1. A six component business model framework.

Source: Own figure based on Morris *et al.* (2005, p. 730).

analysis that bridges different system levels in an boundary-spanning activity system that involves both the content and process of doing business (George and Bock, 2011; Zott et al., 2011). According to Morris et al. (2005) the business model as central construct in entrepreneurship research has six fundamental components at three different levels that are evaluated: the foundation, proprietary and rules level. Especially the foundation level in which the six components are defined allows to compare different business models across different company types because generic decisions are addressed that are essential for every entrepreneur. Thereby, as illustrated in each of the six components (value creation, market factors, internal capability factors, competitive strategy factors, economic factors, personal/investor factors) developed by Morris et al. (2005) addresses a key question that is based on a broad analysis of theoretical and empirical research (e.g., Amit and Zott, 2001; Applegate, 2001; Timmers, 1998; Viscio and Pasternack, 1996).

Methodology and Data

Research approach

To test the hypotheses and in order to generate new findings concerning the business model design of BG companies, we followed step-wise a mixed-methods approach in which both a quantitative survey and qualitative interviews were conducted. According to Tashakkori and Teddlie (2003, p. 5) this approach is called "the third methodological movement", other authors even claimed that the mixed-methods approach is "a new star in the social science sky" (Mayring, 2007, p. 1). We chose this methodical approach for the following reasons: on the one hand, there are already first theories and applicable frameworks concerning in general the business model design and its influence on firm performance (e.g., Zott and Amit, 2007, 2008) that need to be tested for the species of BGs and their international performance. On the other hand, the literature dealing with BGs' business models in conjunction with international performance is still scarce (Hennart, 2014). In a first step, existing business model theory helped us to conceptualize our framework of research and provides an opportunity for a quantitative measurement of our construct. However, this setting only allows testing the hypotheses on a relatively high level of granularity, which promises some new findings concerning the dominant business model design theme of BGs (novelty-versus efficiency-centered). To gain a broader understanding for the configuration of a BG's business model in a more sophisticated way, we dipped deeper in a second step and connected these findings with qualitative in-depth interviews with BG founders and business executives. The outcomes from that endeavor helped us to enrich the dataset, and hence yielded in new propositions.

Sample

Quantitative dataset

In order to test our hypotheses, we investigated the business models and international performance of international operating companies in the German-speaking area. Thereby, we focused on business executives and founders of companies that were established in Germany, Switzerland or Liechtenstein to ensure that cultural context factors, which may lead to different internationalization pathways, could be kept as constant as possible. The founders and business executives at the top management level of a company are in particular responsible for making strategic and international business decisions (Nielsen and Nielsen, 2011). Therefore, to guarantee the reliability of the information provided, we selected CEOs and company founders as key informants who have developed the companies' business models and have general management responsibility for international operations. Additionally, the selected target countries have one of the highest rate of international activities in Europe, and international trade is therefore very important for the national economies. German industry, for example, is highly international, and exports account for approximately 50% of German GDP (Eurostat, 2014). The same structure of the balance of trade applies for the economic area of Switzerland and Liechtenstein, which have signed a customs treaty in 1924 and share the Swiss Franc as official currency (Schuessler *et al.*, 2014). Because of the high similarities between Switzerland and Liechtenstein in terms of culture, economic conditions and international trade, we decided to aggregate both countries to one target area.

We randomly sent an email with an invitation to the online-based survey to 2000 respondents in Germany and 1000 respondents in Switzerland and Liechtenstein. After the survey was completed, in total 174 responses from Germany (response rate of 8.7%) and 78 responses from Switzerland and Liechtenstein (response rate of 7.8%) were received. The response rates were comparable, and even slightly higher, to similar studies in the German-speaking area. Out of these returns we could generate a final sample of 252 questionnaires which were completely filled out. Thereby, the companies that are included in the final sample represent fairly well the company population of the target countries. Regarding the firm size, the number of employees ranged from just a couple of persons to 70,000 persons, while the percentage of international sales varied from just a few to almost 100%. The main industries of the companies involved manufacturing, services and retail. Due to the measurement of the three BG's key dimensions (speed, scope, scale), we could easily separate the sample into two groups: in total 46 companies could be identified as BGs and 206 companies have followed a

Table 2. Sample characteristics.

	Germany	Switzerland and Liechtenstein
Number of companies surveyed	174	78
Industry/Sector		
Agriculture and forestry	0.6	1.3
Manufacturing	23.5	33.3
Construction	8.6	9.0
Wholesale and retail trade	19.0	11.5
Transportation and storage	4.0	6.4
Financial and insurance activities	2.9	1.3
Professional, scientific and technical activities	21.2	20.5
Accommodation and food service activities	1.8	—
Information and communication activities	11.5	15.4
Education	3.4	—
Other	3.5	1.3
Firm Size (Employees)		
<10	40.8	33.4
<50	34.9	35.9
<250	17.8	11.5
>249	6.5	19.2
Internationalization Pathway		
Traditional approach	79.9	85.9
Born global	20.1	14.1

Note: Distribution in percentage.

traditional internationalization pathway. The overall BG ratio of 18.3% represents very well the assumed proportion of BGs among international companies in German-speaking countries (Mandl and Celikel-Esser, 2012). The descriptive characteristics of the companies in the sample are summarized in Table 2.

Qualitative dataset

The selected companies in our qualitative data set had their headquarters in Germany (nine cases) or Switzerland (two cases), ranged in firm size from two to 270 employees and generated an annual turnover from EUR 0.05 million to EUR 24 million in total. The percentage of international sales varied from 50 to 100%. This diversity of companies in terms of size and international performance was included in the sample to comply with Eisenhardt (1989) suggestions that multiple-case studies should include a large variety of cases. The descriptive characteristics of the companies in the sample are summarized in Table 3.

Table 3. Born globals characteristics.

Case	Home country	Industry	Year of foundation	No. of employees	Annual turnover (€ m.)	Scale — Share of int. sales	Scope — Number of foreign markets served	Speed — Year of market entry in Neighbor markets	Speed — Year of market entry in EU markets	Speed — Year of market entry in Int. markets	Form of financing
A	CH	Manufacturer/Software	2010	12	0.25	50%	8	2	4	1	Crowd-funding
B	D	Service Provider/HR development	2002	270	20	50%	29	1	3	5	Profit retention
C	D	Service Provider/Market Research	1993	2	0.75	65%	8	1	1	1	Profit retention
D	D	Retailer/Steel Industry	1998	5	1	100%	8	2	1	2	Profit retention
E	D	Service Provider/Logistics	1986	8	4	90%	100	1	1	1	Profit retention
F	D	Retailer/E-commerce	2011	7	0.7	80%	15	1	2	3	Equity increase
G	D	Manufacturer/Sports industry	2002	20	9	80%	25	1	2	3	Profit retention
H	CH	Service Provider/Biotech	1998	10	1	90%	15	2	5	3	Profit retention
I	D	Service Provider/Logistics	2007	21	24	100%	4	1	1	1	External investor
J	D	Manufacturer/Medical Devices	2002	7	15	95%	45	2	5	2	Profit retention
K	D	Manufacturer/E-Commerce	2011	6	0.05	60%	8	3	3	1	External investor

Five companies were engaged in service industries, four companies came from the manufacturing industries and two operated in retail. Only two of the 11 companies were in the business-to-customer sector, the majority focused on business clients. Furthermore, the majority of the companies was financed by profit retention (seven cases), two of them were funded by external investors, and one company each raised money from equity increase or crowd funding. The youngest companies included were founded in 2011 and the oldest in 1986. However, most of the companies were established in the late 1990s and at the beginning of the 21st century (nine cases). Interestingly, this period of time goes along with the first emergence of scientific theories dealing with the phenomenon of BGs (Cesinger *et al.*, 2012). The number of foreign markets served ranged from four to 100 markets. All of the companies had been active in international markets (that dispose of a high psychic, cultural and geographic distance) within the first five years after foundation. Five companies even jumped in their first year directly into markets that are very far from their home country. Nevertheless, all companies had been engaged in rapid internationalization within the first five years after founding, had conquered at least four foreign markets and had generated more than 25% from international sales, qualifying them as BGs.

Data collection

To gain relevant data concerning the companies' business models, we were faced with the challenge of finding an appropriate concept which can be operationalized. The discussion about the difficulty of obtaining objective measures of business model design is still an ongoing process in the scientific community (Dess and Robinson, 1984). As it was neither our goal to develop a new conceptualization of business model design nor to extend the framework further by including new subcategories, we rather followed the attempt of Zott and Amit (2007, 2008) and built composite scales for each business model design theme (novelty and efficiency) and identified and measured relevant items in a survey instrument. In the next step, our approach differed from that of Zott and Amit (2007, 2008), who first conducted a sample selection by itself and then took a third party (research assistants) in charge to rate and fill in the survey instrument for assigned sample companies. To avoid potential researchers' bias, we decided to address business executives and founders of international companies directly. In order to keep the full sample as representative as possible we also avoided a preselection of BGs by ourselves, but rather implemented variables to measure the three BG dimensions in the survey, and hence could separate the full sample afterwards into two groups. The data were self-reported by the business executives and company owners, which we invited via email to participate in our online-based survey. The invitation email was randomly sent to potential respondents out of a sample that was gathered from several databases like Hoppenstedt and Schober, chamber of commerce business lists from all of the respective countries and commercial broker lists. The selection criteria for the database search were the country of origin (Germany, Switzerland or Liechtenstein), involvement in international business ("European and international markets"), industry ("all sectors") and firm size ("SME and large companies"). The data were collected from May to June 2015.

Measures and variables

To operationalize our core research construct of the business model design, we referred to the approach of Zott and Amit (2007, 2008), but modified it in terms of suitability for a primary survey. Following the assumption that entrepreneurs are faced with two alternatives when it comes to create wealth (imitation versus innovation), we derived two sets of questions dealing with the business model design themes: efficiency- and novelty-centered (Aldrich, 1999; Porter, 1980; Zott, 2003). The efficiency-centered theme goes back to the transaction cost perspective (Milgrom and Roberts, 1992; Williamson, 1975, 1983), which refers to the design of economic transactions. As Zott and Amit (2007) claimed, "Efficiency-centered design refers to the measures that firms may take to achieve transaction efficiency through their business models." (p. 185), we transferred their items into questions and statements, which can be answered directly by business executives and company founders. The novelty-centered theme based on the core concept of entrepreneurship, which describes the process of wealth creation as connection of previously unconnected parties, as linkage of transaction participants in new ways or as design of new transaction mechanisms (Kirzner, 1978; Rialp *et al.*, 2005; Schumpeter, 1934; Zott and Amit, 2007, 2008). Therefore, we centered our questions and statements around the design themes, referring to the sub-items of Zott and Amit (2007). In the following we describe the development of the variables and their measurement.

Independent variables

To measure the three BG dimensions (speed, scope, scale) we transferred the definitional corridors (Cesinger *et al.*, 2012; Kuivalainen *et al.*, 2007; Madsen *et al.*, 2012) into questions, asking the entrepreneurs about their own impressions. By asking whether or not the beginning of the internationalization process was in the first five years after the company's founding (coded as "1" for yes, "0" for no), asking for the total number of markets in which the company was already active and asking for the proportion of the total turnover, which was international generated. Additionally, we asked the respondents in which markets their companies were already active. For this purpose we presented them three options that map the concept of psychic, cultural and geographic distance (Kraus *et al.*, 2015): neighboring markets, European markets and international markets.

Following the approach of Zott and Amit (2007), we selected two independent variables of business model design: novelty and efficiency. We also adopted their 13 sub-items as measures of design novelty and 13 sub-items as measures of design efficiency and transferred them to questions and statements. The expression of each of these sub-items in a given business model was measured using Likert-type scales and coded into a standardized score. After coding, we aggregated the item scores for each design theme into an overall score for the composite scale using equal weights (Mendelson, 2000). This process yielded distinct quantitative measures of the extent to which each business model in the sample leveraged novelty and efficiency as design themes.

Dependent variables

A company's international performance can be measured through the DOI that combines several parameters to a composite index. In order to employ an overall and objective measurement that is applicable for every type of company, we used the TNi which is based on the World Investment Report by the UNCTAD (1995) which focuses on the relationship between the activities in the domestic and foreign country for any particular company and represents the first and one of the most sophisticated multi-dimensional DOI indicators (Dörrenbächer, 2000; Ietto-Gillies, 1998). To operationalize the measurements, we developed three sub-items that asked for the company's current ratio of foreign assets to total assets, foreign sales to total sales and foreign employment to total employment. The calculation of the TNi was conducted through the average of the three ratios (Ietto-Gillies, 1998) while the objective measures of these variables were provided by the respondents.

Control variables

To avoid unobserved influences, we included further factors that might have an impact on the company's international performance as control variables in the analyses. On the *firm level* we used firm age, firm size and business sector. Firm size was measured as total number of employees. The variable can be seen as a proxy for a company's internationalization power in relation to its competitors: the larger the company, the greater its potential to conquer new markets, and hence, the higher its international performance. We controlled for the variable business sector using a dummy ("1" for service, "0" for industry related business sectors). On the *individual level*, we included the variables education and prior experience in international business. For measuring the personal level of education we used an ascending scale from one to six ("1" for low level education up to "6" for the highest qualification). Prior experience in international business was measured in years, following the assumption that international experienced persons tend to be more open for foreign cultures and overcome more likely psychic distance, and therefore are more determined to start business activities abroad (Sousa and Lages, 2011). The inclusion of these control variables strengthen the claim that our analysis captures the impact of distinct business model design configuration on a company's international performance, in opposite to the effects of firm or individual factors.

Network intensity is associated with a higher international firm performance (Danis *et al.*, 2009), therefore we included that variable as mediator. For the measurement we used a well-received construct that asks for the amount of network partners and the intensity of the exchange among each other (Kotabe *et al.*, 2003; Paulraj and Chen, 2007). After the measurement using a Likert-type scale, we aggregated the scores to an overall network score.

Case study selection and data generation

After finishing the online-based survey and the evaluation of the quantitative data set we could identify 46 companies in total that meet all requirements of the BG definition. Out of these companies we chose 11 cases (23.9%), which differ the most in terms of industry, firm size and age, as supposed by Eisenhardt (1989) who argues that the best results are achieved when the cases selected are not typical but extreme cases.

Our qualitative dataset is based on evidence from these 11 cases that had followed a BG internationalization pattern, and was collected in in-depth, semi-structured interviews with the business executives or founders of the companies. Following the approach of John (1984), who argued for the selection of valuable informants, the choice of this respondent group is based on the argument that people in these positions are the most knowledgeable regarding the configuration of the business model and the internationalization process. Additionally, an analysis of the public and internal material provided by the companies was used to acquire a comprehensive understanding of the cases and their specific business model characteristics. These multiple information sources served as a useful means of data triangulation (Chryssochoidis, 1996) and yielded a great amount of useful background information that provided invaluable insights into the issues raised by the study.

To perform the interviews in a semi-structured way we developed a guideline based on the integrative framework of Morris et al. (2005) that supports characterizing a company's business model by unveiling the decision-making process in the background. This framework seems to us "reasonably simple, logical, measurable, comprehensive, operationally meaningful" (p. 729) and applicable to companies in general but serves also the needs of the individual entrepreneur. Following this approach, we first took the six business model components with its leading questions on the foundation level of decision-making and translated them into German. Afterwards, we developed sub-questions for each business model component which support gaining knowledge about the specific theme. For each sub-question we offered answer options that were derived from the conceptual framework of Morris et al. (2005). At the end of the question set we asked the respondents for their perception: from their point of view, what is the most important business model component for the company's international performance? Additionally, the respondents were asked for detailed information concerning the market entry process when internationalization has started. The questions dealt with the form of internationalization: first, target countries, reasons for the internationalization, form of financing and enabling/aggravating factors. Finally, we asked for other factors that might have had an influence on the international performance (e.g., network intensity, prior experiences, etc.). By doing this, we took great care of the correct translation of the meaning into the German language and performed a *pretest* to ensure that all questions were understandable for the respondents.

For the analysis of the interviews, a coding system was developed that integrates the key aspects of the business model design and international performance. Useful quotations from the interviewees were translated from German into English. The data generated using this procedure were subjected to content analysis — a valid and widely employed method of developing an objective and systematic description of the manifest content of qualitative archival data (Bartunek et al., 2006; Sydserff and Weetman, 2002). In particular, we conducted a meaning-orientated analysis which requires the researcher to focus on underlying themes and to interpret the findings accordingly (Sydserff and Weetman, 2002). This was preferred to the more form-oriented approach (typically requiring routine, counting of words and some form of objective, computerized procedures) because it better suits studies in which patterns of linkage and attribution between the observed data and a number of key research questions are sought (Sydserff and Weetman, 2002), as was the case in our study. This meaning-orientated analysis is more amenable to an issue-by-issue presentation approach, as it allows for judicious use of exact quotes and vignettes from the case firms to further illuminate the explored research questions (Miles and Huberman, 1994).

Results

Quantitative analysis

Descriptive statistics

Table 4 provides an overview of the sample data we used in this study. For reasons of clarity we divided the full sample ($n = 252$) into two groups: companies that had followed a traditional internationalization approach (BG0) and BG companies (BG1). The BG1 group disposes an average TNi of 43.98%, which is more than twice as high as the average of the BG0 group (21.45%). Also, the average values for the business model design themes tend to be higher in the BG1 group: the efficiency-centered design (BME) is 2.955 versus 2.878 and the novelty-centered design (BMN) is 2.296 versus 1.995. Furthermore, the BG1 group also has a greater average value in terms of the network intensity (Net), as already observed in several previous studies (Acedo and Jones, 2007; Autio, 2005; Freeman et al., 2006; Hennart, 2014; Moen, 2002; Oviatt and McDougall, 1994). The average age of the BG0 group is 34.4 years, whereas the BG1 group is clearly younger (16.46 years). These finding corresponds well with the first emergence of the BG phenomenon in academia in the 1990s (Cesinger et al., 2012). The average firm size (number of employees) of the BG1 group (mean of 510) is smaller than in the BG0 group (mean of 829). The business sectors (coded as $0 =$ industry, $1=$ service) of the BG1 group tend to be more service oriented (mean of 0.52). Interestingly, the individual level of education as well as the previous international experience of the BG1 entrepreneurs is lower than in the BG0 group which supports the argumentation by Hennart (2014) that individual factors of the entrepreneur do not allow sufficient presumptions concerning the internationalization pathway.

Table 5 lists the Pearson correlations among the variables used in regressions. The correlation between an efficiency-centered business model (BME) and international performance (TNi) is significant (0.248), the same applies for the novelty-centered business model (BMN), even though the correlation is lower (0.171). Also the correlation between network intensity (Net) and the international performance is significant and relatively high (0.402). As expected, firm age (0.165)

Table 4. Descriptive statistics.

Variable name (acronym)	Mean		Median		Standard deviation		Min		Max	
	BG0	BG1	BG0	BG1	BG0	BG1	BG0	BG1	BG0	BG1
TNi	21.499	43.980	16.700	40.000	20.694	20.803	0.000	8.000	100.00	96.700
BME	2.878	2.955	2.920	3.000	0.625	0.544	0.310	1.850	4.000	3.850
BMN	1.995	2.296	2.030	2.335	0.770	0.753	0.150	0.380	3.850	3.640
Net	2.728	3.304	2.875	3.415	0.916	0.604	0.000	1.830	4.000	4.000
Firm age	31.4	16.46	21.0	12.0	49.3	19.4	1	1	515	117
Firm size	829	510	15	12	6146	2541	0	1	70000	17000
Sector	0.46	0.52	0.00	1.00	0.50	0.50	0	0	1	1
Education	4.30	3.98	4.50	4.00	1.298	1.468	1	1	6	6
Experience	4.02	3.43	1.00	1.00	7.04	5.47	0	0	39	24

Table 5. Pearson correlation.

	TNi	BME	BMN	Net	Firm age	Firm size	Sector	Education	Experience
TNi	1.000								
BME	0.248**	1.000							
BMN	0.171**	0.321**	1.000						
Net	0.402**	0.276**	0.349**	1.000					
Firm age	0.165**	0.053	−0.083	0.034	1.000				
Firm size	0.279**	0.095	−0.005	0.116	0.296**	1.000			
Sector	−0.015	−0.046	−0.074	−0.006	−0.031	−0.082	1.000		
Education	−0.040	−0.211**	−0.057	−0.041	−0.078	−0.064	0.025	1.000	
Experience	0.140*	0.075	0.079	0.107	−0.085	−0.015	0.117	−0.034	1.000

Notes: *$0.01 \leq p < 0.05$, **$p < 0.01$.

and firm size (0.279) correlate positively with international performance, the same applies for international experience (0.140).

Hypotheses tested

Following the approach of Zott and Amit (2007), we measured the strength of the items of the independent variables (BME, BMN) and the mediator variable (NET) using five-point Likert-type scales, which we coded into a standardized score. After coding, we aggregated the item scores for each scale into overall indices using equal weights (Mendelson, 2000). This process yielded distinct quantitative measures of business model design and network intensity. The dependent variable (TNi) was also used as index, consisting of the three ratios of foreign sales, assets and employees. All calculations were executed using the statistic software *R*. We validated reliability of our measures by using squared multiple correlations (also known as Guttman's Lambda 6) and Cronbach's standardized Alpha (in parentheses), which was 0.83 (0.80) for the business model efficiency measure, 0.88 (0.85) for the business model novelty measure, 0.81 (0.80) for the network intensity and 0.83 (0.78) for the Transnationality index. Since squared multiple correlations account for specific factors while Alpha does not (Revelle and Zinbarg, 2009), present correlations between sub-dimensions of the indices do not endanger overall reliability of our multi-item measures. Convergent and discriminant validity was confirmed for all four measures because explained common variance of the underlying factor was always greater >0.50 (0.54 for BME, 0.62 for BMN, 0.80 for Net and 0.69 for TNi) and exceeded all squared correlations with another measure (Fornell and Larcker, 1981). To assess the validity of our models, we tested for principle requirements (severe violations of normal distribution residuals, autocorrelation, multicollinearity and heteroscedasticity). However, applying a Breusch-Pagan test, heteroscedasticity was of concern for linear regressions. Thus, we decided to use Huber's (1981) robust linear models instead, which accounts for heteroscedasticity. Finally, for a complete analysis of the mediational hypothesis, we followed Preacher and Hayes (2008) and applied bootstrapping to test mediations.

The four models are defined as follows ("+" indicates main effects, "*" interaction effects):

(1) BME = BMN + BG
(2) BMN = BME + BG
(3) Net = BME + BMN + BG + BME*BG + BMN*BG
(4) TNi = Net + BME + BMN + BG + BME*BG + BMN*BG

The outcomes of the robust regression analysis are depicted in Table 6 and described subsequently.

A basic premise of this study is that a BG's business model design differs from that of a company which has followed a traditional internationalization pathway, and hence a BG company generates a higher international performance. In order to test our hypotheses, we analyzed the interactions within our proposed framework stepwise.

Table 6. Robust regression results.

		Dependent variable			
Variables	Hypotheses	Model 1: BME Value (std. err.)	Model 2: BMN Value (std. err.)	Model 3: Net Value (std. err.)	Model 4: TNi Value (std. err.)
Intercept		2.43***	0.83**	1.36***	−9.65
		(0.10)	(0.24)	(0.25)	(5.98)
BG	H1	−0.03	0.28*	1.09±	31.36*
		(0.10)	(0.13)	(0.65)	(15.17)
BMN	H2	0.24***		0.36***	−0.45
		(0.05)		(0.07)	(1.69)
BME	H4		0.41***	0.26***	5.01**
			(0.08)	(0.08)	(2.00)
Net	H3, H5				5.62***
Firm age		0.00	−0.00	0.00	0.07**
Firm size		0.00	0.00	0.00	0.00***
Sector		0.02	−0.17	−0.00	−0.72
Education		−0.09**	0.01	0.02	0.87
Experience		0.00	0.00	0.01	0.36*

Notes: One tailed: ± $p < 0.05$; two tailed: *$0.01 \leq p < 0.05$, **$p < 0.01$, ***$p < 0.001$.

Effects on business model efficiency and business model novelty

Model 1 and 2 show that business model novelty ($p = 0.000$, $\beta = 0.3072$) has a high positive significant influence on business model efficiency. And vice versa business model efficiency ($p = 0.000$, $\beta = 0.3201$) influences business model novelty in a high positive significant way. These findings illustrate the interrelation between the two design themes, thus they do not compete with each other, but mutually reinforce one another. BG does not have a significant influence on business model efficiency, but has a positive significant influence ($p = 0.03$, $\beta = 0.1384$) on business model novelty. These findings support H1 because BGs tend to have a slightly higher value for business model efficiency (not significant) and have a significant higher value for business model novelty.

Effects on network intensity

Concerning the variable network intensity, model 3 shows that BG ($p = 0.092$, $\beta = 0.4725$) has a positive significant influence that is relatively high. These findings correspond very well with a body of literature that emphasizes the role of networks for BG companies (Bouncken et al., 2015b; Freeman et al., 2006; Hennart, 2014; Madsen and Servais, 1997; Oviatt and McDougall, 2005). They not only have a larger number of partners on a national level and abroad, but they also exchange more intense than others. Also business model efficiency ($p = 0.003$, $\beta = 0.1739$) and business model novelty ($p = 0.000$, $\beta = 0.3115$) have a high positive significant influence on a company's network intensity. These findings are very interesting and will be outlined in the next step.

Effects on international performance

As already assumed because of the great differences in terms of the TNi averages, the positive significant influence of BG ($p = 0.04$, $\beta = 0.5411$) on international performance has a strong effect and supports the assumption. Business model efficiency ($p = 0.013$, $\beta = 0.1364$) also has a positive significant influence on TNi, but business model novelty does not show significant effects. Network intensity ($p = 0.000$, $\beta = 0.2245$) has a positive significant influence on international performance, and thus supports the following assumption: both business model design themes have a positive effect on network intensity and therefore this variable seems to have a mediating role. We explored this observation by running a separate test for the mediation effect of network intensity.

Table 7 depicts the results and shows that the mediating effect on business model efficiency is positive significant for both paths indirect ($p = 0.02$) and direct ($p = 0.00$). The total effect on TNi is relatively strong ($p = 0.00$) and the

Table 7. Mediation effect of network intensity.

Variables		Estimate	Lower 95% CI	Upper 95% CI
BME	Indirect effect	1.754*	0.254	3.383
	Direct effect	5.413*	1.577	9.585
	Total effect	7.167**	3.229	11.490
	Proportion mediated	0.245*	0.038	0.580
BMN	Indirect effect	2.447**	1.089	4.092
	Direct effect	0.163	−3.402	3.442
	Total effect	2.610	−0.698	5.771
	Proportion mediated	0.938[1]	−5.404	9.480

Notes: *$0.01 \leq p < 0.05$, **$p < 0.001$; 1: indicates a nonsignificant difference from total mediation.

proportion mediated, that is the share of the indirect effect compared to the total effect (Iacobucci et al., 2007), is 0.245. Therefore, H4 can be fully supported, whereas H5 is partially supported because only a significant indirect mediating effect was found (Zhao et al., 2010). Also the mediating effect on business model novelty is highly significant, with a significant indirect effect ($p = 0.00$) and a nonsignificant direct effect ($p = 0.95$). Since the mediated proportion is 0.938 and only the indirect effect of novelty is significant, we found partial support for H2. Due to the high mediation of business model novelty through network intensity, H3 can be fully supported. In a nutshell, most of the business model efficiency effect on international performance is direct (75.5%) and in total stronger ($\beta = 0.1364$) than the total effect of business model novelty ($\beta = -0.0154$). Most of the business model novelty effect on international performance is (due to the high mediation effect of network intensity) indirect.

Multiple case study: Findings and propositions

Business model profiling

First of all, each of the 11 cases can clearly be identified as having followed a BG internationalization pathway according to the definitional corridor of BG companies as measured by the three key dimensions (Cesinger et al., 2012; Kuivalainen et al., 2007; Madsen et al., 2012): all companies started rapid internationalization in the first five years after founding, generated more than 25% of their total revenue from international sales in at least four foreign markets and all of them jumped straight into markets that dispose a relative high degree of psychic, cultural and geographic distance. Interestingly, nine of the 11 companies started internationalization very traditionally with exports as form of market entry. Especially for small entrepreneurial firms, which are mostly lacking in resources, this is the only option to expand their business in foreign countries in a fast and efficient way (Madsen and Servais, 1997). The analysis of the interviews and the supplementary material revealed that the design of the BGs' business models is highly interrelated with their internationalization pathway, and thus led to superior international performance. These findings go along with the works of Zott and Amit (2007, 2008), who showed a positive relationship between the business model design and firm performance. On the highest level of granularity, both an efficiency- and a novelty-centered business model design can be seen as the key driver for firm performance (Zott and Amit, 2007, 2008). Overall, a strong focus on an efficiency-centered design theme could clearly be identified in nine of the 11 cases (A, C, D, E, F, H, I, J, K). However, in two cases (B, G) the business model designs were more novelty-centered. As our focus laid on the analysis of the configuration of BGs' business models, we concentrated on the specific features of the six components developed by Morris et al. (2005) in the following.

Component 1 – value creation

The first business model component deals with questions concerning the offering and the value creation of a company. In six cases the companies offered a heavy mix of products and services which were strongly individualized to customer needs and focused on a narrow product portfolio. Especially the high degree of individualization was described as a very important and unique feature of a company. The founder of case firm A stated: *"We consider ourselves to be the tailor-made suit of distribution. [...] We exactly manufacture the product to fit the requirements of our customers. [...] Our product portfolio is narrow because we specialize, we do not want to run around with the famous vendor's tray."* The founder of case firm B described the product portfolio as a *"Lego kit"* that includes *"individual elements that are standardized, but from which we can build very different things"*. The strong focus on individualization corresponds very well with the findings of Knight and Cavusgil (1996), who had described a greater demand for specialized or customized products as one of the trends that have led to the emergence of BGs. Most of the case firms were highly involved in the production processes and do it in-house to keep full control over the quality and their know-how. The founder of case firm H particularly emphasized: *"Of course, we manufacture the technology sensitive parts by ourselves. We must protect the technology!"* The founder of case firm G underlined the importance of a high production depth with the statement: *"Otherwise, we may not be the best."* Producing leading edge technology products can be seen as one of the key features of BG firms, they compete on quality and create value through innovative technology and product design (Rennie, 1993). In this context, we gain new insights about the BGs *international innovativeness* and how it is designed because mostly BGs are at the leading technological edge and develop and introduce new ideas including their processes, products or services to international markets (Knight and Kim, 2009). The majority (eight cases) of the case firms distribute the products directly to the customers. The reasons are twofold: first, in most of the cases direct delivery is faster and more efficient than using intermediaries; secondly, direct customer relations offer the possibility to actively push sales. The founder of case firm A declared his sales team as *"aggressive proactive"* and used, for example, guerilla marketing strategies at exhibitions to fix *"Dates, in which we can present our product and generate an order."*, as it is typically for young entrepreneurial firms to interact with their customers. This leads us to our first proposition:

> **P1:** Typical BG companies offer access to products and services that are strongly individualized and are sold directly to customers. Their product portfolio is more likely narrow and the depth of the in-house production is high.

Component 2 – market factors

Dealing with component two of the business model framework, we asked the respondents for whom the company creates value. Nine of the 11 case firms serve mainly business-to-business (B2B) customers, which are also the end customers in most of the cases. Not surprisingly, ten companies focused their business mainly on international markets. McDougall *et al.* (1994) described a BG's organizational culture attributes as international entrepreneurial orientated which fosters them to leap into international markets because of unique entrepreneurial competences and outlook. Thus, we gain new insights about the BGs *international orientation* and *international market orientation* because the main vision is about decisions to enter foreign markets and both shape the international business competence (Knight and Kim, 2009). The target countries are various and the international focus changes sometimes; depending on the current business conditions the case companies act in a dynamic and flexible way: *"Within our business we thoroughly observe, which part of the world prospers and which has problems."* (case firm H). Only case firm A named the national market as its current focus, although it generates already 50% of the revenues internationally: *"This year we only behave nationally proactive, abroad we serve only incoming requests."*

Due to the high level of specialization, most of the case firms labeled their market as niche. In some cases this limitation is caused by the small number of customers, in other cases the product itself represents the limiting factor because it only fits a special purpose: *"The market is broad in terms of potential customers, but more likely a niche in terms of limited product application possibilities"* (case firm B). These findings underpin a body of literature that claims niche markets are the natural habitat of BG firms (Andersson and Wictor, 2003; Hennart, 2014; Knight and Cavusgil, 2004; Rennie, 1993; Weerawardena et al., 2007). Therefore, BGs have to cooperate with international dispersed but specialized suppliers and customers in a more efficient way due to their domain-specific familiarity which presents the BGs international market orientation (Cavusgil and Knight, 2015; Hennart, 2014; Knight and Kim, 2009; Zander et al., 2015). The founder of case firm D claimed that *"you first have to build relationships and trust before you can sell something. Relations and contacts are very important!"* These findings are congruent with Fink and Kraus (2007) who postulate that mutual trust in business networks is a key to internationalization. Also Madsen and Servais (1997) highlighted the important role of networks in which especially BG companies "seek partners who complement their own competences; this is necessary because of their limited resources." (p. 564). Therefore, it is not surprising that the business relations of seven out of the 11 case firms are characterized through a transformational partnership. This leads us to our second proposition:

> **P2:** A BG's target business is B2B-oriented and its focus lies on international niche markets. BGs mainly serve end customers with whom they build strong transformational partnerships.

Component 3 – internal capability factors

This component deals with the core capabilities of a firm, which distinguish it from others. As postulated by Applegate (2001) and Viscio and Pasternack (1996), organizational capabilities lie at the heart of the business model and are the main source of the company's performance advantages (Grant, 1991). Two themes (in eight of 11 cases) dominated the set of cases: sales/marketing and technology/RandD capabilities. The statement of case firm F *"We deal mostly with technologies and their interfaces, taking into account that there are also different preferences in different countries."* illustrates that especially the connection between technology and the different customer preferences play an important role. Also the founder of case firm A supported the duality of sales and technology competences: *"Sales and marketing comes first, the technology comes directly afterwards"*. Especially in cases in which the company has developed rapidly, and therefore some business divisions were literally felt by the wayside, *"Technology covers our backs and has ensured our competitiveness in these times"* (case firm G). These observations support the work of Knight and

Cavusgil (2004) in which they emphasized the critical linkages between BG's key organizational capabilities (technology and innovation) and marketing orientation as a driver of international performance. BG companies typically differentiate from others by product innovation and intensive marketing: "Offering unique products, at least to the extent buyers' special needs are served and direct competitive rivalries are minimized, appears to contribute to superior international performance [...]" (Knight and Cavusgil, 2004, p. 654). This leads us to our third proposition:

> **P3:** The core capabilities of a BG company are its strengths in sales and marketing, as well as technology and RandD.

Component 4 – competitive strategy factors

In component four the core competencies of the company provide the basis for strategic positioning in the global marketplace. Therefore, the founder has to develop competitive strategy factors that differentiate the company from competitors. We asked the respondents to name the most unique selling proposition of their company. Two themes could be shaped from the answers (in eight of 11 cases): low costs/efficiency and innovation/unique products. These findings correspond very well with Knight *et al.* (2004), who took into account Charles Darwin's theorem and claimed that BGs only survive because they "thrive via the efficient and effective use of a limited range of resources and by adapting themselves to the demands of international marketing." (p. 649). The founder of case firm H illustrated how his company gained its whole market position through innovation: *"We are the only ones on the market that have this particular product. We don't even have to pursue business development, we are busy exclusively with the processing of requests."* Technology-based products obviously foster reaching a position as innovation leader in the market, but brings also some challenges with it, depending on the specific international market: *"Many German customers have difficulties to change existing processes (so complicated and nonsensical they are). English customers have less trouble with it"* (case firm A). As counterpart to innovation, efficiency plays an important role when it comes to superiority over competitors: *"Due to the small size, we are faster and more flexible, thus we act more efficient"* (case firm J). Also the founder of case firm I supported the claim that (small) size matters in terms of efficiency: *"Due to the small size, we perform better in certain markets (e.g., Ghana or Brazil) than big competitors, and thereby generate appropriate cost advantages."* These findings are congruent with the body of literature that deals with the advantages of small entrepreneurial firms (Chen and Hambrick, 1995; Nooteboom, 1994), especially when it comes to internationalization (Westhead *et al.*, 2001). This leads us to our fourth proposition:

> **P4:** A BG company stands out against the competition either through high efficiency or innovation; its image is based on speed and its products and services are of high quality.

Component 5 – economic factors

Researching the economic model of a company allows insights into its revenue mechanism; it combines the operating leverage, the volume, the margin and the revenue model. As typical for young entrepreneurial companies, the case firms tended to reach rather lower volumes (six of 11 cases). In most cases not the market opportunity was the limiting factor, but the internal capacity of the company to handle higher demand. The founder of case firm J justified its low volume with the words: *"Our company is a small one. We have not the same power as the big corporate groups."* But the small firms were aware of their typical strengths: *"We may be much smaller than our competitors, but we are much stronger in the niche!"* (Case firm G). This statement illustrates that it is typically not the aim of BG companies to compete with big players, but rather to "serve narrow niche markets that may not interest larger firms." (Knight *et al.*, 2004, p. 653). Ten of the 11 case firms have variable price models and mixed revenue sources. On the one hand, strong individualized products foster a variable pricing model, on the other hand, most case firms combine physical products with services and this leads to several revenue sources or even to lock-in effects. *"At the beginning we need to customize the specific product that causes one-time costs, and after that we require an annual license fee"* (case firm A). The founder of case firm J emphasized the possibility to adapt prices, depending on the target country: *"We have an internal pricelist that is fixed, but is adapted to each country and region."* These kind of international arbitrage opportunities allow the companies to generate relative high margins (six of 11 cases). Although some of the case firms are strongly technology-oriented, which is mostly associated with high fixed costs, their operating leverage is relatively low. Obviously, the case firms use their network of suppliers efficiently and thereby minimize the risk of high cost blocks. This strategy is typical for entrepreneurial startup companies. Due to the lack of resources they have to build strong relationships with partners (national or abroad) to scale their own possibilities and to overcome the liability of smallness (Moen and Servais, 2002). This leads us to our fifth proposition:

> **P5:** Due to a combination of products and services BGs have mixed revenue sources which generate more likely low volumes.

They are able to achieve high margins and keep their operating leverage low.

Component 6 – personal/investor factors

The last of the business model components deals with the entrepreneur's time, scope and size ambitions. "Differences among venture types have important implications for competitive strategy, firm architecture, resource management, creation of internal competencies, and economic performance." (Morris *et al.*, 2005, p. 730). Overall, the integrated business model has to fit the goals of the entrepreneur, his ambitions can reach from ensuring subsistence, generating stable income, reaching growth or pursuing short term speculation. As it is typical for entrepreneurial startup companies, and especially for BGs, most of the case companies followed a growth model (seven of 11 cases). Their growth strategies are diverse and change cyclically: *"In the second year we have focused on expansion of the product range, in the third year we have concentrated exclusively on expanding the markets"* (case firm F). Most of the case firms expressed their strong growth ambitions relatively clear: *"During the next three years we want to achieve a growth rate of at least 1000%!"* (case firm K), or *"Our company will grow by 10–15% during the next five years!"* (case firm J). Although these growth rates are very ambitious, most of the case firms believe that they will handle it by growing organically. *"The main goal is to expand further through organic growth — we do not intend to grow through acquisitions. With that strategy we achieve currently an annual growth rate of 30–35%, and we are trying to maintain this rate. We are able to achieve this goal only via the international market – in Germany we can also note a strong growth, but the international share is accumulating faster."* This statement of case firm's B founder illustrates the importance of the international markets for the growth strategies of a BG oriented company. This leads us to our sixth proposition:

P6: BGs pursue a clear growth strategy via international markets.

Developing a conceptual framework for a born global business model archetype

As it was our aim to shed more light on the configuration of BGs' business models, we sum up the findings gained in the analysis of the case firms. Having the "Building Theories from Case Study Research" guideline by Eisenhardt (1989) in mind, we decided to develop a theoretical archetype of a BG business model by combining our formulated propositions. We chose this approach because a business model framework relies on its inherent logic in which the specific components build upon one another. By using the framework of Morris et al. (2005) as vignette, we overlaid all business models of the case firms in order to find similarities (see Table 8). Based on the six developed propositions we suggest the following conceptual framework for a BG business model configuration as archetype:

> *Typical BG companies offer access to products and services that are strongly individualized and are sold directly to customers. The product portfolio is more likely narrow and the depth of the in-house production is high. A BG's target business is B2B oriented and its focus lies on international niche markets. BGs mainly serve end customers with whom they build strong transformational partnerships. The core capabilities of a BG company are its strengths in sales and marketing as well as technology and RandD. A BG company stands out against the competition through either high efficiency or innovation. Its image is based on speed, and its products and services are of high quality. Due to a combination of products and services they have mixed revenue sources which generate more likely low volumes. BGs are able to achieve high margins and keep their operating leverage low. BGs pursue a clear growth strategy via international markets.*

This conceptual framework of the BG business model archetype is derived from the investigated cases and their developed propositions analogous to each business model component which gets presented in detail in the following Table 8.

Additionally, we asked the interviewees to name one of the components that had the most influence on their internationalization success. Component 3, which describes the company's internal capability factors, was the most frequently mentioned key driver for international performance (six of 11 cases). *"Our capabilities represent the most important factor in our business!"* the owner of case firm D stated. Case firm I emphasized how important capability factors are in terms of global awareness: *"We were mainly perceived by the major customers through our expertise."* Component 1, dealing with value creation, was named five times as the most important driver for international performance. The founder of case firm B explained his international success as follows: *"In our market we face the challenge that every client demands a certain amount of customization. [The modular principle] allows us to obtain a high flexibility and a high level of profitability at the same time. The way we generate value is very efficient, and thus provides us an advantage."* The respondent of case firm F added: *"In the end, value creation is everything! This stands out against all other components: we*

Table 8. Conceptual framework of a Born Global business model archetype linked with the cases.

		Configuration	Cases observed	No. of cases
P1	Component 1: Factors related to offering	Sell products and services	A, B, C, H, J, K	6
		Strongly individualized offering	A, B, C, D, E, H	6
		Narrow product portfolio	A, C, G, I, J, K	6
		High production depth	A, B, C, E, G, H, I, K	8
		Offer access to products and services	B, C, D, E, F, I, J, K	8
		In-house production	A, B, C, E, F, G, H, I, K	9
		Direct distribution	A, B, C, D, E, F, H, J	8
P2	Component 2: Market factors	B2B	A, B, C, E, G, H, I, J, K	9
		International	B, C, D, E, F, G, H, I, J, K	10
		End customers	A, B, E, F, G, J	6
		Niche market	B, D, E, F, G, I, K	7
		Transformational partnership	C, D, E, G, H, J, K	7
P3	Component 3: Internal capability factors	Sales and marketing/ Technology and RandD	A, D, F, G, H, I, J, K	8
P4	Component 4: Competitive strategy factors	Low costs/efficiency OR Innovation/ unique products	A, B, D, F, G, H, I, J	8
		Image of speed	B, C, F, J, K	5
		Quality of products and services	C, E, G, H, I, J	6
P5	Component 5: Economic factors	Mixed revenue source	A, B, C, D, E, F, H, I, J, K	10
		Low operating leverage	A, F, G, H, I, J, K	7
		Low volumes	A, C, D, G, I, J	6
		High margins	B, C, F, G, H, K	6
P6	Component 6: Growth/ exit factors	Growth model	A, B, C, F, G, J, K	7

have a high value creation, because the competitive situation in other countries is less distinct and because the markets are more attractive."

These findings are very interesting because the BG entrepreneurs obviously know exactly which components of their company's business model support international performance, and thus they can focus all their attention on them.

In contrast to the claim of Hennart (2014), who describes BG companies as accidental internationalists, we formulate our last proposition as follows:

> **P7:** BGs' entrepreneurs are aware of the decisive business model components, and thus they focus specifically on the companies' capabilities and value creation to enhance their international performance.

Discussion and Conclusion

Summary of findings

The central thesis of our study is the notion that the business model design of BGs influences their international performance. To the best of our knowledge, this is the first systematic large-scale empirical study of BGs' business models. To enrich the findings which were gathered throughout the quantitative analysis, we additionally performed a multiple case study to shed more light on the specific business model components of BG companies. We disagree with research suggesting that the key to BGs international performance lies on the individual-level and argue to focus the firm-level as central and primary cause for their specific nature. By bridging the outcomes of both research approaches using a mixed method approach and taking a closer look at the results which yielded in a conceptual framework of a BG business model archetype, we can draw two major conclusions.

First, on the highest level of granularity, which deals with the overall design theme, BGs tend to have a more efficiency-centered business model design. This supports the line of reasoning by Zott and Amit (2007) who constitute that an efficient business model is related to higher performance when resources are scarce which, according to the revealed literature, is a common condition of BGs due to their liability of smallness. We assume that two defining elements of BGs contribute to their focus on efficiency. On the one hand, the highly globalized context, which brings with it great uncertainty, rapid changes and fast pace, forces BGs to focus on fast and efficient transactions. On the other hand, BGs have to handle their inherent liability of smallness by using their scarce resources in a very efficient way to leverage the outcomes. Obviously BGs' managers consider social networks as an efficient means of helping to go international more rapidly and profitably (Zhou et al., 2007) and they use them as a kind of mediator. These observations are consistent with previous research on BGs (Fan and Phan, 2007; Madsen and Servais, 1997; Oviatt and McDougall, 2005). The specific configuration of the BGs' business model components supports this assumption. They concentrate on a narrow product portfolio that is distributed directly and mainly to B2B customers in niche markets which are characterized by a high level of professionalism and close partnerships that allow fast and efficient transactions, due to a high level of mutual trust (Fink and Kraus, 2007). Additionally, BGs center their competitive strategy factors around an image of speed and use their suppliers to keep the operating leverage low. These observations support Hennart's (2014) approach which claims that "the speed with which firms can develop their international sales, and hence the probability that they will be INVs/BGs, depends on the business model they are implementing" (p. 129). Interestingly, BGs also have significant higher values in the novelty-centered business model design theme than companies that have followed a traditional internationalization pathway. As shown by the analysis of the cases, this is either caused by BGs' product strategy focusing on innovative and unique products, or by the establishment of novel transactional activities (e.g., non-equity market entry modes). Furthermore, BGs have developed novel and innovative strategies to interact with their customers in terms of offering easy access to their products and services, and thus reducing the time and costs for transactions substantially. These findings imply that efficiency and novelty are no opponents, but can be rather seen as two faders on a panel that controls a company's business model, as already stated by Zott and Amit (2007), "they complement, but do not replace" (p. 195).

Second, business model design matters for the international performance of BGs, but as revealed by our quantitative analysis, the interaction between the two design themes is complex. Efficiency as well as novelty are mediated through network intensity. Whereas the business model efficiency-centered design has a significant direct influence on international performance and the novelty-centered design has a significant indirect effect. These findings give partially support to scholars that emphasize the role of networks for the international performance of BGs (e.g., Oviatt and McDougall, 1994, 2005), but contribute also to a deeper understanding of their specific function. Due to high network intensity, BGs dispose of a wide range of national and international partners, but they mainly use them to foster novelty driven aspects of the business model (e.g., by exploiting easy access to technology). The benefit of networks is obviously less when it comes to efficiency-centered business model components. These findings are in line with the observations of Zucchella *et al.* (2007) and Nummela *et al.* (2004), who even noticed that firms with more business partnerships are actually slower to internationalize. Although it was not our focus, the analysis of the individual factors of the BG entrepreneur (e.g., level of education or prior international experience) do not allow any conclusions on the firm's international success, and thus strengthen the hypotheses from Hennart (2014): "foreign sales may occur even when the firm's founders do not have an international orientation, experience, or education." (p. 128). On the other hand, we disagree with the aspect suggested by Hagen *et al.* (2014) that BGs are accidental internationalists and present our seventh proposition that BGs are aware of the decisive business model components, and thus they focus specifically on the companies' capabilities and value creation to enhance their international performance.

In a nutshell, we followed the methodological call from Cavusgil and Knight (2015) and applied a mixed method approach by examining stepwise the business model configuration of BGs to deduce new aspects regarding their accelerated internationalization by providing a conceptual framework of a BG business model archetype.

Practical and theoretical implications

Inspired by Hennart's (2014) call for further research on the specific characteristics of BGs' business models that lead to fast internationalization, we developed the business model frameworks of Zott and Amit (2007) and Morris *et al.* (2005) further and adapted them to a BG context yielding into a comprehensive conceptual framework of a BG business model archetype that illustrates the important design themes and components to help both researchers and practitioners in their endeavor to investigate the phenomenon of BGs.

Our work can contribute to international entrepreneurship research in two respects. First, our study shows that capturing BGs' business model configuration as variables (as suggested by Zott and Amit (2007)) helps to measure design themes in a quantitative approach and provides the opportunity to assess their influence on international performance. In conjunction with a multiple case study approach, we offer deeper insights into a BG's business model design that may contribute to a better understanding of BGs' success factors. Second, we believe that the perspective of the BGs' business models, their design themes and our propositions concerning the features of the specific components broadens the discussion concerning the triggers of a BG internationalization pathway and opens up new promising research opportunities. On the individual-level we found no support for the hypotheses that put the entrepreneur in the center to explain the rapid internationalization of BGs (Andersson, 2011; Bingham and Davis, 2012; Cohen and Levinthal, 1990; Freeman and Cavusgil, 2007; Grant, 1996; Knight and Cavusgil, 2004; Park and Rhee, 2012; Roberts *et al.*, 2012). Also on the national-level the attempts of explanation are not sufficient, because small economies like Liechtenstein as well as big economies like Germany can yield a growing number of BG companies (Almor, 2013; Fan and Phan, 2007; Gabrielsson and Manek Kirpalani, 2004; Knight and Cavusgil, 1996; Madsen and Servais, 1997; Oviatt and McDougall, 1994; Schuessler *et al.*, 2014).

Instead, we give a conceptual framework of a BG business model archetype in Table 8 and argue for a perspective that lays the focus on the firm-level according to the presented insights: first, a business model design is per se a theory of the firm that contains hypotheses based on assumptions about the world. Second, these hypotheses have to be tested in business practice on the firm-level. In the context of a globalized world, BGs show their innovative nature as they have obviously developed a worldview, respectively a business model design that fits best to this setting. The focus on efficiency helps them to cope with their liability of smallness and supports rapid actions. Novelty-based aspects of a BG's business model brings with it innovation in terms of new products or new combinations following the Schumpeterian definition of entrepreneurship.

One group, the so called *"innovative born globals"* succeed in globalized markets through their high technological as well as RandD intensity characterizing their innovative nature leading to high specialized products (Amshoff et al., 2015; Trkman et al., 2015). Based on their product differentiation strategy which gets reflected in their business models they achieve a higher export performance the more they innovate (Gudiksen, 2015; Pels and Kidd, 2015). In a broader view, future research can orientate on the BG business model archetype with its components presented in the propositions (e.g., market, capability factors) to have a guiding map for research.

In addition to these implications for research, our findings also hold some practical implications. As our findings suggest, business model design matters for a BG's international performance and can be configured accordingly. Having an efficiency-centered design can help to cope with the complexity and rapid changes of today's world by acting in a fast and efficient way. However, including novel aspects into the business model does not necessarily mean to have the most revolutionary innovation on the market, but rather focuses on novel strategies to interact with customers in terms of offering easy access to products and services. Our proposed archetype framework can support entrepreneurs as well as managers with valuable insights into a business model configuration in a more practical way and shows what needs to be considered when planning a BG business. They may use it as a guide to assess or develop their own company's business model in order to enhance international sales. In times of extensive globalization and increasing competition, especially young entrepreneurial firms are forced to think internationally right from the start in order to establish a strong market position. Focusing on an efficiency-centered business model can help them to overcome their liability of smallness, or even to overcome big competitors by concentrating on niche markets.

Limitations of research

Finally, it is important to mention some limitations in our research that might be related to the collection of our data and the interpretation of the results.

We included business executives and founders of BG companies in our research who, according to previous research, are the organizations' strategy decision-makers (Nielsen and Nielsen, 2011). However, as other executive members (e.g., board members) decide major decisions for business model strategies, the results cannot be applied to the opinions and decisions of absolutely all organizational members because they are based on the entrepreneurs' own impression of the company. Furthermore, we cannot eliminate possible loss of information, as we asked the respondents for their firms' business models at the time of foundation, but most of the companies were already in a later stage of their life cycle. The only way to handle this kind of bias might be a longitudinal study approach that follows potential BG firms right from their inception. Another possible limitation is that we focused our investigations only on companies based in Germany, Switzerland and Liechtenstein. Therefore, these findings are generalizable only to some other countries. Nonetheless, generalizations of these findings can be applicable in countries that have similar economic conditions and dispose of analog structural characteristics and export rates comparable to those in Germany, Switzerland and Liechtenstein.

Another potential limitation is related to the measurement. As already noted by Zott and Amit (2007), "the measurement of business model design themes may not have captured all the lines of a firm's business that have revenue potential, and therefore might not explain all the variation in the dependent variable due to business model design themes." (p. 195). There is still an ongoing discussion about capturing a company's business model and generating valid and quantifiable data is one of the main issues (George and Bock, 2011; Hagen et al., 2014; Onetti et al., 2012). For this reason, we enriched our data by performing a multiple case study with the aim to reconcile the results. Although the findings do not allow to draw generalizable conclusions about the role of business model designs in the broader population of firms, we might inspire future research to develop business model measurements further.

Recommendations for future research

In line with our findings and propositions that yield in the conceptual framework of the BG business model archetype, some interesting research directions need to be further explored in order to clarify to what extent the process of BGs' rapid internationalization is influenced by the specific characteristics of business models, especially in different industry sectors and comparing fast and slow firms within this specific industry. In order to increase the generalizability of the results gained in our study, it would be necessary to test the frameworks by conducting the analysis in other countries. Furthermore, when the economic conditions change dramatically, firms face the challenge of reassessing their set of activities and they need to decide between which ones to continue and which ones not (Chesbrough, 2010; Rask, 2014; Trimi and Berbegal-Mirabent, 2012). Therefore, also from a methodological point of view, longitudinal studies looking at whether or not BGs change their business model design during their life cycle leading to the concept of *business model innovation* would be instructive. Additionally, we propose to also include the theoretical derivate of BGs, referred to as *born again globals* (Bell, 1995; Bell *et al.*, 2001), in further research projects to gain a deeper understanding of the influence of business model design on international performance.

Appendix A. Measurements for research constructs in the quantitative research.

Research construct	Measurement	References
Born Global	(1) Was your company already involved in international activities in the first five years after the foundation?	Kuivalainen *et al.* (2007)
	(2) Please indicate the approximate total number of international markets in which your company was active in the first five years after the foundation.	Madsen *et al.* (2012)
	(3) In which markets was your company already active in the first five years after the foundation?	Cesinger *et al.* (2012)
	(4) What proportion of the total turnover did your company internationally achieve within the first five years after its foundation?	
Business Model Design	*Efficiency-centered*	
	(1) We reduce storage costs for all actors within our business model.	Zott and Amit (2007, 2008)
	(2) The execution of transactions is simple from the user's perspective.	
	(3) When doing business, our business model ensures a low error rate.	
	(4) Other than the above-mentioned costs are reduced for all actors within our business model (marketing and selling expenses, transaction and processing costs, communication costs, etc.).	
	(5) Our business model is scalable (it can handle both a small and a large number of transactions).	

(*Continued*)

Appendix A. (*Continued*)

Research construct	Measurement	References
	(6) Our business model allows all actors to make informed decisions.	
	(7) We keep the execution of transactions transparent: all information flows and their use, services and goods are traceable.	
	(8) When doing business all actors are provided with relevant information, so an imbalance of knowledge in terms of product quality and nature is reduced.	
	(9) When doing business, we provide all actors with information about each other.	
	(10) We offer access to a large range of products, services, information and other actors.	
	(11) Our business model enables efficient order processing (e.g., through a coordinated process between the various divisions).	
	(12) Our business model allows fast transactions.	
	(13) Overall, our business model provides a high efficiency in the processing of transactions.	
	Novelty-centered	
	(1) Our business model provides new combinations of products, services and information.	
	(2) Our business model brings new actors together.	
	(3) We offer actors new incentives for the execution of transactions.	
	(4) Our business model allows access to an unprecedented variety and number of actors and/or products.	
	(5) Our business model combines actors in entirely new ways with business opportunities.	
	(6) The richness (quality and depth) of several connections between the actors is new.	
	(7) Number of patents, which keeps our company for individual components of our business model (e.g., for product innovations or processes).	
	(8) Extent to which our business model is based on trade secrets and/or copyrights.	

(*Continued*)

Appendix A. (*Continued*)

Research construct	Measurement	References
	(9) Does your business claim to have a pioneering role due to the business model?	
	(10) Our company has continuously introduced innovations to the business model.	
	(11) There are competing business models with the potential to overtake our own business model.	
	(12) There are other important aspects of our business model, which make it novel.	
	(13) Overall, our business model is novel.	
International Performance	(1) What is the percentage of international sales in total sales?	Ietto-Gillies (1998)
	(2) What is the percentage of international assets in total assets?	
	(3) What is the percentage of employees who primarily work abroad?	
Network Intensity	(1) How many people form your current international network?	Kotabe *et al.* (2003); Paulraj and Chen (2007)
	(2) I regularly exchange views with my network partners.	
	(3) I maintain close links with my network partners.	
	(4) There is an informal exchange between my network partners and me.	

Appendix B. Interview guide in the qualitative research.

Business model	Questions/Offerings
Component 1 value creation	(1) What kind of products does your company mainly sell? offering: primarily products/primarily services/heavy mix
	(2) How do you rate the degree of standardization of your products? offering: standardized/some customization/high customization
	(3) What kind of product portfolio does your company have? offering: broad line/medium breadth/narrow line
	(4) Which production depth does your company have? offering: deep lines/medium depth/shallow lines
	(5) What do you provide your customers regarding the product offer? offering: access to product/product itself/product bundled with other firm's product
	(6) How does your company obtain its end-products? offering: internal manufacturing or service delivery/outsourcing/licensing/reselling/value added reselling
	(7) How does your company sell its products? offering: direct distribution/indirect distribution (if indirect: single or multichannel)
Component 2 market factors	(1) In which segment is your primary business located? offering: b-to-b/b-to-c/both
	(2) In which markets do you mainly operate? offering: local/regional/national/international
	(3) What kind of customers do you serve mainly? offering: upstream supplier/downstream supplier/government/institutional/wholesaler/retailer/service provider/final consumer
	(4) How big is the market you serve? offering: broad or general market/multiple segment/niche market
	(5) On what foundation are your business relationships mainly based on? offering: transactional/relational
Component 3 internal capability factors	(1) What are the (main) competences of your company? offering: production/operating systems OR selling/marketing OR information management/mining/packaging OR technology/RandD/creative or innovative capability/intellectual OR financial transactions/arbitrage OR supply chain management OR networking/resource leveraging
Component 4 competitive strategy factors	(1) In what way does your company clearly differentiate itself from competitors? offering: image OR products/services OR innovation leadership OR low cost/efficiency OR intimate customer relationship/experience

(*Continued*)

Appendix B. (*Continued*)

Business model	Questions/Offerings
	(2) In which area does your company especially have a positive image? offering: operational excellence/consistency/dependability/speed
	(3) Which attributes are connected with your products/services? offering: quality/selection/features/availability
Component 5 economic factors	(1) How are your company's pricing policy and revenue streams designed? offering: fixed/mixed/flexible
	(2) What is the ratio of fixed costs to variable costs? offering: high/medium/low
	(3) How high do you estimate the sales volume of your company compared to the industry? offering: high/medium/low
	(4) What margin does your company achieve on average? offering: high/medium/low
Component 6 personal/investor factors	(1) What are your time, scope, and size ambitions? offering: subsistence model/income model/growth model/speculative model

Source: Based on Morris *et al.* (2005).

References

Acedo, FJ and MV Jones (2007). Speed of internationalization and entrepreneurial cognition: Insights and a comparison between international new ventures, exporters and domestic firms. *Journal of World Business*, 42(3), 236–252.

Afuah, A and CL Tucci (2001). *Internet Business Models and Strategies*. New York: McGraw-Hill.

Aldrich, H (1999). *Organizations Evolving*. Thousand Oaks: SAGE Publications.

Almor, T (2013). Conceptualizing paths of growth for technology-based born-global firms originating in a small-population advanced economy. *International Studies of Management and Organization*, 43(2), 56–78.

Alt, R and H-D Zimmermann (2001). Introduction to special section-business models. *Electronic Markets-The International Journal*, 11(1), 1019–6781.

Amit, R and C Zott (2001). Value creation in e-business. *Strategic Management Journal*, 22(6–7), 493.

Amit, R and C Zott (2012). Creating value through business model innovation. *MIT Sloan Management Review*, 53(3), 41–49.

Amshoff, B, C Dülme, J Echterfeld and J Gausemeier (2015). Business model patterns for disruptive technologies. *International Journal of Innovation Management*, 19(3), 1.

Andersson, S (2011). International entrepreneurship, born globals and the theory of effectuation. *Journal of Small Business and Enterprise Development*, 18(3), 627–643.
Andersson, S and I Wictor (2003). Innovative internationalisation in new firms: Born globals — the swedish case. *Journal of International Entrepreneurship*, 1(3), 249–276.
Applegate, LM (2001). Emerging e-business models. *Harvard Business Review*, 79(1), 79–87.
Autio, E (2005). Creative tension: The significance of Ben Oviatt's and Patricia McDougall's article 'toward a theory of international new ventures. *Journal of International Business Studies*, 36(1), 9–19.
Autio, E, HJ Sapienza and JG Almeida (2000). Effects of age at entry, knowledge intensity, and imitability on international growth. *Academy of Management Journal*, 43(5), 909–924.
Barney, J (1991). Firm resources and sustained competitive advantage. *Journal of Management*, 17(1), 99–120.
Bartunek, JM, SL Rynes and RD Ireland (2006). What makes management research interesting, and why does it matter? *Academy of Management Journal*, 49(1), 9–15.
Bell, J (1995). The internationalization of small computer software firms. *European Journal of Marketing*, 29(8), 60–75.
Bell, J, R McNaughton and S Young (2001). 'Born-again global' firms: An extension to the 'born global' phenomenon. *Journal of International Management*, 7(3), 173–189.
Bell, J, R McNaughton, S Young and D Crick (2003). Towards an integrative model of small firm internationalisation. *Journal of International Entrepreneurship*, 1(4), 339–362.
Belso-Martínez, JA (2006). Do industrial districts influence export performance and export intensity? Evidence for Spanish SMEs' internationalization process. *European Planning Studies*, 14(6), 791–810.
Bingham, CB and JP Davis (2012). Learning sequences: Their existence, effect, and evolution. *Academy of Management Journal*, 55(3), 611–641.
Bouncken, R, A Brem and S Kraus (2015a). Multi-cultural teams as sources for creativity and innovation: The role of cultural diversity on team performance. *International Journal of Innovation Management*, 20(1), 1–34.
Bouncken, RB, M Muench and S Kraus (2015b). Born globals: Investigating the influence of their business models on rapid internationalization. *The International Business and Economics Research Journal*, 14(2), 247–256.
Brem, A and F Freitag (2015). Internationalisation of new product development and research and development: Results from a multiple case study on companies with innovation processes in Germany and India. *International Journal of Innovation Management*, 19(1), 1–32.
Carayannis, EG, S Sindakis and C Walter (2015). Business model innovation as lever of organizational sustainability. *Journal of Technology Transfer*, 40(1), 85–104.
Casadesus-Masanell, R and JE Ricart (2010). From strategy to business models and onto tactics. *Long Range Planning*, 43(2–3), 195–215.

Cavusgil, ST and G Knight (2015). The born global firm: An entrepreneurial and capabilities perspective on early and rapid internationalization. *Journal of International Business Studies*, 46(1), 3–16.

Cesinger, B, M Fink, TK Madsen and S Kraus (2012). Rapidly internationalizing ventures: How definitions can bridge the gap across contexts. *Management Decision*, 50(10), 1816–1842.

Chen, M-J and DC Hambrick (1995). Speed, stealth, and selective attack: How small firms differ from large firms in competitive behavior. *The Academy of Management Journal*, 38(2), 453–482.

Chesbrough, H and S Rosenbloom (2002). The role of the business model in capturing value from innovation: Evidence from Xerox Corporation's technology spinoff companies. *Industrial and Corporate Change*, 11(3), 529–555.

Chesbrough, HW (2010). Business model innovation: Opportunities and barriers. *Long Range Planning*, 43(2–3), 354–363.

Chetty, S and C Campbell-Hunt (2004). A strategic approach to internationalization: A traditional versus a "born-global" approach. *Journal of International Marketing*, 12(1), 57–81.

Chroneer, D, J Johansson and M Malmstrom (2015). Business model management typologies-cognitive mapping of business model landscapes. *International Journal of Business and Management*, 10(3), 67–80.

Chryssochoidis, GM (1996). Successful exporting. *Journal of Global Marketing*, 10(1), 7–31.

Cohen, WM and DA Levinthal (1990). Absorptive capacity: A new perspective on learning and innovation. *Administrative Science Quarterly*, 35(1), 128–152.

Coviello, N (2015). Re-thinking research on born globals. *Journal of International Business Studies*, 46(1), 17–26.

Danis, WM, DS Chiaburu and MA Lyles (2009). The impact of managerial networking intensity and market-based strategies on firm growth during institutional upheaval: A study of small and medium-sized enterprises in a transition economy. *Journal of International Business Studies*, 41(2), 287–307.

DaSilva, CM and P Trkman (2014). Business model: What it is and what it is not. *Long Range Planning*, 47(6), 379–389.

Dess, G and R Robinson (1984). Measuring organizational performance in the absence of objective measures: The case of the privately-held firm and conglomerate business unit. *Strategic Management Journal*, 5(3), 265–273.

Dhanaraj, C and PW Beamish (2003). A resource-based approach to the study of export performance. *Journal of Small Business Management*, 41(3), 242–261.

Dörrenbächer, C (2000). Measuring corporate internationalisation. *Intereconomics*, 35(3), 119–126.

Dunford, R, I Palmer and J Benveniste (2010). Business model replication for early and rapid internationalisation: The ING direct experience. *Long Range Planning*, 43(5–6), 655–674.

Eisenhardt, KM (1989). Building theories from case study research, *Academy of Management Review*, 14(4), 532–550.

Eisenhardt, KM and JA Martin (2000). Dynamic capabilities: What are they? *Strategic Management Journal*, 21(10–11), 1105–1121.

Eurostat (2014). *International Trade, Investment and Employment as Indicators of Economic Globalization*. Luxembourg: European Commission.

Evers, N (2010). Factors influencing the internationalisation of new ventures in the Irish aquaculture industry: An exploratory study. *Journal of International Entrepreneurship*, 8(4), 392–416.

Fan, T and P Phan (2007). International new ventures: Revisiting the influences behind the 'born-global' firm. *Journal of International Business Studies*, 38(7), 1113–1131.

Fink, M, R Harms and S Kraus (2008). Cooperative internationalization of SMEs: Self-commitment as a success factor for international entrepreneurship. *European Management Journal*, 26(6), 429–440.

Fink, M and S Kraus (2007). Mutual trust as a key to internationalization of SMEs. *Management Research News*, 30(9), 674–688.

Fornell, C and DF Larcker (1981). Evaluating structural equation models with unobservable variables and measurement error. *Journal of Marketing Research*, 18(1), 39–50.

Frankenberger, K, T Weiblen, M Csik and O Gassmann (2013). The 4i-framework of business model innovation: A structured view on process phases and challenges. *International Journal of Product Development*, 18(3), 249–273.

Freeman, S and ST Cavusgil (2007). Toward a typology of commitment states among managers of born-global firms: A study of accelerated internationalization. *Journal of International Marketing*, 15(4), 1–40.

Freeman, S, R Edwards and B Schroder (2006). How smaller born-global firms use networks and alliances to overcome constraints to rapid internationalization. *Journal of International Marketing*, 14(3), 33–63.

Gabrielsson, M and VH Manek Kirpalani (2004). Born globals: How to reach new business space rapidly. *International Business Review*, 13(5), 555–571.

Gabrielsson, P, M Gabrielsson and T Seppäl (2012). Marketing strategies for foreign expansion of companies originating in small and open economies: The consequences of strategic fit and performance. *Journal of International Marketing*, 20(2), 25–48.

George, G and AJ Bock (2011). The business model in practice and its implications for entrepreneurship research. *Entrepreneurship: Theory and Practice*, 35(1), 83–111.

Glaum, M and M-J Oesterle (2007). 40 years of research on internationalization and firm performance: More questions than answers? *Management International Review*, 47(3), 307–317.

Grant, RM (1991). The resource-based theory of competitive advantage: Implications for strategy formulation. *California Management Review*, 33(3), 114–135.

Grant, RM (1996). Prospering in dynamically-competitive environments: Organizational capability as knowledge integration. *Organization Science*, 7(4), 375–387.

Gray, B and A Farminer (2014). And no birds sing-reviving the romance with international entrepreneurship. *Journal of International Entrepreneurship*, 12(2), 115–128.

Gudiksen, S (2015). Business model design games: Rules and procedures to challenge assumptions and elicit surprises. *Creativity and Innovation Management*, 24(2), 307–322.

Hagen, B, S Denicolai and A Zucchella (2014). International entrepreneurship at the crossroads between innovation and internationalization. *Journal of International Entrepreneurship*, 12(2), 111–114.

Hennart, J-F (2014). The accidental internationalists: A theory of born globals. *Entrepreneurship: Theory and Practice*, 38(1), 117–135.

Huber, PJ (1981). *Robust Statistics*. New York: Wiley.

Iacobucci, D, N Saldanha and X Deng (2007). A meditation on mediation: Evidence that structural equations models perform better than regressions. *Journal of Consumer Psychology*, 17(2), 139–153.

Ietto-Gillies, G (1998). Different conceptual frameworks for the assessment of the degree of internationalization: An empirical analysis of various indices for the top 100 TNCs. *Transnational Corporations*, 7(1), 17–39.

Jantunen, A, N Nummela, K Puumalainen and S Saarenketo (2008). Strategic orientations of born globals — Do they really matter? *Journal of World Business*, 43(2), 158–170.

John, G (1984). An empirical investigation of some antecedents of opportunism in a marketing channel, *Journal of Marketing Research*, 21(3), 278–289.

Jones, MV, N Coviello and YK Tang (2011). International entrepreneurship research (1989–2009): A domain ontology and thematic analysis. *Journal of Business Venturing*, 26(6), 632–659.

Kim, SK and S Min (2015). Business model innovation performance: When does adding a new business model benefit an incumbent? *Strategic Entrepreneurship Journal*, 9(1), 34.

Kirzner, I (1978). *Competition and Entrepreneurship*. Chicago: University of Chicago Press.

Knight, G, TK Madsen and P Servais (2004). An inquiry into born-global firms in Europe and the USA. *International Marketing Review*, 21(6), 645–665.

Knight, GA and ST Cavusgil (1996). The born global firm: A challenge to traditional internationalization theory. In *Advances in International Marketing*, S Tamer Cavusgil and T Koed Madsen (Eds.).

Knight, GA and ST Cavusgil (2004). Innovation, organizational capabilities, and the born-global firm. *Journal of International Business Studies*, 35(2), 124–141.

Knight, GA and D Kim (2009). International business competence and the contemporary firm. *Journal of International Business Studies*, 40(2), 255–273.

Kotabe, M, X Martin and H Domoto (2003). Gaining from vertical partnerships: Knowledge transfer, relationship duration, and supplier performance improvement in the U.S. and Japanese automotive industries. *Strategic Management Journal*, 24(4), 293–316.

Kraus, S, F Meier, F Eggers, R Bouncken and F Schuessler (2015). Standardisation vs. Adaption: A conjoint experiment on the influence of psychic, cultural and geographical distance on international marketing mix decisions. *European Journal of International Management*.

Kuivalainen, O, S Sundqvist, S Saarenketo and R McNaughton (2012). Internationalization patterns of small and medium-sized enterprises. *International Marketing Review*, 29(5), 448–465.

Kuivalainen, O, S Sundqvist and P Servais (2007). Firms' degree of born-globalness, international entrepreneurial orientation and export performance. *Journal of World Business*, 42(3), 253–267.

Laanti, R, M Gabrielsson and P Gabrielsson (2007). The globalization strategies of business-to-business born global firms in the wireless technology industry. *Industrial Marketing Management*, 38(8), 1104–1117.

Madsen, TK, S Kraus and M O'Dwyer (2012). International entrepreneurship and SME internationalisation: An introduction and overview. *International Journal of Entrepreneurship and Small Business*, 12(2), 131–135.

Madsen, TK and P Servais (1997). The internationalization of born globals: An evolutionary process? *International Business Review*, 6(6), 561–583.

Mahdjour, S and S Fischer (2014). International corporate entrepreneurship with born global spin along ventures – a cross case analysis of telekom innovation laboratories' venture portfolio. *International Journal of Innovation Management*, 18(3), 1–18.

Mandl, I and F Celikel-Esser (2012). *Born Global: The Potential of Job Creation in New International Businesses*. Luxembourg: Publications Office of the European Union.

Mason, K and M Spring (2011). The sites and practices of business models. *Industrial Marketing Management*, 40(6), 1032–1041.

Mayring, P (2007). *Qualitative Inhaltsanalyse: Grundlagen und Techniken*. Weinheim: Beltz.

McAuley, A (1999). Entrepreneurial instant exporters in the Scottish arts and craft sector. *Journal of International Marketing*, 7(4), 67–82.

McDougall, PP (1989). International versus domestic entrepreneurship: New venture strategic behavior and industry structure. *Journal of Business Venturing*, 4(6), 387–400.

McDougall, PP and BM Oviatt (1996). New venture internationalization, strategic change, and performance: A follow-up study. *Journal of Business Venturing*, 11(1), 23.

McDougall, PP, S Shane and BM Oviatt (1994). Explaining the formation of international new ventures: The limits of theories from international business research. *Journal of Business Venturing*, 9(6), 469–487.

Mendelson, H (2000). Organizational architecture and success in the information technology industry. *Management Science*, 46(4), 513–529.

Miles, M and A Huberman (1994). *Qualitative Data Analysis*, 2nd edn. Thousand Oaks: SAGE Publications.

Milgrom, P and J Roberts (1992). *Economics, Organization and Management*. Englewood Cliffs: Prentice-Hall.

Miller, D (1996). Configurations revisited. *Strategic Management Journal*, 17(7), 505–512.

Moen, Ø (2002). The born globals: A new generation of small European exporters. *International Marketing Review*, 19(2), 156–175.

Moen, Ø and P Servais (2002). Born global or gradual global? Examining the export behavior of small and medium-sized enterprises. *Journal of International Marketing*, 10(3), 49–72.

Morris, M, M Schindehutte and J Allen (2005). The entrepreneur's business model: Toward a unified perspective. *Journal of Business Research*, 58(6), 726–735.

Nair, S, H Paulose, M Palacios and J Tafur (2013). Service orientation: Effectuating business model innovation. *Service Industries Journal*, 33(9/10), 958–975.

Nielsen, BB and S Nielsen (2011). The role of top management team international orientation in international strategic decision-making: The choice of foreign entry mode. *Journal of World Business*, 46(2), 185–193.

Nooteboom, B (1994). Innovation and diffusion in small firms: Theory and evidence. *Small Business Economics*, 6(5), 327–347.

Nummela, N, S Saarenketo and K Puumalainen (2004). Rapidly with a rifle or more slowly with a shotgun? Stretching the company boundaries of internationalising ICT firms. *Journal of International Entrepreneurship*, 2(4), 275–288.

Onetti, A, A Zucchella, M Jones and P McDougall-Covin (2012). Internationalization, innovation and entrepreneurship: Business models for new technology-based firms. *Journal of Management and Governance*, 16(3), 337–368.

Osterwalder, A and Y Pigneur (2005). Clarifying business models: Origins, present, and future of the concept. *Communications of the Association for Information Systems*, 16(1), 1–25.

Oviatt, BM and PP McDougall (1994). Toward a theory of international new ventures. *Journal of International Business Studies*, 25(1), 45–64.

Oviatt, BM and PP McDougall (1997). Challenges for internationalization process theory: The case of international new ventures. *MIR: Management International Review*, 37(1), 85–99.

Oviatt, BM and PP McDougall (2005). Defining international entrepreneurship and modeling the speed of internationalization. *Entrepreneurship: Theory and Practice*, 29(5), 537–553.

Park, T and J Rhee (2012). Antecedents of knowledge competency and performance in born globals. The moderating effects of absorptive capacity. *Management Decision*, 50(8), 1361–1381.

Paulraj, A and IJ Chen (2007). Environmental uncertainty and strategic supply management: A resource dependence perspective and performance implications. *Journal of Supply Chain Management*, 43(3), 29–42.

Pels, J and TA Kidd (2015). Business model innovation. *International Journal of Pharmaceutical and Healthcare Marketing*, 9(3), 200–218.

Porter, ME (1980). *Competitive Strategy: Techniques for Analyzing Industries and Competitors*. New York: Free Press.

Ramaswamy, K, KG Kroeck and W Renforth (1996). Measuring the degree of internationalization of a firm: A comment. *Journal of International Business Studies*, 27(1), 167–177.

Rask, M (2014). Internationalization through business model innovation: In search of relevant design dimensions and elements. *Journal of International Entrepreneurship*, 12(2), 146–161.

Rennie, MW (1993). Born global. *McKinsey Quarterly*, 4, 45–52.

Revelle, W and R Zinbarg (2009). Coefficients Alpha, beta, omega, and the glb: Comments on sijtsma. *Psychometrika*, 74(1), 145–154.

Rialp, A, J Rialp and GA Knight (2005). The phenomenon of early internationalizing firms: what do we know after a decade (1993–2003) of scientific inquiry? *International Business Review*, 14(2), 147–166.

Richardson, J (2008). The business model: An integrative framework for strategy execution. *Strategic Change*, 17(5–6), 133–144.

Ritala, P, A Golnam and A Wegmann (2014). Coopetition-based business models: The case of Amazon.com. *Industrial Marketing Management*, 43(2), 236–249.

Roberts, N, PS Galluch, M Dinger and V Grover (2012). Absorptive capacity and information systems research: Review, synthesis, and directions for future research. *MIS Quarterly*, 36(2), 625–A626.

Schneider, S and P Spieth (2013). Business model innovation: Towards an integrated future research agenda. *International Journal of Innovation Management*, 17(1), 1–34.

Schuessler, F, MT Schaper and S Kraus (2014). Entrepreneurship in an Alpine micronation: The case of Liechtenstein. *International Journal of Entrepreneurship and Small Business*, 22(1), 106–114.

Schumpeter, JA (1934). *The Theory of Economic Development: An Inquiry into Profits, Capital, Credit, Interest, and the Business Cycle*. Cambridge: Harvard University Press.

Shrader, RC (2001). Collaboration and performance in foreign markets: The case of young high-technology manufacturing firms. *The Academy of Management Journal*, 44(1), 45–60.

Sleuwaegen, L and J Onkelinx (2014). International commitment, post-entry growth and survival of international new ventures. *Journal of Business Venturing*, 29(1), 106–120.

Smith, D (2015). Disrupting the disrupter: Strategic countermeasures to attack the business model of a coercive patent-holding firm. *Technology Innovation Management Review*, 5(5), 5–16.

Solberg, CA (1997). A framework for analysis of strategy development in globalizing markets. *Journal of International Marketing*, 5(1), 9–30.

Sousa, CMP and LF Lages (2011). The PD scale: A measure of psychic distance and its impact on international marketing strategy. *International Marketing Review*, 28(2), 201–222.

Sullivan, D (1994a). Measuring the degree of internationalization of a firm. *Journal of International Business Studies*, 25(2), 325–342.

Sullivan, D (1994b). The "threshold of internationalization:" Replication, extension, and reinterpretation. *Management International Review (MIR)*, 34(2), 165–186.

Sullivan, D (1996). Measuring the degree of internationalization of a firm: A reply. *Journal of International Business Studies*, 27(1), 179–192.

Sydserff, R and P Weetman (2002). Developments in content analysis: A transitivity index and DICTION scores. *Accounting, Auditing and Accountability Journal*, 15(4), 523–545.

Tanev, S, ES Rasmussen, E Zijdemans, ROY Lemminger and LL Svendsen (2015). Lean and global technology start-ups: Linking the two research streams. *International Journal of Innovation Management*, 19(3), 1–41.

Taran, Y, H Boer and P Lindgren (2015). A business model innovation typology. *Decision Sciences*, 46(2), 301–331.

Tashakkori, A and C Teddlie (2003). *Handbook of Mixed Methods in Social and Behavioral Research*. SAGE Publications.

Teece, DJ (2007). Explicating dynamic capabilities: The nature and microfoundations of (sustainable) enterprise performance. *Strategic Management Journal*, 28(13), 1319–1350.

Teece, DJ (2010). Business models, business strategy and innovation. *Long Range Planning*, 43(2–3), 172–194.

Timmers, P (1998). Business models for electronic markets. *Electronic Markets*, 8(2), 3–8.

Tracey, P and O Jarvis (2007). Toward a theory of social venture franchising. *Entrepreneurship: Theory and Practice*, 31(5), 667–685.

Trimi, S and J Berbegal-Mirabent (2012). Business model innovation in entrepreneurship. *International Entrepreneurship and Management Journal*, 8(4), 449–465.

Trkman, P, M Budler and A Groznik (2015). A business model approach to supply chain management. *Supply Chain Management*, 20(6), 587–602.

Turcan, R and A Juho (2014). What happens to international new ventures beyond start-up: An exploratory study. *Journal of International Entrepreneurship*, 12(2), 129–145.

UNCTAD (1995). *World Investment Report 1995. Transnational Corporations and Competitiveness*. Geneva: United Nations.

Viscio, AJ and BA Pasternack (1996). Toward a new business model. *Strategy Business*, 2(1), 25–34.

Weerawardena, J, GS Mort, PW Liesch and G Knight (2007). Conceptualizing accelerated internationalization in the born global firm: A dynamic capabilities perspective. *Journal of World Business*, 42(3), 294–306.

Welch, LS and R Luostarinen (1988). Internationalization: Evolution of a concept. *Journal of General Management*, 14(2), 34–55.

Westhead, P, M Wright and D Ucbasaran (2001). The internationalization of new and small firms: A resource-based view. *Journal of Business Venturing*, 16(4), 333–358.

Williamson, OE (1975). *Markets and Hierarchies: Analysis and Antitrust Implications*. New York: Free Press.

Williamson, OE (1983). Credible commitments: Using hostages to support exchange. *The American Economic Review*, 73(4), 519–540.

Young, S (1987). Business strategy and the internationalization of business: Recent approaches. *Managerial and Decision Economics*, 8(1), 31–40.

Zahra, SA, RD Ireland and MA Hitt (2000). International expansion by new venture firms: International diversity, mode of market entry, technological learning, and performance. *Academy of Management Journal*, 43(5), 925–950.

Zander, I, P McDougall-Covin and EL Rose (2015). Born globals and international business: Evolution of a field of research. *Journal of International Business Studies*, 46(1), 27–35.

Zhao, X, JG Lynch Jr. and Q Chen (2010). Reconsidering baron and kenny: Myths and truths about mediation analysis. *Journal of Consumer Research*, 37(2), 197–206.

Zhou, L, WP Wu and X Luo (2007). Internationalization and the performance of born-global SMEs: the mediating role of social networks. *Journal of International Business Studies*, 38(4), 673–690.

Zott, C (2003). Dynamic capabilities and the emergence of intraindustry differential firm performance: Insights from a simulation study. *Strategic Management Journal*, 24(2), 97–125.

Zott, C and R Amit (2007). Business model design and the performance of entrepreneurial firms. *Organization Science*, 18(2), 181–199.

Zott, C and R Amit (2008). The fit between product market strategy and business model: Implications for firm performance. *Strategic Management Journal*, 29(1), 1–26.

Zott, C and R Amit (2010). Business model design: An activity system perspective. *Long Range Planning*, 43(2–3), 216–226.

Zott, C and R Amit (2013). The business model: A theoretically anchored robust construct for strategic analysis. *Strategic Organization*, 11(4), 403–411.

Zott, C, R Amit and L Massa (2011). The business model: Recent developments and future research. *Journal of Management*, 37(4), 1019–1042.

Zucchella, A, G Palamara and S Denicolai (2007). The drivers of the early internationalization of the firm. *Journal of World Business*, 42(3), 268–280.

Index

A
alliances, 153
antecedents, 5
ATB Antriebstechnik, xvi

B
barriers, 244
born global firm, xiv, xviii, xx, 199–200, 257
BT 21st Century Network, 59, 61
business model, 199

C
case study, 257
China, 97
CITIC, xvi
communication, 7
competition, 129
competitive advantage, 156
competitiveness, xvii, 4
consequences, 5
corporate venturing, 257
cross-border communication, 15

D
Deutsche Telekom (DT), 65
dyadic alliances, 183

E
efficiency-centered, 310
emerging economies, 28
emerging markets, 27
export intensity, 129, 141
external corporate venturing, 258
external platform-based approach, 64

F
focal firm, 155
foreign direct investment (FDI), 28, 65
France Telecom, 65

G
Germany, 27
global corporate ventures, 260
global innovation, xiv, xvi–xvii, xx, 241
global product development, 6
global standardisation, 3
globalization, xv, xx, 243

H
Hofstede, 32

I
ICT services, 59
imitation, 153

incumbent telecom operators, 59–60
India, 27
indigenous innovation, xvi
Information and Communication Technologies (ICT), xvi, 7
infrastructure, 91
innovation, 97
innovation capabilities, xv–xvii, 124
innovation management, 199
innovation partnership, 153
innovation phase, 183
innovation practices, 31
innovation process, 27
innovativeness, 132
intellectual property, xvii, 155
interaction with the headquarters, 124
internal corporate venturing (ICV), 258
internal platform-based approach, 64
international innovation, xiii, xv–xvii, xix, xxi
international new ventures (INVs), 199–200
international performance, 275
international product strategy, 3
internationalization of R&D, 27, 241
Internet Protocol, 59, 155

K
Knowledge, viii, xv, xvii
Knowledge sharing, xvii
KSM castings, xvi

L
L2G start-ups, 199
lean global start-ups (LGS), xvii, 199
lean start-ups (LSs), 199–200
lean, xiv, xvi–xvii, xix, xxi
lean-and-global, 199

Lenovo, xvi
L&G start-ups, 199
local adaptation, 3

M
manufacturing industry, 97
market requirements, 3
Medion, xvi
mitigation, 252
multi-local, adaptation-based, 12
multinational companies, 97
multinational corporations (MNCs), xv, 3
Multi-Protocol Label Switching (MPLS), 61

N
New Product Development (NPD), xvi, 3
Next Generation Network (NGN), 59

O
offshoring, xviii
open innovation, 153
operations, 32
organisation, 32
organisational change, 97
organisational structures, 124

P
Pang Da, xvi
partner location, 183
partner type, 183
partners, 154
partnership portfolios, 153
platform, xiii, xviii, xxi, 3, 59
platform innovation, 59
platform leadership, 59
platform strategies, xvii

platform-based, 12
problem-centred interviews (PCIs), 37
product adaptation, 5

R
R&D expenditure, 142
R&D management, 27
R&D, viii, xiv–xv, xviii–xx, xxi, 10, 27
resource-based view (RBV), 153–154
risk, xvii
Russia, 129

S
Saab, xvi
science and engineering (S&E), 30
small and medium-sized enterprises (SMEs), 241
spin-along, 257
spin-in, 259

spin-off, 259
spin-out, 257, 259
strategy, 32
subsidiary, 97

T
technology entrepreneurship, 199
telecommunications industry, 61
Telekom Innovation Laboratories, 257, 258

U
United Nations (UN), xv

W
Wolong, xvi

Y
Youngman, xvi

Series on Technology Management

(Continuation of series card page)

Vol. 26 *Total Value Development: How to Drive Service Innovation*
by Frank M. Hull (Cass Business School and Fordham University Graduate School of Business-Executive Education, USA) &
Christopher David Storey (University of Sussex, UK)

Vol. 25 *Small Firms as Innovators: From Innovation to Sustainable Growth*
by Helena Forsman (University of Tampere, Finland)

Vol. 24 *The Knowledge Enterprise: Innovation Lessons from Industry Leaders (Second Edition)*
by E. Huizenga (University of Amsterdam, The Netherlands)

Vol. 23 *Open Innovation Research, Management and Practice*
edited by J. Tidd (University of Sussex, UK)

Vol. 22 *Discontinuous Innovation: Learning to Manage the Unexpected*
by P. Augsdörfer (Technische Hochschule Ingolstadt, Germany),
J. Bessant (University of Exeter, UK),
K. Möslein (Universität Erlangen-Nürnberg, Germany),
B. von Stamm (Innovation Leadership Forum, UK) &
F. Piller (RWTH Aachen University, Germany)

Vol. 21 *Workbook for Opening Innovation: Bridging Networked Business, Intellectual Property and Contracting*
by Jaakko Paasi (VTT Technical Research Centre of Finland, Finland),
Katri Valkokari (VTT Technical Research Centre of Finland, Finland),
Henri Hytönen (VTT Technical Research Centre of Finland, Finland),
Laura Huhtilainen (University of Eastern Finland, Finland) &
Soili Nystén-Haarala (University of Eastern Finland, Finland)

Vol. 20 *Bazaar of Opportunities for New Business Development: Bridging Networked Innovation, Intellectual Property and Business*
by Jaakko Paasi (VTT Technical Research Centre of Finland, Finland),
Katri Valkokari (VTT Technical Research Centre of Finland, Finland),
Tuija Rantala (VTT Technical Research Centre of Finland, Finland),
Soili Nystén-Haarala (University of Eastern Finland, Finland),
Nari Lee (University of Eastern Finland, Finland) &
Laura Huhtilainen (University of Eastern Finland, Finland)

Vol. 19 *From Knowledge Management to Strategic Competence: Assessing Technological, Market and Organisational Innovation (Third Edition)*
edited by Joe Tidd (University of Sussex, UK)

Vol. 18 *Perspectives on Supplier Innovation: Theories, Concepts and Empirical Insights on Open Innovation and the Integration of Suppliers*
edited by Alexander Brem (University of Erlangen-Nuremberg, Germany) &
Joe Tidd (University of Sussex, UK)

Vol. 17 *Managing Process Innovation: From Idea Generation to Implementation*
by Thomas Lager (Grenoble Ecole de Management, France)

Vol. 16 *Perspectives on User Innovation*
edited by Stephen Flowers (University of Brighton, UK) &
Flis Henwood (University of Brighton, UK)

Vol. 15 *Gaining Momentum: Managing the Diffusion of Innovations*
edited by Joe Tidd (SPRU, University of Sussex, UK)

Vol. 14 *Innovation and Strategy of Online Games*
by Jong H. Wi (Chung-Ang University, South Korea)

Vol. 13 *Building Innovation Capability in Organizations: An International Cross-Case Perspective*
by Milé Terziovski (University of Melbourne, Australia)

Vol. 12 *Project-Based Organization in the Knowledge-Based Society*
by Mitsuru Kodama (Nihon University, Japan)

Vol. 11 *Involving Customers in New Service Development*
edited by Bo Edvardsson (Karlstad University, Sweden),
Anders Gustafsson (Karlstad University, Sweden),
Per Kristensson (Karlstad University, Sweden),
Peter Magnusson (Karlstad University, Sweden) &
Jonas Matthing (Karlstad University, Sweden)

Vol. 10 *Open Source: A Multidisciplinary Approach*
by Moreno Muffatto (University of Padua, Italy)

Vol. 9 *Service Innovation: Organizational Responses to Technological Opportunities & Market Imperatives*
edited by Joe Tidd (University of Sussex, UK) &
Frank M. Hull (Fordham University, USA)

Vol. 8 *Digital Innovation: Innovation Processes in Virtual Clusters and Digital Regions*
edited by Giuseppina Passiante (University of Lecce, Italy),
Valerio Elia (University of Lecce, Italy) &
Tommaso Massari (University of Lecce, Italy)

Vol. 7 *Innovation Management in the Knowledge Economy*
edited by Ben Dankbaar (University of Nijmegen, The Netherlands)

Vol. 6 *Social Interaction and Organisational Change: Aston Perspectives on Innovation Networks*
edited by Oswald Jones (Aston University, UK),
Steve Conway (Aston University, UK) & Fred Steward (Aston University, UK)

Vol. 5 *R&D Strategy and Organisation: Managing Technical Change in Dynamic Contexts*
by Vittorio Chiesa (Università degli Studi di Milano-Bicocca, Milan, Italy)

Vol. 4 *Japanese Cost Management*
by Yasuhiro Monden (University of Tsukuba, Japan)

Vol. 3 *From Knowledge Management to Strategic Competence: Measuring Technological, Market and Organizational Innovation*
edited by Joe Tidd (SPRU, University of Sussex, UK)

Vol. 2 *The Knowledge Enterprise: Implementation of Intelligent Business Strategies*
by J. Friso den Hertog (MERIT, Maastricht University & Altuïtion bv, 's Hertogenbosch, The Netherlands) &
Edward Huizenga (Altuïtion bv, 's Hertogenbosch, The Netherlands)

Vol. 1 *Engines of Prosperity: Templates for the Information Age*
by Gerardo R. Ungson (University of Oregon, USA) &
John D. Trudel (The Trudel Group, USA)

Printed in the United States
By Bookmasters